Green Energy and Technology

Climate change, environmental impact and the limited natural resources urge scientific research and novel technical solutions. The monograph series Green Energy and Technology serves as a publishing platform for scientific and technological approaches to "green"—i.e. environmentally friendly and sustainable—technologies. While a focus lies on energy and power supply, it also covers "green" solutions in industrial engineering and engineering design. Green Energy and Technology addresses researchers, advanced students, technical consultants as well as decision makers in industries and politics. Hence, the level of presentation spans from instructional to highly technical.

Indexed in Scopus.

Indexed in Ei Compendex.

Sarah Farrukh · Xianfeng Fan · Takeshi Matsuura ·
Syed Shujaat Karim
Editors

Facilitated Transport Membranes (FTMs) for CO_2 Capture: Overview and Future Trends

Springer

Editors
Sarah Farrukh
School of Chemical and Materials Engineering (SCME)
National University of Sciences and Technology (NUST)
Islamabad, Pakistan

Takeshi Matsuura
Department of Chemical and Biological Engineering
University of Ottawa
Ottawa, ON, Canada

Xianfeng Fan
Department of Chemical Engineering
University of Edinburgh
Edinburgh, UK

Syed Shujaat Karim
School of Chemical and Materials Engineering (SCME)
National University of Sciences and Technology (NUST)
Islamabad, Pakistan

ISSN 1865-3529 ISSN 1865-3537 (electronic)
Green Energy and Technology
ISBN 978-3-031-21443-1 ISBN 978-3-031-21444-8 (eBook)
https://doi.org/10.1007/978-3-031-21444-8

This Springer imprint is published by the registered company Springer Nature Switzerland AG
The registered company address is: Gewerbestrasse 11, 6330 Cham, Switzerland

I want to dedicate this work to my parents "Abdul Hafeez," "Jamila Hafeez," husband "Farrukh Hanif," my darling son "M. Sarim Janjua," and my deceased brother "Abdul Mateen" as without their support and guidance, I am unable to achieve anything in my life. I cannot repay my both supervisors, Prof. Dr. Arshad Hussain and Prof. Dr. Xianfeng Fan, who are always beacons of light for me.

—Sarah Farrukh

I dedicate this book to my parents, siblings, friends, and especially to my teachers, whose unwavering support and guidance made this work possible.

—Syed Shujaat Karim

Preface

Membrane technology has seen a wide expansion in its application in the past decades, mainly due to the increase in the understanding of its working principle and the development of progressively better materials, synthesis methods, and operating procedures. One of the most promising candidates among novel membrane separation technologies is the facilitated transport membranes (FTMs). These membranes were first developed in the 1960s for improving CO_2 gas separation performance in membranes by exploiting the reversible acid-base reaction between the acid gases and basic amino groups in various polymers. This technology was only confined to the laboratory scale until recently, when it was realized that there is an urgent need for improving the carbon capture efficiency of emerging technologies to combat climate change as well as the impending energy crisis. This was needed as the current technologies (such as liquid absorption, solid adsorption, and cryogenic separation), although very efficient, required large amounts of energy. Thus, FTMs came up as a very advantageous technology owing mainly due to their high performance efficiency and low energy consumption. Since then, there has been a vast expanse of literature being published on the use of FTMs for several industrial processes such as flue gas separation, biogas upgrading, syngas purification, and natural gas processing. In spite of such an extensive body of literature available on this subject, there is still need to organize and compile as concisely as possible all the data regarding the use of FTMs in the different gas separation processes. This was thought necessary to understand our current standing on this subject and to map out what needs to be done in the future for the improvement of gas separation technology to fulfill our energy needs as well as tackle environmental pollution. As our area of expertise spans all the various CO_2 gas separation technologies and the use of membrane technology in these fields, therefore we are well-equipped to accomplish this task of compiling, summarizing, and understanding the available literature regarding the use of FTMs for CO_2 gas separation processes.

Islamabad, Pakistan Sarah Farrukh
Edinburgh, UK Xianfeng Fan
Ottawa, Canada Takeshi Matsuura
Islamabad, Pakistan Syed Shujaat Karim

Acknowledgements We acknowledge the support of the Membrane for Advanced Research (MEMAR) Lab at the School of Chemical and Materials Engineering (SCME), National University of Sciences and Technology (NUST), Sector H-12, Islamabad, Pakistan. We also appreciate the support of the Norwegian University of Science and Technology (NTNU), NO-7491 Trondheim, Norway for the Chap. 9 titled "Techno-economic Analysis of Facilitated Transport Membranes (FTMs) Based CO_2 Separation Processes". Heartfelt appreciation to Prof. Dr. Xianfeng Fan (University of Edinburgh, UK), Prof. Dr. Takeshi Matsuura (Ottawa University, Canada), Prof. Dr. Arshad Hussain (Pak-Austria Fachhochschule Institute of Applied Sciences and Technology (PAF-IAST), Haripur, Pakistan), Prof. Dr. Mohammad Younas (University of Engineering and Technology (UET) Peshawar, KPK, Pakistan), Dr. Muhammad Ayoub (Universiti Teknologi Petronas (UTP), Perak, Malaysia), and Dr. Huma Shaikh (University of Sindh, Jamshoro, Sindh, Pakistan) for their support.

Contents

Chapter 1
Introduction

Sarah Farrukh, Subhan Ali, Syed Shujaat Karim, Hizba Waheed, and Inamullah Douna

Abstract Carbon dioxide (CO_2) and other greenhouse gas emissions from quickly expanding industry and homes are a major problem. These pollutants are to blame for issues like climate change and global warming. One potential solution for the climate change caused by greenhouse gases is the capture and storage of CO_2. For this purpose, effective separation technologies must be used to remove CO_2 from industrial flue gas streams in-order to reduce the greenhouse effects. In addition to this source, CO_2 is also found in raw natural gas and syngas, which must be removed in order to increase the calorific value of the gaseous fuel. Membrane-based gas separation is an emerging technology that is more energy-efficient, compact, and cost-effective than conventional technologies such as amine absorption, adsorption, and cryogenic distillation processes. This introductory chapter briefly discussed membrane technology designed for the application of CO_2 separation. Initially, the background to membrane separation is provided, with a discussion of fundamental concepts, the classification of membranes, and their transport mechanisms. This is followed by a discussion on the different gas separation applications, CO_2 capture processes, and facilitated transport membranes (FTMs), along with their types, transport phenomena, material characteristics, and benefits of using FTMs in CO_2 separation processes.

S. Farrukh (✉) · S. Ali · S. S. Karim · I. Douna
School of Chemical and Materials Engineering (SCME), National University of Sciences and Technology (NUST), Sector H-12, Islamabad, Pakistan
e-mail: sarah.farrukh@scme.nust.edu.pk

S. Ali
e-mail: subhan.che8scme@student.nust.edu.pk

S. S. Karim
e-mail: sshujaat.che6scme@student.nust.edu.pk

I. Douna
e-mail: inamullah.che6scme@student.nust.edu.pk

H. Waheed
Department of Chemical Engineering, Wah Engineering College (WEC), University of Wah (UOW), Quaid Avenue, Wah Cantt, Pakistan
e-mail: hizba.waheed@uow.edu.pk

S. Farrukh et al. (eds.), *Facilitated Transport Membranes (FTMs) for CO_2 Capture: Overview and Future Trends*, Green Energy and Technology,
https://doi.org/10.1007/978-3-031-21444-8_1

Keywords Facilitated transport membranes (FTMs) · Mobile and fixed carrier · CO_2 separation technologies · Membrane technology · Gas separation applications · Transport mechanisms

1.1 Carbon Dioxide (CO_2) Capture Technologies

Many industries and power plants, such as the cement industry, iron and steel industry, food and tobacco industry, chemical and machine manufacturing sector, and so on, rely heavily on CO_2 capture. Flue gases are created as a result of the processing of fuel, air, and raw materials in the industries indicated above. As seen in Fig. 1.1, it is critical to segregate flue gases in order to cleanse the environment. Aside from that, separating CO_2 from methane gas is a huge part of purifying natural gas. Various techniques, like amine absorption, pressure swing adsorptions, cryogenic distillation, and membrane gas separations, are used on a commercial scale to achieve these separations.

1.1.1 Amine Absorption Technique

During the absorption process through amine, the exhaust gases are wiped with amine-containing solvents and form a CO_2-amine complex. Later stages of the process required the use of heat for the amine and CO_2 separation. This process is a low-temperature process containing a minute CO_2 in the exhaust. However, this process has few challenges, like solvent toxicity, high-rate corrosion of equipment and solvent degradation, which have to be resolved and opens a field for researchers to seek new approaches for gas separation [1].

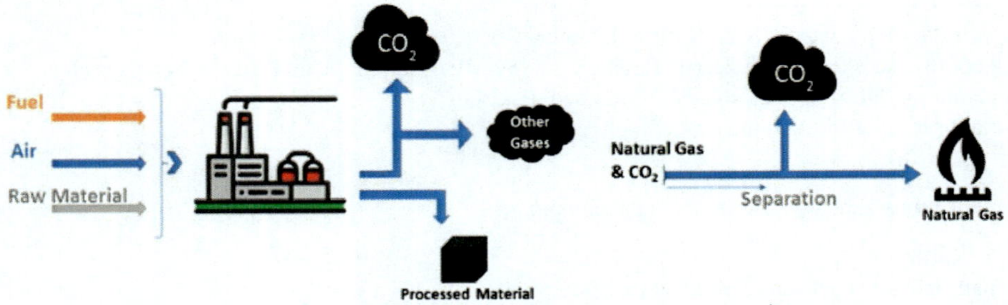

Fig. 1.1 Schematic representation of gas separation from industrial processes and natural gas

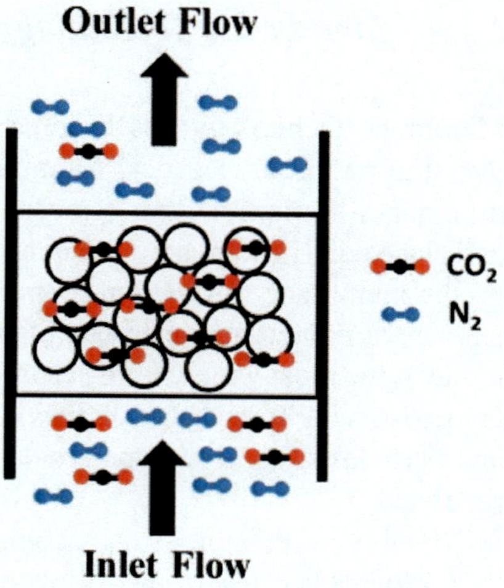

Fig. 1.2 Representation of pressure swing adsorption process [1]

1.1.2 Pressure Swing Adsorption (PSA) Technique

Commercially, pressure swing adsorption (PSA) is a technique that absorbs CO_2 from both natural as well as flue gases under pressure according to the molecular affinity of the adsorbent. Gases mixture is able to pass through a column filled with pellets or powder, revealing a large surface area per total weight of pellets or powder, during the operation. These gases become adhered to or adsorbed on the surface of the pellets as they pass through [1]. By utilizing the appropriate material like pellets, CO_2 may be absorbed more effectively than that of other gases. The PSA techniques's primary drawbacks are the adsorbent's reduced gas adsorption capability and the energy and time spent in the process while regeneration. Figure 1.2 depicts the pressure swing adsorption in diagram form.

1.1.3 Cryogenic Air Separation Technique

To separate N_2, O_2, and liquid products, the cryogenic air separation process is well-thought-out. It requires cooling pure gases until they liquefy, then selectively distilling the components at their respective boiling temperatures to separate them from the air. This technique can be used to make liquid items. In comparison to plants that produce liquid goods, plants that produce gases use less energy. However, this technology consumes a lot more energy, making this technique uneconomical for many commercial sectors.

1.1.4 Membrane Technology

Membrane technology has become more important in a number of applications over the past four decades. Membrane-based separation methods like ultrafiltration, nanofiltration, reverse osmosis, microfiltration, gas separation, pervaporation, and electrodialysis are now frequently employed [2]. All procedures mentioned rely on the membrane to function properly. The capacity of a membrane to restrict the passage of numerous species and its selectivity for a variety of chemicals have made it one of the most widely used separation technologies. In the presence or absence of applied driving forces, membranes act as selective barriers to a variety of molecules and particles [3]. Certain species can pass through the membrane, while others are restricted.

Membrane technology is recommended for gas separation over other current technologies because of the following advantages.

- It is simple to set up and run, more adaptable and cost-effective.
- It is safe for the environment.
- It is adaptable to managing high flow rates, pressures, and feed compositions since they are continuous and can operate in mild ambient conditions.
- It may be tweaked to some extent depending on the work at hand.
- It can easily be combined with the other gas separation techniques to increase operating efficiency.

1.2 Membrane Technology

1.2.1 Historical Background

When Von Wroblewski enumerated Graham's model and described gas permeability via membranes in 1879, the membrane technology became well-known. He proposed that the product of diffusivity and solubility be used to calculate permeability [4]. In membrane gas separation applications, this type is still used. After permeability, H. Kayser defined Henry's law of CO_2 absorption in 1891. Lord Rayleigh studied the permeability of N_2 and Argon in rubber in 1900 after discovering these rules and models.

Loeb and Sourirajan created a cellulose acetate-based asymmetric membrane for reverse osmosis water purification in the twentieth century [5]. This was the first step toward membrane technology being used in both industrial and commercial applications. Scientists and engineers created a variety of laboratory equipment to include membranes in the reverse osmosis process for sea water treatment in the late 1920s. As a result, industrialized countries began to invest in membrane technology research in order to investigate the use of membranes in a variety of industrial and environmental applications. Membranes have been established as one of the finest choices for gas separation through research and development in membrane

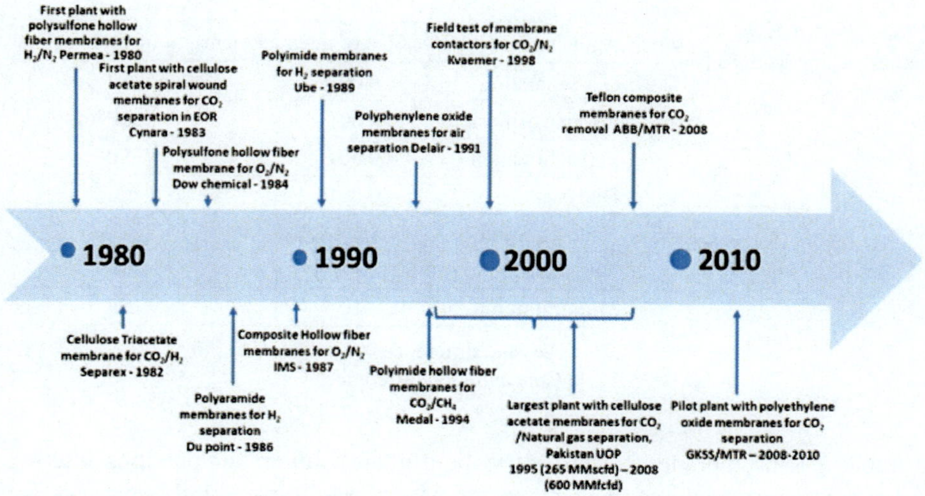

Fig. 1.3 Developments in membrane gas separation in late twentieth century [8]

technology. Prism® was the first commercially available asymmetric membrane, developed by Henis and Tripodi [6]. Scientists later tried to develop new materials and processes for fabricating membranes. Some of the first commercially available manufactured membranes were deployed in a variety of operating units and industries [7]. Commercialization of gas separation membranes took off in the late 1980s and continues to develop, as seen in Fig. 1.3.

1.2.2 *Fundamental Concept*

Under the effect of various gradients, membranes are a semipermeable conduit that allows certain species to travel through it but not others, as seen in Fig. 1.4. Pressure, voltage, or concentration can all be used as driving gradients or forces [9].

Membranes are employed in microfiltration (MF), ultrafiltration (UF), nanofiltration (NF), reverse osmosis (RO), electro-dialysis (ED), dialysis, pervaporation (PV), and gas separation on the basis of the filtrate's size and gradients, as shown

Fig. 1.4 Representation of membrane process [9]

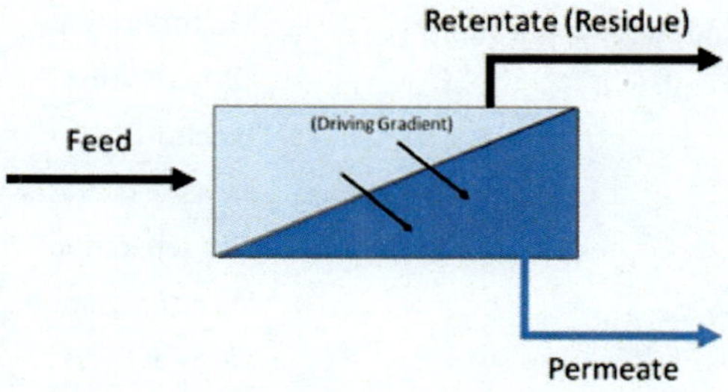

Table 1.1 Different membrane applications [7]

Applications	Size of retentate (μm)	Gradient
Gas separation	< 0.001	ΔP
Reverse osmosis	< 0.001	ΔP
Pervaporization	< 0.001	ΔP
Dialysis	< 0.001	c
Nano filtration	0.0005–0.005	ΔP
Ultrafiltration	0.001–0.1	ΔP
Microfiltration	0.1–10	ΔP

ΔP = Pressure difference
ΔC = Concentration difference

in Table 1.1. Membranes are proven to be a good solution for previous filtering and separation procedures in many sectors, as seen in Table 1.1. Membranes are preferred over older traditional approaches for a variety of reasons, which include their handling simplicity, energy savings, maintenance ease, environmentally friendly nature, excellent compatibility, and long-term performance.

Membrane technology has now replaced a variety of classic industrial processes such as adsorption, absorption, distillation, extraction, and filtration [10]. Membrane separation methods such as microfiltration, ultrafiltration, electro-dialysis, and reverse osmosis have been developed. Table 1.2 shows that the gas separation and pervaporization processes are still in development.

The feed streams in infiltration/separation systems move primarily in two directions toward membranes. The orientation of the membranes determines this direction. The feed stream flows at a 90° angle on the membrane surface in Fig. 1.5 (left). Species that are allowed to pass through membranes flow perpendicular to the membrane surface, whereas those that are rejected flow in the same direction as the membrane surface. Dead end filtration is the name for this type of filtration (see Fig. 1.5).

The pressure differential across the membrane is caused by the feed stream passing tangentially to the membrane surface (Fig. 1.5 right). Crossflow filtration is the name given to this flow mode. The permeate flows normally across the membrane in this

Table 1.2 Status of membrane systems utilized in industrial applications [7]

Process	Status	Remarks
Microfiltration	Developed	Established
Ultrafiltration	Developed	Established
Electro-dialysis	Developed	No major break through
Reverse osmosis	Developed	Seems imminent
Gas separation	Developing	Plants have been installed
Per vaporization	Developing	Major problems exist
Piezo dialysis	To be developed	No systems installed

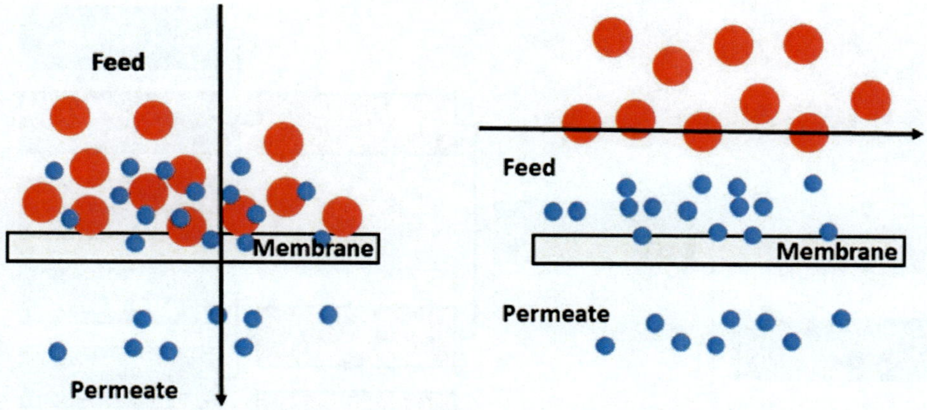

Fig. 1.5 Dead end (left) and cross flow mode (right) of membranes filtration process

situation. Many commercial gas separation technologies employ this mode as well. Shear forces act on the feed stream during cross flow, preventing the formation of retentate on the membrane surface. Cross flow allows permeate to flow continuously without losing focus.

1.2.3 Classification of Membranes

Membrane technology's adaptability may be demonstrated by considering various membrane types. The membrane, as mentioned in the preceding section, is an interface that allows some species to flow through while inhibiting others that are irrelevant or unwelcome. In terms of structure and composition, the membrane that forms the contact might be homogeneous or heterogeneous. It might be dense or porous [1]. Figure 1.6 depicts the various membrane types, which are further detailed below.

(a) **Isotropic Membranes**

Isotropic membranes or symmetric membranes are porous or dense and have a uniform composition structure throughout. The entire membrane thickness determines the barrier to mass transport in these membranes. The permeation rate increases as the membrane thickness decreases [11].

(b) **Dense Membranes**

Dense membranes have a smooth and hole-free surface. Under the influence of pressure, concentration, or an electrical gradient, the permeate travels across these membranes. Filtration by diffusion takes place in a thick membrane. The flowrate of permeate over these membranes controls the separation of distinct species from the incoming stream. Based on size exclusion, dense membranes may separate a combination of feeds. In reverse osmosis, pervaporization, and mostly in gas separation applications, dense membranes are utilized.

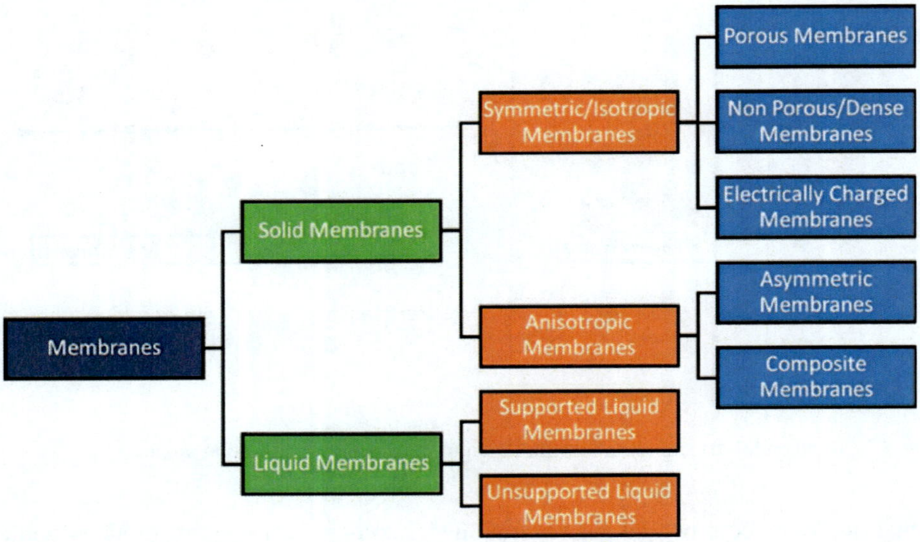

Fig. 1.6 Various membrane types and classification

(c) **Microporous Membranes**

Pores and voids run the length of these membranes' cross section. Several manufacturing processes are used to create these pores inside the membrane framework. Temperature, heat treatments, drying conditions, and other factors impacting membrane production all have a part in the structure of the membrane. A porous membrane is formed when one or more of the aforementioned parameters change. The structure of a microporous membrane is made up of interconnecting pores that are spread throughout the membrane and on its surface. The pores range in size from 0.01 to 10 μm, making them stand out from traditional filters. Membrane-based filtration/separation is dependent on the size of pores and permeant molecules. Small molecules flow through the membrane's numerous pores, whereas large molecules remain in the retentate. Ultrafiltration and nanofiltration both use these membranes.

(d) **Electrical Membranes**

There are two types of membranes: dense and microporous. Electrical membranes, which are charge dependent and carried by the pore of the membrane, can be positive or negative. An anion exchange membrane contains a positive charge on the pore walls, whereas a cation exchange membrane carries a negative charge. In electrodialysis, an electrical membrane is advantageous.

(e) **Anisotropic Membranes**

In the filtering process, the mechanical strength of the membrane is essential. Composition or pressure gradients can harm membranes. The separating layer and thick support layer are joined to generate anisotropic membranes, which enhance mechanical strength. The thickness of homogenous integrated membranes is kept narrow to maximize flow rate. Species will flow quickly through the thin membrane. When the membrane thickness is lowered, however, the membrane's mechanical strength decreases. To solve this problem, a thin separating membrane is cast on a separate strong, thick, and porous support layer, which enhances its mechanical strength and separation efficiency for the filtration process [12]. Asymmetric thin film composite membranes are membranes with a support and separation layer comprised of different materials. These are the ones that are in high demand right now. There is no effect of the thick support layer on the membrane's permeability or selectivity due to its inert and fully porous nature. The Loeb-Sourirajan technique produces membranes with a single membrane material, but various layers of the membrane have varied porosity and pore sizes.

(f) **Metal and Ceramic membranes**

Since the previous decade, metal and ceramic membranes have also been manufactured. Ceramic membranes are formed by combining a metal with a nonmetal, such as oxide, nitride, or carbide. Whereas, sintering of metal particles yields metallic membranes (e.g., tungsten or molybdenum). These microscopic membranes are used in Ultrafiltration and Microfiltration procedures. These membranes are solvent-resistant and mechanically strong. Palladium membranes are employed in H_2 gas separation applications.

(g) **Facilitated Transport membranes (FTMs)**

Facilitated transport membranes (FTMs) are the most modern gas separation membrane technology. In these membranes, the liquid (separating substance) is immobilized in a solid porous support. When flue gases or other species pass through these membranes, there is a reaction between the components of the input stream and the liquid, forming a complex. The diffusion occurs across the membrane when the substance hits the permeate side, and it releases after that. The carrier agent moves back to the feed side after releasing the permeants. Metal separation applications are also possible with these membranes (Fig. 1.7).

1.2.4 Transport Mechanisms

Various morphologies of membranes must have different transport phenomena. In general, two models describe mass transfer across membranes:

- Pore flow model
- Solution diffusion model

The pore flow model is used in porous membranes, whereas the solution diffusion model is used in dense membranes. Permeate flows through membrane pores under a pressure gradient in a pore flow model. Pore flow models include molecular sieving, Knudson diffusion, and Poiseuille flow, in which the molecular size of

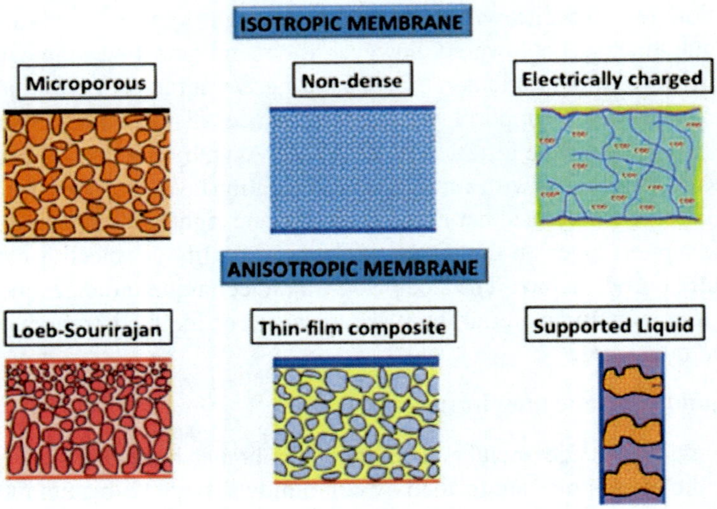

Fig. 1.7 Schematic representation of various types of membranes. (Reprinted with permission from Ref. [13] copyright © 2022 Elsevier)

the permeate and pores is critical in the separation process. Figure 1.8 shows the following processes in porous and dense membranes:

(a) Knudsen Diffusion Mechanism

This form of diffusion takes place in porous membranes with pore sizes ranging from 40 to 100 μm. The mean free path (the distance travelled by a molecule in a membrane before collision) of gas molecules is more than the pore size of the

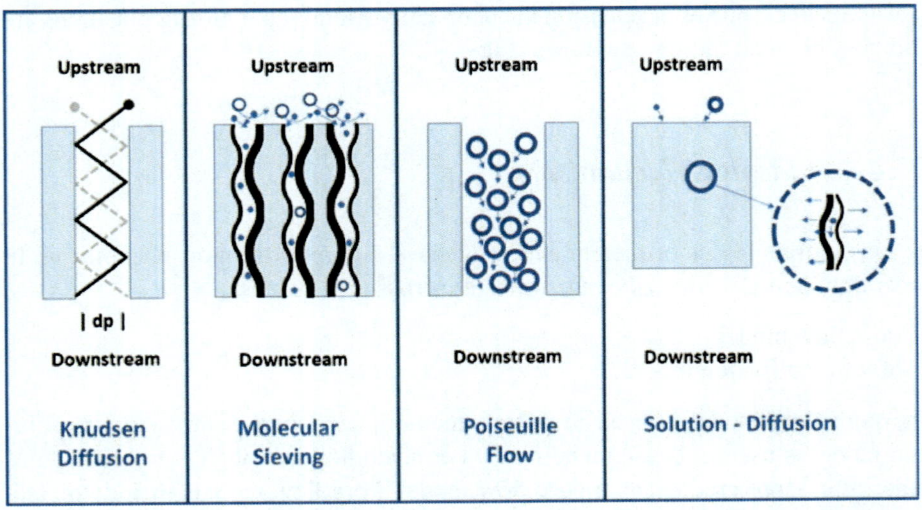

Fig. 1.8 The illustration of transport mechanisms in porous and dense membranes [14]

membrane in this form of diffusion [15]. Because of this, molecules clash with pore walls more than they collides with one other, resulting in a rapid lighter molecules or particles diffusion through these pores. The Knudsen diffusion equation is written as:

$$DK = 0.667\mathrm{rv} = 97\mathrm{r}\sqrt{}\mathrm{TMw} \tag{1.1}$$

In Eq. (1.1),
r is radius of pore
T is temperature and
v is molecular velocity

Gas components diffusion across membrane is independent of one another. Hence, pressure has no effect on the Knudson diffusion (Dk) coefficient. The permeate flowrate is inversely related to the molecular weight of different species, according to Eq. (1.1) [16]. Membrane selectivity diminishes as pore size increases. Membranes with a bigger pore size are a suitable choice for gas separation, but they're also used in the Nanofiltration and Ultrafiltration processes.

(b) **Poiseuille Flow Mechanism**

Membranes with pore sizes greater than 10 μm are subjected to the Poiseulli flow model. The mean free route of molecules in this scenario is less than the pore diameter of membranes. As a result, the molecular flow is convective. When molecules travel across a membrane, they collide with each other. This type is used in membranes that are suitable for Microfiltration applications.

(c) **Molecular Sieving Mechanism**

Membranes with pore sizes smaller than 7 Å are subjected to the molecular sieving phenomenon. Because of the large flow of gases, these membranes are used in gas separation applications [17]. Nowadays, zeolites and carbon membranes use the same transport phenomena to separate gases. The pore width and the molecular size of gas molecules are essential characteristics in gas separation [18]. Various manufacturing procedures and factors can alter the pore diameter of membranes, affecting the molecular sieving transport phenomena [19].

(d) **Solution Diffusion Mechanism**

Solution diffusion is the transport process in dense membranes. This includes reverse osmosis, pervaporization, and gas separation membranes. In the commercial world, dense membranes are used for the purification of gases. In the solution diffusion model, there is an adsorption of molecules of gas molecules on the surface of membrane, which diffuses across it, and subsequently desorb at the permeate side [20].

(e) **Facilitated Transport Mechanism**

Simple diffusion transport combined with one or more reversible chemical reactions that enable the transport of the target species also in the form of a complex are what make up the carrier-mediated facilitated transport of a gas in a membrane. Due to the fact that the differential transport problem is typically nonlinear and lacks an accurate analytical solution, such a process is difficult to quantitatively characterize.

Numerous researchers have created analytical solutions to the issue based on certain simplifications or approximations, such as the achievement of the local equilibrium condition, equality of the carrier and complex diffusivities, and an excess of the carrier relative to the solute. The suggested ideas and models will be looked into and discussed in more detail in the following sections.

The generic formulation of the assisted transport problem is given here in terms of mathematical differential equations that express the mass balance for the relevant species. Three species—solute, carrier, and their reaction product—will coexist in a single chemical reaction that will serve as the basis for the treatment (also called a complex). In a rectangular geometry, for each species *i* involved, we can write the continuity equation: [21]

$$\begin{aligned}\frac{DC_i}{Dt} &= \left(\frac{\partial C_i}{\partial t} + v_x\frac{\partial C_i}{\partial x} + v_y\frac{\partial C_i}{\partial y} + v_z\frac{\partial C_i}{\partial z}\right)\\ &= -\left(\frac{\partial J_{ix}}{\partial x} + \frac{\partial J_{iy}}{\partial y} + \frac{\partial J_{iz}}{\partial z}\right) - r_i + r_i' \end{aligned} \tag{1.2}$$

where,

- C = concentration
- J = flux
- R = dissipative term
- r' = generative term
- v = velocity
- x, y, z = Cartesian directions
- t = time, to be considered with consistent units.

1.2.5 Membrane Gas Separation Applications

Membranes for gas separation applications were initially commercialized in the late twentieth century. Breakthroughs in membrane technology have been observed over time. As a result of this breakthrough, the use of membranes in gas separation applications has increased. A number of gas separation applications have employed membranes, some of which are described below.

(a) **Air Separation**

The air composition is 78% nitrogen (N_2), 21% oxygen (O_2), and 1% additional gases. The N_2 and O_2 separation has several industrial applications. These two gases are purified to be used in a wide range of applications. The membrane's separated nitrogen is used to preserve vegetables and fruits. Membrane technology is responsible for roughly 30% of the total N_2 utilized in industry [22]. On the other end, O_2 is mostly employed for medical applications. In industrial sector, O_2 is also used for the coal treatment. In this case, O_2 enhanced air aids coal combustion process. Furthermore, around 50% of O_2 separation occurs by using the membranes. However, the separation rate is not that much high in this process.

(b) **Separation of H_2 from Hydrocarbons**

The gases containing light hydrocarbons are produced whenever we use hydrogen gas in the refining sector, which has to be evacuated from the system [23]. Hydrogen recovery from hydrocarbons is necessary from the purge stream because the recovered H_2 will be about 20% of the total H_2 produced by the plant. Hydrogen separation through membranes is considered better than other techniques.

(c) **H_2 and N_2 Separation**

The production of ammonia is done at high temperatures and pressures. In the production process, Syngas is utilized. During the synthesis process, a possibility exists that 100% conversion of ammonia does not take place, so unreacted gases are utilized again in the reactor. The purge gas is also used to control the amount of inert gases in the process. The purge gas contains mostly H_2 and N_2 gas molecules, which must be separated for the recycling. Membrane technology is considered a better option to separate above-mentioned gases than cryogenic distillation, adsorption, and other conventional techniques.

(d) **Natural Gas Sweetening Process—(CO_2/CH_4) Separation**

Natural gas is an extremely necessary fuel for industrial and domestic applications. The natural gas reservoirs are predominantly composed of CH_4 as well as CO_2, H_2S, N_2, water, and heavy hydrocarbons. Separating CH_4 is essential. Earlier, this separation was performed by amine absorption, but the method' corrosiveness, massive space needs, and environmental unfriendliness encouraged scientists to investigate new technologies. Membrane technology has lately been presented as a better option for purifying natural gas [24].

(e) **Flue Gas Separation (CO_2/N_2)**

With the growing industrial development, the issue of CO_2 emission into the environment has alarmingly increased. The cement industry, power plants, fossil fuel and biomass combustion, all emit large amounts of flue gases. Most of the exhaust or flue gas is made up of CO_2, N_2, and hydrocarbons. To capture CO_2 from flue gases, membranes have been implanted in industries.

1.3 Global Warming and CO_2 Capture

Global warming is a serious problem that many environmentalists, engineers, and scientists are concerned about. The earth's temperature and CO_2 content have doubled over the past 650,000 years, according to ice core research in Antarctica. Annual CO_2 concentration increases of 1% are observed in Hawaii. It is also predicted that CO_2 concentrations will reach 600 ppm by 2050. It is suggested that CO_2 concentrations can be stabilized in the range of 450–750 ppm by limiting human activities that cause CO_2 emissions [25, 26].

CO_2 emissions from human activities are the primary cause of global warming. These operations mostly entail the burning of fossil fuels in power generation plants to provide enough energy for industry, transportation, domestic, and commercial use. In this context, Fig. 1.9 represents the annual CO_2 emissions from the fossil fuels, and the growth in their usage over time. The cement and petroleum industries as well as the natural gas reserves, are the primary sources of CO_2. If these companies adopt effective emission reduction measures, they can cut CO_2 emissions by about 45%.

For a wide range of industries, CO_2 capture from natural and exhaust gases is critical. As previously said, the rapid expansion of businesses and population growth have increased CO_2 gas emissions into the atmosphere. Controlling the quantity of CO_2 gas in the atmosphere is critical for reducing global warming. CO_2 concentrations around the globe were predicted to be 28,051 million metric tons (MMT) in 2005, and they are expected to climb to 42,325 MMT by 2030. Furthermore, natural gas reserves include multiple gases like CO_2, H_2S, N_2, water, and heavy hydrocarbons. Purifying natural gas from other gases to fulfil pipeline criteria is a top priority [28].

1.3.1 CO_2 Capture Processes

Cost-effective strategies for controlling CO_2 emissions from industrial operations, where CO_2 is created by the burning of fossil fuels, and from natural gas reserves are currently being investigated. A wide range of procedures were investigated. However, three approaches are used. Figure 1.10 depicts these processes in schematic form.

(a) **Post-Combustion CO_2 Capture**

Flue gases are created as a byproduct of combustion operations in industries (such as cement). CO_2 and N_2 gases make up the majority of the flue gas. To capture CO_2 from exhaust gases in post-combustion processing, many approaches like chemical absorption (MEA, MDEA, etc.), physical absorption (Selexol, Rectisol, etc.), physical adsorption (molecular sieves, metal organic frameworks, etc.), and membrane technology can be utilized. Membrane technology is considered as a viable approach for CO_2 capturing since it saves energy, space, and is easily scalable.

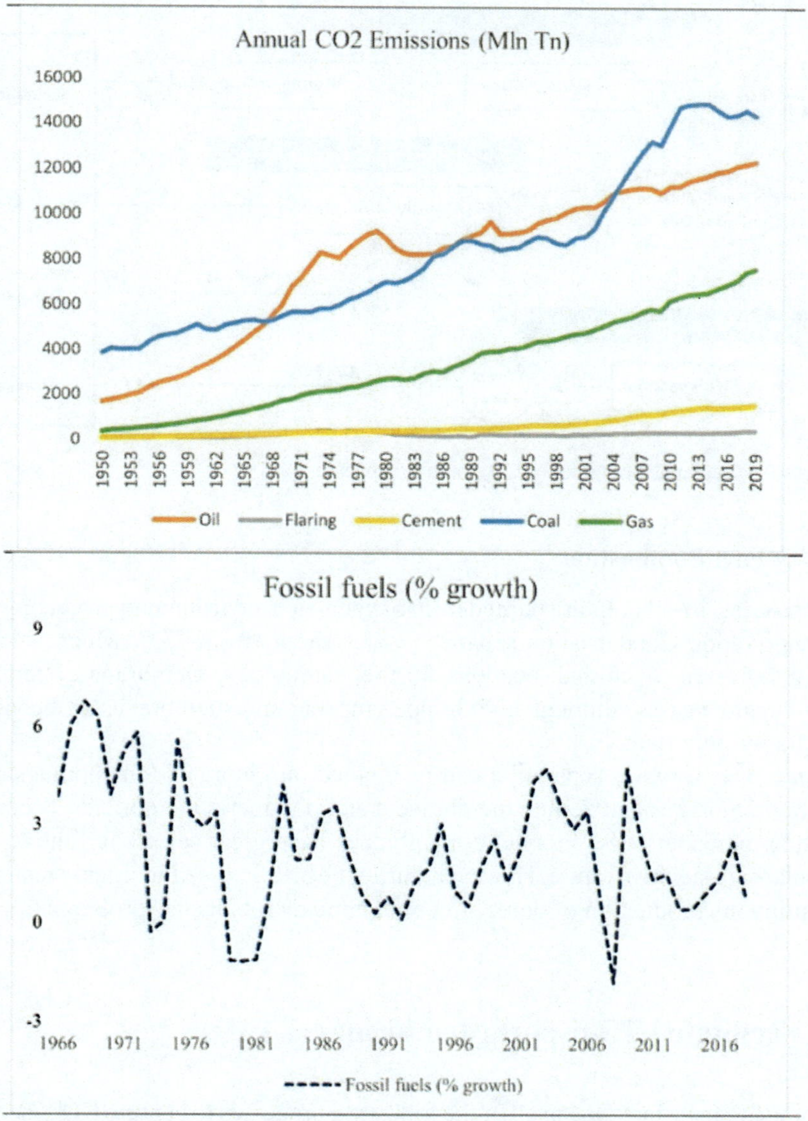

Fig. 1.9 CO_2 emission from fossil fuels and the growth in their usage over time. Reused with the permission of Ref. [27]

(b) **Pre–Combustion CO_2 Capture**

Natural gas is transformed into a gaseous mixture of CO and H_2 in order to capture CO_2. Syngas is the name given to this gaseous combination. CO_2 may be produced via the CO–shift reaction at temperatures ranging from 180 °C to 550 °C, and CO_2 gas can be separated from H_2 and stored. This H_2 can also be used in the production of fuel, chemicals, and biomass. This approach works best with polymeric, porous ceramic, and proton conducting membranes.

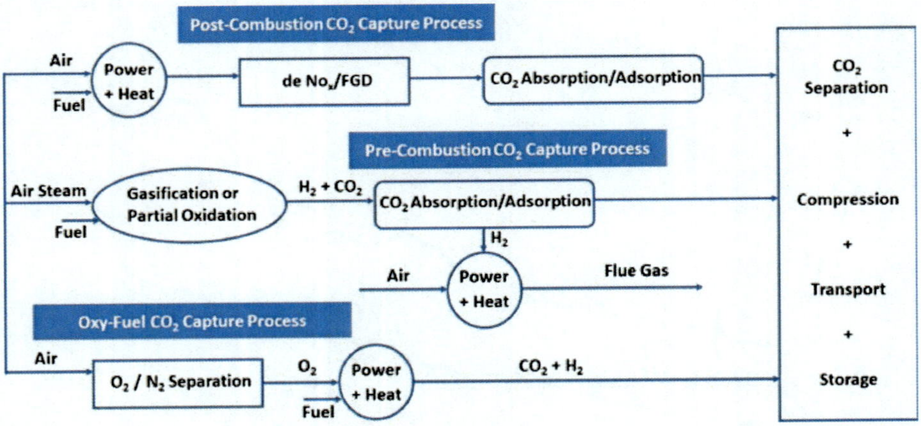

Fig. 1.10 The schematic diagram of CO_2 capture processes [29]

(c) **Oxy Fuel Combustion**

In this process, fossil fuels are burned in an oxygen-rich environment, producing CO_2 and water vapor. Condensation separates water vapors from CO_2, which is subsequently delivered to storage locations. In this technology, membrane systems are used to separate gases, although high-temperature ceramic membranes are becoming increasingly important.

Membranes for CO_2 separation can be divided into inorganic membranes (such as silica membranes and zeolite membranes) and polymeric membranes, according to membrane technology. Inorganic membranes have great selectivity due to their molecular size sieving feature. However, due to the brittleness of inorganic materials, the continuous production of defect-free inorganic membranes is problematic.

1.4 Facilitated Transport Membranes (FTMs)

Membranes can be installed in all of the following processes to capture CO_2. Because of its great energy efficiency, compact size, and operational simplicity, membrane separation is a potential technique for CO_2 separation and capture. However, today's conventional polymeric membranes, like cellulose acetate and polyimide, have limited CO_2 permeability and selectivity as they are reliant on a solution-diffusion process and mostly on the tiny size variations of the penetrants to achieve separation. Furthermore, these membranes are unsuitable for high-temperature and high-humidity operation, which is necessary for direct CO_2 capture from exhaust gases in power plants [30]. Facilitated transport membranes are an important group of CO_2 separation membranes among all the approaches covered due to the high CO_2 permeability across these membranes, especially when the gas is humid, as it is in flue gas after burning, and their unique ability to selectively separate target species (e.g., ions and tiny molecules) [1].

The term "facilitated transport" (FT) describes a coupled transport mechanism composed of pure physical diffusion and reversible chemical interactions between the targeted solute (A) and the carrier (C). At the feed side of the membrane, the solute interacts, and the reaction product (AC), also known as "complex" in FT jargon, diffuses across the thickness. The reversed chemical reaction releases the solute on the downstream side. According to Fick's law, unreacted solutes and other dissolved species (which have no chemical interaction with the carrier) simply disperse throughout the system. Because of this carrier-mediated transport, the membranes function better in terms of solute flux and separation capabilities.

1.4.1 Types of FTMs

FTMs are usually classified into two types:

1. Mobile carrier (MC)-FTMs
2. Fixed site carrier (FSC)-FTMs

(a) **Mobile Carrier Facilitated Transport Membranes (MC-FTMs)**

The mobile carrier has completed free movement across the membrane. Supported liquid membrane (SLM) or immobilized liquid membrane (ILM) are other names for this type of membrane. In these membranes, On the feed side, CO_2 initially diffused on the membrane's surface, then the carrier interacts with CO_2 molecules to formed the reversible bonded reaction product, which flows across the membrane due to its high selectivity, which provides a rapid channel for CO_2 molecules. In the final step, the CO_2-carrier bond breaks down and the CO_2 is then liberated on the permeate side at a lower pressure. However, the other gas species only follows solution diffusion mechanism, which enhances the efficiency of CO_2 separation process.

(b) **Fixed-site Carrier Facilitated Transport Membranes (FSC-FTMs)**

The fixed carrier is linked covalently to the polymer's backbone and has restricted mobility. The CO_2 molecule binds to one carrier site before moving on to the next one until it makes it to the permeate side. These two types of CO_2 transportation mechanisms in FTMs are depicted in Fig. 1.11.

1.4.2 Transport Phenomena in Facilitated Transport Membranes (FTMs)

Due to the natural flexibility of polymer materials, polymeric membranes are far more advantageous in terms of ease of scaling up. In a recent review study published

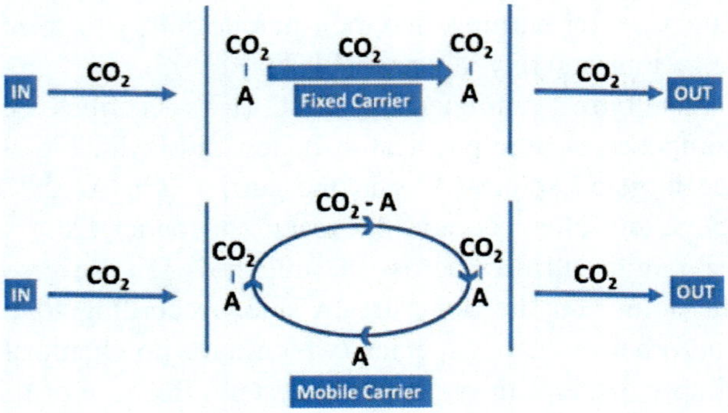

Fig. 1.11 Illustration of CO_2 transport in facilitated transport membranes (FTMs) showing two types of mechanism: mobile carrier (top) and fixed carrier (bottom) [31]

in 2015 by Kai and Duan [32], the many types of membranes for CO_2 capture are summarized, including polymeric membranes, inorganic membranes, ionic liquid membranes, and facilitated transport membranes.

Typically, the solution-diffusion mechanism is used for gas transport through polymeric membranes [33]. Gas molecules in such membranes first dissolve in the membranes at the feed side and membrane interface before diffusing across the membrane to the permeate side. Gas solubility and diffusivity combine to provide permeability [33]. The Robeson's upper bound-controlled trade-off between permeability and selectivity is the primary drawback of solution-diffusion membranes [34].

However, inside FTMs, things are different. The "carrier" inside these membranes can reversibly react with CO_2 molecules, which facilitates CO_2 transport across the membrane. When CO_2 and the carrier interact at the feed-side interface of the membrane, a CO_2-carrier reaction product is created. This product diffuses along its concentration gradient to the permeate side of the membrane. The CO_2 is then released, when the CO_2-carrier reaction product breaks down due to a lower partial pressure of CO_2 on the permeate side. This regenerates the carrier, which can then react with another CO_2 molecule on the feed side. As a result, CO_2 transport in these membranes is dependent on the facilitated transport and solution diffusion coupled mechanisms. Whereas the other non-reactive gases, including N_2, H_2, CH_4, and CO, only rely on the pure diffusion process. Because of these factors, FTMs demonstrated significantly higher perm-selectivity than other types of membranes, which generally exceeded the membrane's revised Robeson's upper bound [35]. The following equation can be used to express the steady state CO_2 flux across a FTM [36].

$$N_A = \frac{D_{AB}\left(C_{AB,1m} - C_{AB,m2}\right)}{l} + \frac{D_A\left(D_{A,1} - D_{A,2}\right)}{l} \tag{1.3}$$

N_A denotes the total flow of CO_2 through a membrane that facilitates transport at steady state. The diffusion coefficients of CO_2 and the CO_2-carrier reaction product inside a facilitated transport membrane are DAB and DA, respectively. The CO_2-carrier reaction product concentrations at the feed-side and permeate-side interfaces of the membrane are $C_{AB,\,1m}$ and $C_{AB,\,m2}$, respectively. The amounts of CO_2 that are dissolved at the feed-side and permeate-side interfaces of the membrane are denoted as $C_{A,\,1}$ and $C_{A,\,2}$, respectively.

When compared to other technologies, the utilization of FTMs is more advantageous at low CO_2 pressure streams, such as, in the case flue gas from coal-fired power plants, where the CO_2 concentration is just 14 vol.%. The "carrier saturation" phenomenon [33], which happens when CO_2 partial pressure or concentration is high enough that CO_2 may react with all of the carrier molecules present in the membrane, can be used to explain this phenomena. This is the highest level of facilitation that the available carrier molecules can achieve. Any further rise in CO_2 partial pressure will only result in a somewhat larger flow through the solution-diffusion mechanism than through the facilitated transport mechanism. On the other hand, when the CO_2 partial pressure is low, only a small portion of the carrier molecules present in the membrane can react with the CO_2; as a result, there are more carrier molecules present in the membrane for reaction with the CO_2, leading to significantly increased transport of CO_2 molecules.

The small, mobile molecules are referred to as "mobile carriers" in these FTMs, whereas functional groups linked to polymer backbones are referred to as "fixed-site carriers". The product of the reaction between CO_2 and a mobile carrier can easily diffuses across the membrane due to its higher perm-selectivity. For a fixed-site carrier, CO_2 transport is guided by the "hopping" process [36]. The CO_2 and non-reactive gas transport mechanisms inside the FTMs are depicted in Fig. 2.2. Due to its polymeric composition and inherent stability, the fixed-site carriers can boost the stability of the membrane. On the other hand, mobile carriers can freely move within the membrane matrix; therefore, MC-FTMs exhibited a higher CO_2 flux as compared to the FSC-FTMs. However, these membranes are unstable than the FSC-FTMs. To overcome this problem and ensure that the mobile carrier remains in the membrane, a low volatility substance, such as amino acid salt, should be used. Furthermore, good membrane stability and a high CO_2 transport rate can be achieve by using a combination of fixed-site and mobile carriers in the membrane matrix.

The complete transport phenomena of FTM including the CO_2 carrier chemistry and the models are discussed in Chap 2.

1.4.3 Membrane Materials

Generally, amine molecules are present in CO_2-selective polymers, which are mainly utilized for CO_2/CH_4, CO_2/N_2, and CO_2/H_2 separation processes. Whether these carriers are attached to the polymer backbone or freely diffused in the polymer matrix, amine molecules have a higher affinity for CO_2 molecules, which results in higher perm-selective carbon capturing efficiency as compared to the other types of membranes.

(a) **Amine-containing fixed-site carriers**

Several polymers containing amino groups either in the polymer backbone or as a pendent group have been used as the fixed-site carriers to facilitate CO_2 transport. Polyvinyl amine (PVAm), poly (allylamine) (PAA) [9], poly (ethyleneimine)(PEI) [37], chitosan, and Poly (amidoamine) (PAMAM) dendrimer [38] with a reducing density of amine groups are some of the most often utilized polyamines.

(b) **Amine-containing mobile carriers**

When compared to membranes that only have fixed-site carriers, adding a mobile carrier can:

(1) Maximize the FTM's total amine content.
(2) Add additional polar moieties to the membrane to minimize the physisorption of inert gases like N_2 and H_2.
(3) Increase the light gas diffusion coefficients by disrupting polymer chain packing.

The most commonly used polymers to host the mobile carrier are PVA, either pristine or cross-linked, and PVAm. These polymers can establish hydrogen bonds with small molecule amines due to this factor they have good miscibility even at large mobile carrier loadings. Additionally, the ability of PVA and PVAm to form good membranes also enables the easy fabrication of ultrathin FTMs using a solution casting or dip coating method.

1.4.4 Advantages and Characteristics of FTMs for CO_2 Separation Applications

Facilitated Transport Membranes (FTMs) are the new approach in the field of membrane technology, which contains mobile and fixed carriers that enhance both the separation capabilities and the flux of CO_2 molecules across the membrane. FTMs have multiple properties, which makes them unique and better than other polymeric membranes. Some of these characteristics are briefly described below:

- The membrane separation process offers several advantages over classic separation methods such as cryogenic rectification and adsorption, including operating flexibility, simplicity of scale-up, no phase shift, and no regeneration. It is possible to construct the membrane separation system as an integrated skid-mounted structure that is easy to carry and move in the isolated wilderness [39].
- Membrane technology is thought to be one of the most effective CO_2 capture systems. Mixed matrix and facilitated transport membranes have attracted a huge amount of interest in recent years among the other types of membranes due to their high CO_2 perm-selectivity across the membranes matrix, especially when there is humidity in the gas, such as in the flue gas that is produced during the combustion process [40].

- Because of their unique ability to selectively separate target species (e.g., ions/tiny molecules), FTMs have been widely used in a number of industrial applications [38].
- These membranes are increasingly being used in a variety of commercial applications, including biological molecule separation, trace chemical detection, wastewater treatment, and gas separation processes [41, 42].
- FTMs with varied transport properties can also be combined in a hybrid membrane configuration to achieve 99% H_2 recovery at an electricity cost of $118.5/MWh, which is 12.5% less than the Selexol method [43].
- FTMs typically offer more selective CO_2 separation as compared to the other types of membranes due to the combined solution-diffusion and facilitated transport mechanisms [44, 45].
- Supported ionic liquid membranes (SILMs), which are made up of ILs and polymer membranes, have been produced and shown to be effective CO_2 separation candidates [46].
- The reversible interactions between carriers and target molecules (CO_2) in a facilitated transport mechanism lead to enhanced separation performance, allowing the permeability and selectivity to increase at the same time [47] (As represented in Fig. 1.11).
- Despite high concentrations of acidic impurities, membranes with mobile carriers showed good resistance to performance degradation when subjected to feed gas including SO_x and NO_x [48].
- Using plate-and-frame and hollow-fiber modules, many pilot demonstration systems have been used in the power plants and cement factories [49].
- The FTMs presented here are the only ones that can have high CO_2 permeance while also being selective against hydrogen (H_2), nitrogen (N_2), methane (CH_4), and carbon monoxide (CO), which do not react with an amine carrier [32].
- To lower the separation energy barrier, FTMs having a homogeneous distribution of facilitated transport carriers are extremely desirable [50].
- Despite this, it is still difficult to produce uniform carrier dispersion inside membranes due to the disorganized and random assembly of membrane matrix materials/blocks, [50] and it is still challenging to resolve the trade-off issue between membrane stability and its separation efficiency [38].

1.5 Conclusion

Facilitated transport membranes (FTMs) have been the focus of extensive research since more than 40 years ago, but no practical commercial technology has emerged. Academic researchers are interested in these membranes because they enable remarkable separations using only basic laboratory tools. Because these membranes have obvious potential economic and technological advantages for a variety of significant separation challenges, the possibilities for FTMs for gas separation are greater. But there are challenging technological issues that have to be resolved before these membranes can be developed on an industrial scale, which include, high-performance membranes with a minimum lifespan of one year and ideally many years are required for industrial operations. Furthermore, industrial-scale FTMs and their system development requires access to cutting-edge membrane technology that is often unavailable in academic settings as well as a dedication to a protracted development program that only few businesses are prepared to undertake.

References

1. Yi L (2006) Development of mixed matrix membranes for gas separation application. National University Singapore
2. Roelofs KS, Hirth T, Schiestel T (2010) Sulfonated poly (ether ether ketone)-based silica nanocomposite membranes for direct ethanol fuel cells. J Membr Sci 346(1):215–226
3. Wang L (2012) Cyclic membrane gas separation processes. Université de Lorraine
4. v. Wroblewski S (1879) Ueber die natur der absorption der gase. Ann Phys 244(9):29–52
5. Loeb S, Sourirajan S (1962) Sea water demineralization by means of an osmotic membrane. ACS Publications
6. Henis JM, Tripodi MK (1981) Composite hollow fiber membranes for gas separation: the resistance model approach. J Membr Sci 8(3):233–246
7. Baker RW (2002) Future directions of membrane gas separation technology. Ind Eng Chem Res 41(6):1393–1411
8. Bernardo P, Clarizia G (2013) 30 years of membrane technology for gas separation. Chem Eng Trans 32:1999–2004
9. Koros WJ, Fleming GK (1993) Membrane-based gas separation. J Membr Sci 83(1):1–80
10. Spillman RW, Sherwin MB (1990) Gas separation membranes: the first decade. Chemtech 20(6):378–384
11. Tian M, Yuan S, Decaesstecker F, Zhu J, Volodine A, Van der Bruggen B (2020) One-step fabrication of isotropic poly (vinylidene fluoride) membranes for direct contact membrane distillation (DCMD). Desalination 477:114265
12. Li Y, Cao C, Chung TS, Pramoda KP (2004) Fabrication of dual-layer polyethersulfone (PES) hollow fiber membranes with an ultrathin dense-selective layer for gas separation. J Membr Sci 245(1–2):53–60
13. Birmod RP, Lade VG, Shambharkar RD (2021) Responsive membranes for wastewater treatment. In: Handbook of nanomaterials for wastewater treatment. Elsevier, pp 673–697
14. Baker R, Cussler E, Eykamp W, Koros W, Riley R, Strathmann H (1991) Recent developments and future directions. Membr Sep Syst
15. Hosseini SS (2009) Membranes and materials for separation and purification of hydrogen and natural gas
16. Uhlhorn R, Keizer K, Burggraaf AJ (1989) Gas and surface diffusion in modified γ-alumina systems. J Membr Sci 46(2–3):225–241
17. Koresh JE, Soffer A (1987) The carbon molecular sieve membranes. General properties and the permeability of CH_4/H_2 mixture 22(2–3):973–982
18. Fuertes A, Nevskaia D, Centeno T, materials M (1999) Carbon composite membranes from Matrimid® and Kapton® polyimides for gas separation. Microporous Mesoporous Mater 33(1–3):115–125
19. Chung TS, Jiang LY, Li Y, Kulprathipanja S (2007) Mixed matrix membranes (MMMs) comprising organic polymers with dispersed inorganic fillers for gas separation. Prog Polym Sci 32(4):483–507
20. Koros W, Fleming G, Jordan S, Kim T, Hoehn H (1988) Polymeric membrane materials for solution-diffusion based permeation separations. Prog Polym Sci 13(4):339–401

21. Rea R, De Angelis MG, Baschetti MG (2019) Models for facilitated transport membranes: a review. Membranes 9(2):26
22. Norris DA, Puri N, Sinko PJ (1998) The effect of physical barriers and properties on the oral absorption of particulates. Adv Drug Deliv Rev 34(2–3):135–154
23. Bernardo P, Jansen J (2015) Polymeric membranes for the purification of hydrogen. In: Compendium of hydrogen energy. Elsevier, pp 419–443
24. Bitterlich S, Kesting RE, Fritsche AK (1995) Polymeric gas separation membranes. Wiley, New York, Chichester. ISBN 0-471-56931-3, 416 Seiten, Preis:£ 66, Wiley Online Library
25. Van Vuuren DP, Edmonds J, Kainuma M, Riahi K, Thomson A, Hibbard K, Hurtt GC, Kram T, Krey V, Lamarque JF (2011) The representative concentration pathways: an overview. Clim Change 109(1):5–31
26. Marland G, Andres R, Boden TA (1994) Global, regional, and national CO_2 emissions. Trends 93: Compendium Data Glob Change 505–584
27. Guo X, Liang C, Umar M, Mirza NJTF, Change S (2022) The impact of fossil fuel divestments and energy transitions on mutual funds performance. Technol Forecast Soc Change 176:121429
28. Washim Uddin M, Hägg M-B (2012) Natural gas sweetening—the effect on CO_2–CH_4 separation after exposing a facilitated transport membrane to hydrogen sulfide and higher hydrocarbons. J Membr Sci 423–424:143–149. https://doi.org/10.1016/j.memsci.2012.08.010
29. Leung DY, Caramanna G, Maroto-Valer MM (2014) An overview of current status of carbon dioxide capture and storage technologies. Renew Sustain Energy Rev 39:426–443
30. Huang J, Zou J, Ho WW, Research EC (2008) Carbon dioxide capture using a CO_2-selective facilitated transport membrane. Ind Eng Chem Res 47(4):1261–1267
31. Doong S (2012) Membranes, adsorbent materials and solvent-based materials for syngas and hydrogen separation. In: Functional materials for sustainable energy applications. Elsevier, pp 179–216
32. Kai T, Duan S (2014) CO_2 capture by membrane. In: Chen W-Y, Suzuki T, Lackner M (eds) Handbook of climate change mitigation and adaptation. Springer New York, New York, NY, pp 1–28 https://doi.org/10.1007/978-1-4614-6431-0_84-1
33. Tong Z, Vakharia VK, Gasda M, Ho WW (2015) Water vapor and CO2 transport through amine-containing facilitated transport membranes. React Funct Polym 86:111–116
34. Robeson LM (1991) Correlation of separation factor versus permeability for polymeric membranes. J Membr Sci 62(2):165–185
35. Gottschlich D, Roberts D, Way JJ (1988) A theoretical comparison of facilitated transport and solution-diffusion membrane modules for gas separation. Gas Sep Purif 2(2):65–71
36. Zou J, Ho WW (2006) CO_2-selective polymeric membranes containing amines in crosslinked poly (vinyl alcohol). J Membr Sci 286(1–2):310–321
37. Uddin MW, Hägg MB (2012) Natural gas sweetening—the effect on CO_2–CH_4 separation after exposing a facilitated transport membrane to hydrogen sulfide and higher hydrocarbons. J Membr Sci 423:143–149
38. Wang D, Liu F, Zhang X, Wu M, Wang F, Liu J, Wang J, Liu Q, Zeng H (2021) A Janus facilitated transport membrane with asymmetric surface wettability and dense/porous structure: enabling high stability and separation efficiency. J Membr Sci 626:119183
39. Zhang C, Sheng M, Hu Y, Yuan Y, Kang Y, Sun X, Wang T, Li Q, Zhao X, Wang ZJ (2021) Efficient facilitated transport polymer membrane for CO_2/CH_4 separation from oilfield associated gas. Membranes 11(2):118
40. Rafiq S, Deng L, Hägg MB (2016) Role of facilitated transport membranes and composite membranes for efficient CO_2 capture–a review. ChemBioEng Rev 3(2):68–85
41. Gao J, Qiu Y, Hou B, Zhang Q, Zhang X (2018) Treatment of wastewater containing nickel by complexation-ultrafiltration using sodium polyacrylate and the stability of PAA-Ni complex in the shear field. Chem Eng J 334:1878–1885
42. Ma J, Stevens GW, Mumford KA (2018) The performance of diphenyldichlorosilane coated ammonium exchange zeolite and its application in the combination of adsorption and biodegradation of hydrocarbon contaminated ground water. Chem Eng J 347:415-423

43. Han Y, Ho WW (2021) Facilitated transport membranes for H_2 purification from coal-derived syngas: a techno-economic analysis. J Membr Sci 636:119549
44. Li P, Wang Z, Qiao Z, Liu Y, Cao X, Li W, Wang J, Wang S (2015) Recent developments in membranes for efficient hydrogen purification. J Membr Sci 495:130–168
45. Ansaloni L, Zhao Y, Jung BT, Ramasubramanian K, Baschetti MG, Ho WW (2015) Facilitated transport membranes containing amino-functionalized multi-walled carbon nanotubes for high-pressure CO_2 separations. J Membr Sci 490:18–28
46. Tu Z, Liu P, Zhang X, Shi M, Zhang Z, Luo S, Zhang L, Wu Y, Hu X (2021) Highly-selective separation of CO_2 from N_2 or CH_4 in task-specific ionic liquid membranes: facilitated transport and salting-out effect. Sep Purif Technol 254:117621
47. Kojabad ME, Babaluo A, Tavakoli AJS (2021) A novel semi-mobile carrier facilitated transport membrane containing aniline/poly (ether-block-amide) for CO_2/N_2 separation: molecular simulation and experimental study. Sep Purif Technol 266:118494
48. Janakiram S, Santinelli F, Costi R, Lindbråthen A, Nardelli GM, Milkowski K, Ansaloni L, Deng L (2021) Field trial of hollow fiber modules of hybrid facilitated transport membranes for flue gas CO_2 capture in cement industry. Chem Eng J 413:127405
49. He X (2021) Polyvinylamine-based facilitated transport membranes for post-combustion CO_2 capture: challenges and perspectives from materials to processes. Engineering 7(1):124–131
50. Yang L, Yang H, Wu H, Zhang L, Ma H, Liu Y, Wu Y, Ren Y, Wu X, Jiang Z (2021) COF membranes with uniform and exchangeable facilitated transport carriers for efficient carbon capture. J Mater Chem 9 (21):12636–12643

Chapter 2
Facilitated Transport Membranes (FTMs) Gas Transport Models and Reaction Mechanism

Syed Shujaat Karim, Takeshi Matsuura, Arshad Hussain, and Sarah Farrukh

Abstract The remarkable CO_2 capture capability of the facilitated transport membranes (FTMs) has captivated both academia and the commercial sectors, which have the potential to overcome the current trade-off constraint of the dense polymeric membranes that solely depends on the solution diffusion mechanism. Different synthetic carriers have been used in membranes by researchers for the rapid transport of selective gas molecules via the carrier mediation process, which demonstrate a similar mechanism of gas transport that occurs in natural biological membranes. However, it is difficult to understand the facilitated transport processes without a thorough theoretical investigation. Therefore, this chapter mainly focused on the comprehensive overview of the facilitated transport reaction mechanism, its chemistries, and gas transport models. In which facilitated transport reaction mechanisms are classified on the basis of different types and nature of carriers and their four main reaction chemistries are explored in detail. Furthermore, the gas transport model and its approximate analytical solutions of both mobile carriers (MC) and fixed-site carriers (FSC) for the facilitated transport of CO_2 gas molecules are also thoroughly discussed.

S. S. Karim · S. Farrukh (✉)
School of Chemical and Materials Engineering (SCME), National University of Sciences and Technology (NUST), Sector H-12, Islamabad, Pakistan
e-mail: sarah.farrukh@scme.nust.edu.pk

S. S. Karim
e-mail: sshujaat.che6scme@student.nust.edu.pk

T. Matsuura
Department of Chemical and Biological Engineering, University of Ottawa, 161 Louis Pasteur Street, Ottawa, ON K1N 6N5, Canada
e-mail: mastuura@uottawa.ca

A. Hussain
Department of Chemical, Mechanical, Materials, and Mining Engineering, Pak-Austria Fachhochschule, Institute of Applied Sciences and Technology, Haripur, KPK, Pakistan
e-mail: arshad.hussain@fcm3.paf-iast.edu.pk

S. Farrukh et al. (eds.), *Facilitated Transport Membranes (FTMs) for CO_2 Capture: Overview and Future Trends*, Green Energy and Technology,
https://doi.org/10.1007/978-3-031-21444-8_2

Keywords Facilitated transport · Reaction mechanisms · Gas permeation models · Reaction chemistries · Membrane separation · CO_2 capture

2.1 FTMs and Its Gas Transport Mechanism

Gas transport through polymeric membranes generally follows the solution-diffusion (S-D) mechanism [1]. In the S-D mechanism, gas molecules first dissolve in the membrane at the feed side, diffuse across the membrane due to the concentration gradient according to Fick's Law and then are desorbed at the permeate side under a lower pressure [2]. The key shortcoming of polymeric membranes is the trade-off relationship between permeability and selectivity, typically demonstrated by the Robeson's upper bound [3, 4]. In contrast, the membranes in which the gas transport takes place by facilitated transport in combination with the S-D mechanism are called facilitated transport membranes (FTMs). Herein, the "carrier" inside the membrane matrix additionally contributes to the transport of the selective molecules across the membrane [5]. Figure 2.1 shows the performance of FTMs that surpasses the Robeson upper bound line [6].

FTMs are based on a selective reversible reaction between the incorporated carrier agents and the desired permeating component. In such membranes, the carrier can either be covalently bonded to the polymer backbone as a fixed site carriers or freely move as a mobile carrier that is confined in the membrane matrix, which then reacts with the permeant, forming a complex and diffuses due to its concentration gradient across the membrane [7]. In case of fixed-site carrier, the permeant transport is

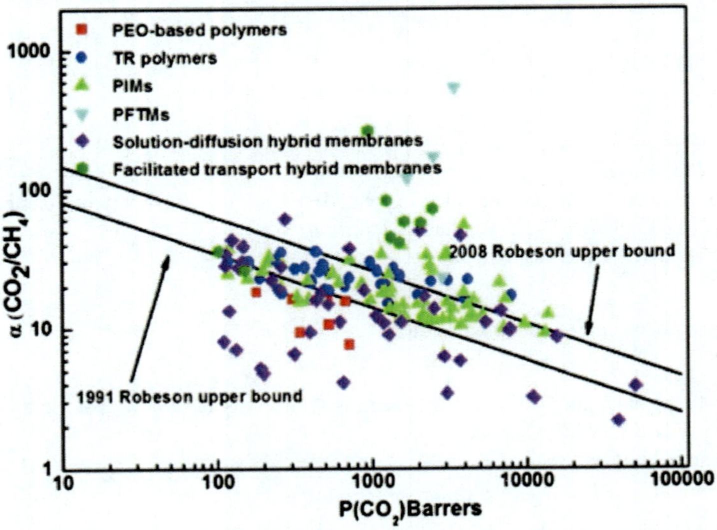

Fig. 2.1 Robeson's upper bound lines illustrating the CO_2/CH_4 selectivity data for different membrane materials versus CO_2 permeability (Reproduced from Ref. [6] with permission from Elsevier Inc.)

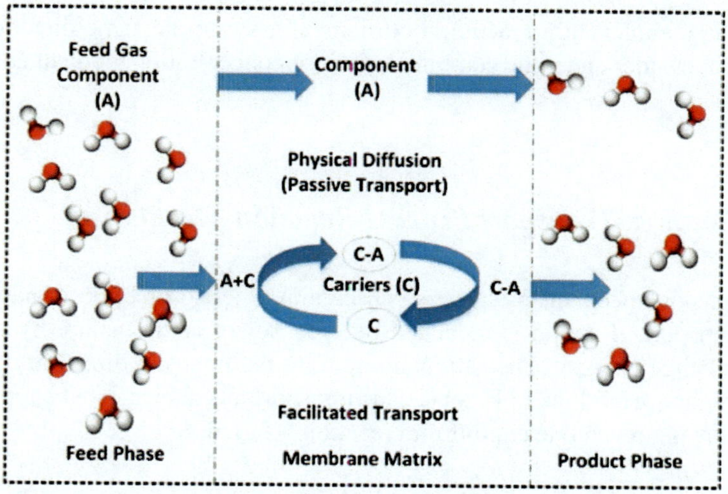

Fig. 2.2 Illustration of FTMs gas transport mechanism [14]

governed by the hopping mechanism [8]. Permeant is then released at the permeate side, regenerating the carrier, which can react with another permeant molecule on the feed side. Therefore, a large part of the permeant is transported by the carriers inside the membrane, in addition to the physical S-D mechanism (non-reactive molecular transport), which leads to the superior gas perm-selectivity of membrane [7–10]. The mechanism of the permeant transport in FTMs is schematically illustrated in Fig. 2.2. The general reversible reaction scheme in FTMs is represented in the Eq. (2.1) [11–13].

$$A(Component) + C(Carrier) \rightleftharpoons AC(solute - carrier\ complex) \tag{2.1}$$

When only one species is transported in FTMs, it is called the simple facilitated transport (SFT). Whereas, if more than one species are able to associate with the carrier, it is called coupled facilitated transport (CFT). In CFT there are two transport system reaction mechanisms, as discussed in the following sections [15, 16].

2.1.1 Co-transport Process Reaction Mechanism

This type of transport mechanism is observed when the carrier is electrically neutral. The feed phase is comprised of both cation and anion (A^{m+} and mX^-), which are reversibly associated with carrier (C). Their interactions with the carrier are represented by the following equilibrium reaction [15, 16].

$$\left(A^{m+} + mX^-\right) \text{“}Components\text{”} + C \text{“}Carrier\text{”} \rightleftharpoons \left(AC^{m+}, mX^-\right) \text{“}complexes\text{”} \tag{2.2}$$

After the complexation reaction, both ions migrate in the same direction (feed side to product side) under the combined effect of concentration gradient and carrier mediation.

2.1.2 *Counter-Transport Process Reaction Mechanism*

In this transport mechanism, carriers can be either acidic or basic in nature. The counter transport of cation is mediated by a carrier of acidic nature by a cation-proton exchange process. The cations migrate in the opposite direction of proton, from the higher pH to lower pH zone. This transport mechanism can be represented by the following reversible equilibrium reaction [15, 16].

$$A^{m+} + mHC \rightleftharpoons AC^{m+} + mH^{+} \tag{2.3}$$

whereas, the anion-based counter transport in the membrane matrix takes place by the interaction of the basic agent (anion B^-) and the carrier C^+ at the feed-membrane interface forming neutral species (B^-C^+). This neutral specie diffuses across the membrane and reaches the membrane-product interface, where the anion B^- is released due to the reversible reaction [15, 16]. The cationic carrier (C^+), interacts with another anion X^- present on the product side forming XC, which diffuses across the membrane but in the opposite direction of B^- towards the feed side. In this case the reaction balance can be expressed as:

$$B^{-} + XC \rightleftharpoons (B^{-}C^{+}) + X^{-} \tag{2.4}$$

This (counter transport) reversible reaction is governed by the interaction between the substrate and carrier, concentration gradient of anion X and electro-neutrality of the feed and product phase.

2.2 Types of Carrier Mediated FTMs Gas Transport Mechanism

In-order to elucidate the gas transport phenomena and separation process of FTMs, several different models have been proposed. These models are based on three main mechanisms as illustrated in Fig. 2.3 [16–19].

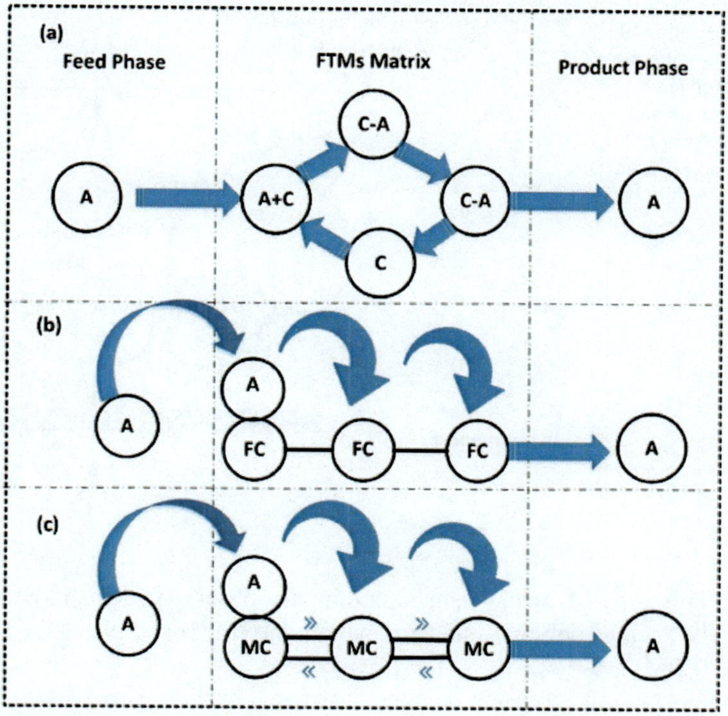

Fig. 2.3 Illustration of different FTMs gas transport mechanism. **a** Component transport by vehicle mechanism. **b** Component jumping on fixed sites mechanism. **c** Component jumping on mobile sites mechanism [16–19]

2.2.1 Component Transport by Vehicle Mechanism (Fig. 2.3a)

In this mechanism carrier serves as a "ferry boat" for the species A forming a complex (A-C) through a reversible reaction. The complex diffuses across the membrane towards the permeate side, where A is released. Then the carrier C migrates back to the feed side for capturing another species A. This type of mechanism is observed in the liquid based FTMs [16–19].

2.2.2 Component Jumping on Fixed Sites Mechanism (Fig. 2.3b)

In this mechanism, species A successively migrates across the membrane matrix by repeating the binding and unbinding with several interaction sites also known as the fixed site. One of such mechanisms is illustrated in Fig. 2.4, where the permeant CO_2 is bound and unbound with primary amine site attached to the polymer backbone

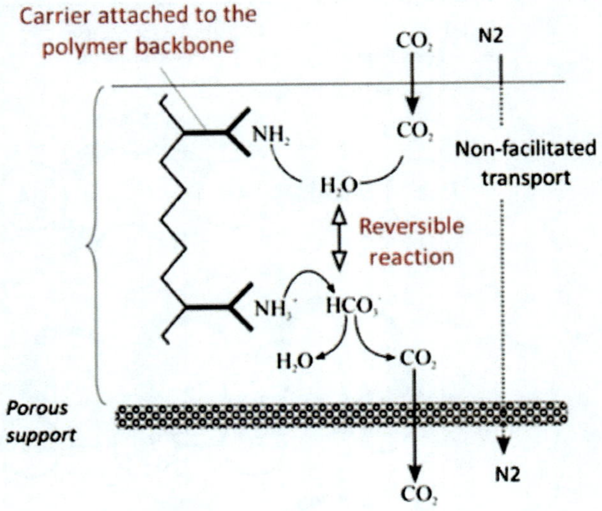

Fig. 2.4 Illustration of component jumping on facilitated fixed sites transport mechanism (Reproduced from Ref. [5] with permission from MDPI journals)

via water. This type of mechanism is commonly observed in Fixed Site Carrier (FSC) FTMs, such as polymer inclusion membrane (PIM) and plasticized polymer membrane (PPM) [16–19].

2.2.3 Component Jumping on Mobile Sites Mechanism (Fig. 2.3c)

In this mechanism, species A successively moves across the membrane matrix by binding to mobile site carriers present in the membrane matrix as depicted in Fig. 2.3c. This type of mechanism is commonly observed in mobile carrier (MC) FTMs such as supported liquid membranes (SLMs) [16–19].

2.3 Facilitated Transport Reaction (FTR) Chemistry

In a reversible process of FTMs system, carriers are usually associated with the desired permeant such as hydrogen (H_2), carbon dioxide (CO_2), carbon monoxide (CO), hydrogen sulfide (H_2S), oxygen (O_2) molecules etc. FTR theories can be applied on a wider range of desired permeants which include both molecules and ions. FTR theories are generally focused on a four (4) different types of reversible reaction as mentioned below [19].

2.3.1 Facilitated Proton Transport (FPT) Reaction

FPT reversible reaction is a kind of acid–base reaction that can be explained by Brønsted-Lowry theory. According to this theory, an acid is a species that dissociates to release a proton (H^+) while a base is a species that accepts the proton (H^+). However, a group of substances (e.g. water) which can act as an acid in one reaction and as a base in another reaction are called amphoteric. Brønsted-Lowry concept is useful in explaining proton transfer and applicable on a reversible reaction in a non-aqueous phase [19–21].

This reaction facilitates the transport of both proton (H^+) and small molecules having Brønsted acidity and basicity, such as carbon dioxide (CO_2), ammonia (NH_3), hydrogen sulfide (H_2S) etc. FPT reaction is a multi-step process between the membrane carrier and the desired permeant for the selective transport, therefore it is necessary to complete the cycle of FPT reversible reaction by releasing the permeant from the carrier due to the breakdown of temporary bond [19–21]. In this type of facilitated transport process, both forward and reverse reaction of the acid–base equilibrium process are involved in proton mediation. The proton transfer reaction can be generally represented by the following chemical equation [19–21].

$$HX + B^- \rightleftharpoons BH + X^- \tag{2.5}$$

where, HX and B^- are Brønsted acid and base pair, respectively. In the above equilibrium reaction, a conjugated base (X^-) is formed by the removal of a proton (H^+) from Brønsted acid (HX). Similarly, a conjugated acid (BH) is formed by the addition of H^+ to the Brønsted base (B^-) [19–21].

In proton exchange membranes, most commonly used membrane materials have a functional acidic group of sulfonic ($-SO_3H$) or phosphoric ($-PO_3H_2$), which requires a Brønsted base (e.g. amine or azole group etc.) to facilitate the transfer of proton. Water is, in general, an important proton carrier that may act both in the vehicle and the proton hopping mechanism. It also functions as a bridge for proton transport between an acid and a base when its mean free path distance exceeds the distance between the proton donating and receiving groups in an acid–base reaction. Another feature is that it helps in the dissociation of acidic groups and forms a continuous network of hydrogen-bonding for successful proton transfer process. Some proton exchange membrane (PEM) materials contain basic functional groups, such as polybenzimidazole (PBI). Those materials, having azole groups in particular, are also likely to function in an anhydrous condition, which is necessary for fuel cell membranes that operate at high temperature and low humidity conditions [19–23].

2.3.2 *Facilitated Transport Nucleophilic Addition Reaction (NAR)*

A nucleophilic addition reaction (NAR) is a type of chemical addition reaction, in which a nucleophile reacts with an electron deficient species (electrophile) having double or triple bond that breaks down to form a sigma bond. This type of reaction usually occurs at the carbon atom having double bond (e.g. C=O). In this reaction, electron cloud is displaced toward the oxygen atom thus making the positively charged carbon atom vulnerable to be reacted by a nucleophilic agent, for example: hydroxyl (OH^-) and amino group etc. In this way, NAR facilitates the transport of small molecules (e.g. carbon dioxide (CO_2) gas molecules) [19, 24, 25]. Typically, amines are often-used carriers in the development of CO_2 selective FTMs. These amines act as a Lewis base and can be covalently bonded to the polymer chains of the membrane materials. As CO_2 molecules contain an electron-deficient carbon center, which gives CO_2 molecules its reactivity to serve as an electrophile. Thus, CO_2 molecules act as a Lewis acid that can undergo a NAR process by reacting with a Lewis base.

For example, when primary amines reacts with CO_2 a zwitterion is formed as shown in the following equation [7, 26].

$$\mathrm{CO_2} + R - \mathrm{NH_2} \rightleftharpoons R - \mathrm{NH_2^+} - \mathrm{COO^-} \tag{2.6}$$

Then, intermolecular proton transfer rapidly occurs in the zwitterion, which results in the formation of carbamic acid having an acidic proton [7, 26].

$$R - \mathrm{NH_2^+} - \mathrm{COO^-} \rightleftharpoons R - \mathrm{NH} - \mathrm{COOH} \tag{2.7}$$

This carbamic acid is readily deprotonated by another amine to form a carbamate and ammonium ions [7, 26].

$$R - \mathrm{NH} - \mathrm{COOH} + R - \mathrm{NH_2} \rightleftharpoons R - \mathrm{NH} - \mathrm{COO^-} + R - \mathrm{NH_3^+} \tag{2.8}$$

This shows that, for a mole of CO_2 molecule, two moles of primary amine are required for the CO_2 transport process. Note that carbamate can be further hydrolyzed to produce a bicarbonate and a free amine, especially when pH is low. However, due to the good stability of carbamates, the hydrolysis reaction is not favorable in this process. The carbamate stability is usually enhanced by the rotational restriction of C–N bond surrounded by a bulky component, which hinders the hydrolyzation of carbamate [7, 26, 27].

Similar steps are followed by the secondary amine as shown by the equations,

$$\mathrm{CO_2} + R' - \mathrm{NH} - R'' \rightleftharpoons R'R'' - \mathrm{NH^+} - \mathrm{COO^-} \rightleftharpoons R'R'' - N - \mathrm{COOH} \tag{2.9}$$

$$R'R'' - N - \mathrm{COOH} + R' - \mathrm{NH} - R'' \rightleftharpoons R'R'' - N - \mathrm{COO}^- + R'R'' - \mathrm{NH}_2^+ \quad (2.10)$$

$$R'R'' - N - \mathrm{COO}^- + \mathrm{H_2O} \rightleftharpoons R' - \mathrm{NH} - R'' + \mathrm{HCO_3^-} \quad (2.11)$$

The last step, hydrolyzation of carbamate, is often prevented by steric hindrance due to the presence of bulky R′ and R″ groups. According to the above mechanism, the transfer of one mole of CO_2 requires only one mole of secondary amine, leading to higher capacity for CO_2 transfer. However, it should be noted that the sterically hindered amine also slows down CO_2 uptake by Eq. (2.9) due to rotational restriction and collision orientation [7, 26, 27].

As for the tertiary amine, it lacks the liable active proton (H^+), which prevents the formation of zwitterion and carbamate ion. Thus, tertiary amine cannot take part in the NAR process. Instead, water (H_2O) can act as a carrier by initially reacting with CO_2 to form carbonic acid (H_3CO_2) and then reacting with tertiary amine group [7, 26, 27].

$$\mathrm{CO_2} + \mathrm{H_2O} \rightleftharpoons \mathrm{H_2CO_3} \quad (2.12)$$

$$\mathrm{H_2CO_3} + R'R'' - N - R''' \rightleftharpoons \mathrm{H_2CO_3^-} + R'R'' - \mathrm{NH}^+ - R''' \quad (2.13)$$

In this case, one mole of tertiary amine is required for the transport of one mole of CO_2 molecule. However, due to slow carbonic acid formation, overall reaction kinetics of tertiary amine process is slow as compared to primary or secondary amine [7, 26, 27].

2.3.3 *Facilitated Transport Pi (π) Complexation Reaction*

This type of facilitated transport depends on the Pi (π) interaction bonding which occurs due to the attraction between the metal-ion (e.g. in MOFs with active metal sites) and the electron-rich (π) system. In case of CO_2 molecule, a chemical structure having asymmetric double bonds can connect the C and O atoms. A positively charged C atom and negatively charged O atoms result from the shift of electron cloud towards O atoms. This enables the CO_2 molecule for Pi (π) complexation reaction by forming a π bond with metal ions [28, 29].

According to Wang et al. [29], incorporation of Zn-poly(N-vinylimidazole)(PVI)@carbon nanotube (CNT) in the polymeric matrix of polyimide (PI) increased the CO_2 gas permeability of FTMs up-to 93% with CO_2/CH_4 selectivity of 71% at dry state condition. In these membranes, Zn^{2+} ions are bound on the CNT surface via association with imidazole groups. As there are more Zn^{2+} ions than imidazole groups, they can interact with CO_2 molecules in a reversible manner via complexation between the unoccupied s orbitals in Zn^{2+} ions and the delocalized π bond in CO_2 molecules. Thus, Zn^{2+} ion acts as carrier for the facilitated transport of CO_2 gas molecules in the membrane [29–32].

Similarly, Peng et al. [33], loaded PVI@CNT with four different type of divalent metal ions; copper (Cu^{2+}), ferrous iron (Fe^{2+}), calcium (Ca^{2+}) and magnesium (Mg^{2+}). These nano-particles were incorporated in the PI membrane as fillers for the enhanced CO_2/CH_4 separation. According to their study, alkaline earth metal ions such as Ca^{2+} and Mg^{2+} have a more stable extra-nuclear electron configuration, forming a strong coordination bond with the CO_2 gas molecules. Transition metal ions on the other hand, have a greater nuclear charge number, a smaller atomic radius, and a non-filled 3d orbital making their electron orbits more complex and enabling them more susceptible to a coordination reaction with CO_2 molecules than the alkaline earth group metal ions.

It indicates that transition metal ions have higher affinity with CO_2 and can serve as an effective CO_2 carrier as compared to the alkaline earth metal-ions for enhanced membrane-based CO_2 gas separation. The alkaline earth group metal ion based FTMs showed no or small enhancement of facilitated transport ability, while transition metal ions Cu^{2+} and Fe^{2+} based FTMs exhibited permeability increment of CO_2 gas up-to 89 and 87%, respectively. They suggested that higher electronegativity of metal ion does not necessarily mean higher FTMs performance. According to the recent studies, Zn^{2+} has the most suitable electronegativity for the facilitated transport of CO_2 gas molecules. The smaller electronegativity difference between the divalent metal ions and the Zn^{2+} metal ion imparts higher capability of facilitated transport of CO_2 gas molecules [33].

2.3.4 Facilitated Transport Via Electrochemical Reaction

Facilitated transport of CO_2 molecules can also occur through the electrochemical reaction. If the membrane materials have the characteristics of conducting both electrons and ions, the electrochemical reaction process, which generally occurs on both sides of the membrane, can be used to initiate the facilitated transport of CO_2 molecules. This kind of membranes usually operate at elevated temperature (e.g. above 600 °C) in a continuous manner. The Oxygen (O_2) Transport membranes (OTMs) are the best example of this type of membranes, as they are designed for the commercial production of both pure O_2 gas and synthesis gas. The main advantage of OTMs is the high O_2 flux due to the internal rapid oxide-ion and electron based facilitated transport, which leads to a considerably simpler and energy efficient reactor design [19, 34, 35].

Applying the similar principle of OTMs for the CO_2 selective membrane, two types of mixed conductor designs can be utilized, which are (a) mixed carbonate-ion and oxide-ion conductor (MOCC) and (b) mixed carbonate-ion and electron conductor (MECC). These membranes are suitable for the CO_2 separation of pre-combustion and post-combustion processes respectively. In the case of MOCC, CO_2 extraction from a syn-gas stream is preferable since CO_3^{2-} and O^{2-} electrochemical gradients occur in the opposite direction and CO_3^{2-} charge is balanced by O^{2-} in the membrane. On the other hand, MECC is better suited for CO_2 separation from a flue-gas stream because the electro-chemical gradients of CO_3^{2-} and electron (e^-) are present in the opposite direction, and CO_3^{2-} is charge-balanced by e^- rather than O^{2-} ions. These mixed conductor membranes can be considered as mobile-carrier membranes for the selective facilitated transport of CO_2 molecules [19, 35–37].

2.4 Gas Transport Model for FTMs

As discussed earlier in Sect. 2.1, FTMs utilize complexing agents (carriers) to form a reversible bond with a selective component (e.g. CO_2 molecules) of the feed gas mixture stream. This reversible reaction can be generally expressed by Eq. (2.1), in which three species: solute (A), carrier (C) and carrier-solute complex (AC) of the reaction process are co-exist. The gas transport through FTMs system occurs by the combination of carrier-mediated and S-D mechanism, which results in higher perm-selectivity towards the selective gas component (e.g. CO_2) as compared to other components of the gas mixture. The mass transfer of each component in FTMs system can be described by considering a general continuity equation for one dimensional transport of flat-plane geometry system [38].

$$\frac{dC_i}{dt} + v_i \frac{dC_i}{dx} = -\left(\frac{dJ_i}{dx}\right) - D_i + G_i' \tag{2.14}$$

where, C_i is the concentration, v_i is the convective flow velocity and J_i indicates the solute flux of component i. The dissipative and generative terms are expressed by D_i and G_i', respectively. Ignoring the convective flow term and at the steady state, a general chemical Eq. (2.1) of FTM system ($A + C \rightleftharpoons AC$) is taken into consideration to formulate the equation for each components of the FTMs system as expressed below:

(i) **For selective gas component (A)**

$$-\left({dJ_A}/{dx}\right) - k_f C_A C_C + k_r C_{AC} = 0 \tag{2.15a}$$

(ii) **For complexing agent as carrier (C)**

$$-\left({dJ_C}/{dx}\right) - k_f C_A C_C + k_r C_{AC} = 0 \tag{2.15b}$$

(iii) **For solute-carrier complex as product (AC)**

$$-\left({dJ_{AC}}/{dx}\right) + k_f C_A C_C - k_r C_{AC} = 0 \tag{2.15c}$$

where, k_f and k_r indicate the forward and reverse reaction rate constant. In-order to solve these equations, various boundary conditions are required, which can be given by the following assumptions.

- Only feed component (species) can diffuse to the permeate side.
- The carrier and its complexes are considered to remain confined within the FTMs matrix.
- The feed gas component concentration at the interface should be in equilibrium with the gas in bulk phase. If this condition is not applicable, external mass transfer resistance needs to be considered as reported by RD Noble et al. [39].

As for the carrier (C) and solute-carrier complex (AC), they are confined in the membrane matrix and their concentrations can be mathematically expressed by the following integral equation:

$$\int_0^L (C_C + C_{AC})dx = C_T L \tag{2.16}$$

In the above equation, the term C_T and L represent the initial concentration of carrier and thickness of membrane, respectively. As for determining the fluxes (J) of each component of the FTMs system, Fick's law is considered. The expression for the fluxes of solute (J_A), carrier (J_C) and solute-carrier complex (J_{AC}) are given by:

$$J_A = -D_A \frac{dC_A}{dx} \tag{2.17a}$$

$$J_C = -D_C \frac{dC_C}{dx} \tag{2.17b}$$

$$J_{AC} = -D_{AC} \frac{dC_{AC}}{dx} \tag{2.17c}$$

Substituting the fluxes in Eqs. (2.17a)–(2.17c) for those in Eqs. (2.15a)–(2.15c):

$$D_A \frac{d^2C_A}{dx^2} - k_f C_A C_C + k_r C_{AC} = 0 \tag{2.18a}$$

$$D_C \frac{d^2C_C}{dx^2} - k_f C_A C_C + k_r C_{AC} = 0 \tag{2.18b}$$

$$D_{AC} \frac{d^2C_{AC}}{dx^2} - k_f C_A C_C + k_r C_{AC} = 0 \tag{2.18c}$$

These differential equations are solved with the boundary conditions:

$$x = 0, \quad C_A = C_A^0, \tag{2.19a}$$

$$x = L, \quad C_A = C_A^L, \tag{2.19b}$$

And,

$$x = 0, \frac{\mathrm{d}C_{AC}}{\mathrm{d}x} = \frac{\mathrm{d}C_C}{\mathrm{d}x} = 0 \tag{2.19c}$$

$$x = L, \frac{\mathrm{d}C_{AC}}{\mathrm{d}x} = \frac{\mathrm{d}C_C}{\mathrm{d}x} = 0 \tag{2.19d}$$

For determining the total flux of solute, Eqs. (2.18a) and (2.18c), are added to obtain the following equation:

$$D_A \frac{\mathrm{d}^2 C_A}{\mathrm{d}x^2} + D_{AC} \frac{\mathrm{d}^2 C_{AC}}{\mathrm{d}x^2} = 0 \tag{2.20}$$

Then, by integrating Eq. (2.20) twice, Eq. (2.21) is obtained.

$$D_A C_A + D_{AC} C_{AC} = C_1 x + C_2 \tag{2.21}$$

where, both C_1 and C_2 are constants of integration, in which C_1 indicates the total flux of solute A. Similarly, from the sum of Eqs. (2.18b) and (2.18c) and by integrating it twice, the equation for total flux of complexing agent (carrier) is obtained.

$$D_A \frac{\mathrm{d}^2 C_C}{\mathrm{d}x^2} + D_{AC} \frac{\mathrm{d}^2 C_{AC}}{\mathrm{d}x^2} = 0 \tag{2.22}$$

$$D_C C_C + D_{AC} C_{AC} = C_3 x + C_4 \tag{2.23}$$

where, C_3 and C_4 are constants of integration, in which C_3 indicate the total flux of complexing agent (carrier).

Differentiating Eq. (2.23),

$$D_C \frac{\mathrm{d}C_C}{\mathrm{d}x} + D_{AC} \frac{\mathrm{d}C_{AC}}{\mathrm{d}x} = C_3 \tag{2.24}$$

From Eqs. (2.19c) and (2.19d),

$$D_C \frac{\mathrm{d}C_C}{\mathrm{d}x} + D_{AC} \frac{\mathrm{d}C_{AC}}{\mathrm{d}x} = 0 \text{ at } x = 0 \text{ and } L \tag{2.25}$$

Since $D_C \frac{\mathrm{d}C_C}{\mathrm{d}x} + D_{AC} \frac{\mathrm{d}C_{AC}}{\mathrm{d}x}$ is a continuous function of x at $0 \le x \le L$, C_3 should be equal to 0.

Therefore,

$$D_C C_C + D_{AC} C_{AC} = C_4 \quad (2.26)$$

Mobile carrier (MC) and fixed-site (FSC) FTM systems are the two most common forms of carrier-based FTM systems. In Fig. 2.5 that depicts various modes of solute transport, the center represents, MC FTM, where reactive carriers can freely move and actively transport specific gas component across the membrane through reversible reaction. The reactive carriers for carbon sequestration can be amine group, carbonate and carboxylate group etc. These (MC) FTMs are also known as support liquid membrane (SLM) or immobilized liquid membrane (ILM). Whereas, in FSC FTMs (Fig. 2.5 left) reactive carriers are held fixed by covalent linkage in the polymeric membrane matrix. The fixed-site carriers can only vibrate in a confined space, while waiting for selective gas molecules (e.g. CO_2) to arrive for the transport.

According to the second assumption, the mobile carrier and solute-carrier complex are confined to the membrane matrix. If their diffusivities ($D_C = D_{AC} = D$) are considered to be equal, the Eq. (2.26) can be simplified as:

$$D(C_C + C_{AC}) = C_4 \quad (2.27)$$

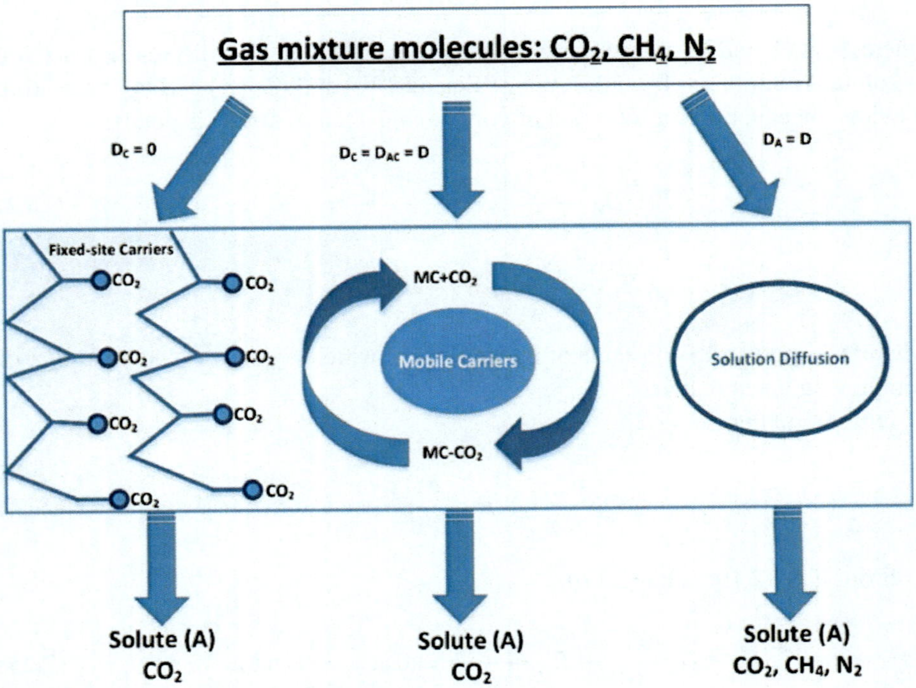

Fig. 2.5 Illustration of different gas transport mechanisms that occur inside the FTMs [40]

Hence,

$$C_C + C_{AC} = {}^{C_4}\!\big/_{D} = C_5 \tag{2.28}$$

where, $C_5 = C_T$ from Eq. (2.16), therefore,

$$C_C + C_{AC} = C_T \tag{2.29}$$

In the case of FSC FTMs, the diffusion coefficient (D_C) can be considered as zero ("0") as there is no mobility of carrier in the facilitated transport process. While (D_{AC}) is the effective diffusion coefficient that is associated with the jumping mechanism of solute (A) as discussed in Sect. 2.2.2. Solute is transported in a facilitated transport process using a carrier mediation mechanism in combination with an S-D mechanism (see Fig. 2.5 right). As a result, the sum of the fluxes of these two mechanisms equals the total flux of the selective solute (permeant), as expressed below:

$$J_A + J_{AC} = -D_A\frac{\mathrm{d}C_A}{\mathrm{d}x} - D_{AC}\frac{\mathrm{d}C_{AC}}{\mathrm{d}x} \tag{2.30}$$

The term facilitation factor (F) was introduced to compare the effects of facilitated transport reactions to the S-D mechanism, which measures the improved selectivity that comes with selective facilitated transport of the key component of a gas mixture. It is the ratio of solute (A) total flux within the membrane to the S-D flux as expressed below:

$$F = \frac{J_A}{J_A^{(S\text{-}D)}} = \frac{-D_A\frac{\mathrm{d}C_A}{\mathrm{d}x} - D_{AC}\frac{\mathrm{d}C_{AC}}{\mathrm{d}x}}{-D_A\frac{\mathrm{d}C_A}{\mathrm{d}x}} \tag{2.31}$$

If no reaction occurs between the key component (solute A) and the carrier, then F is equal to 1. This implies that the transport process is only dependent on the S-D mechanism.

The general formulation of FTM process for gas transport is described in the set of differential equations from Eqs. (2.18a)–(2.18c), which express a steady-state non-linear diffusion reaction problem. These equations can be simply solved if some conditions are satisfied, which include the assumption of chemical equilibrium to be reached among the species within the FTMs matrix and there is negligible effect of chemical reaction in the pure S-D process. Olander [41] used these assumptions to solve the facilitation transport equations. In his work, chemical equilibrium constant was expressed as:

$$K_{eq} = \frac{C_{AC}}{C_A C_C} \tag{2.32}$$

In the case of MC FTMs, both the carrier and its complex diffusivity are considered equivalent. As a result, by using Eq. (2.32) as a chemical equilibrium constant, the facilitation factor can be rewritten as:

$$F = 1 + \frac{K_{eq} C_T}{(1 + K_{eq} C_A^0)(1 + K_{eq} C_A^L)} \frac{D_{AC}}{D_A} \tag{2.33}$$

This approximation of the MC-based facilitated transport system can also be validated by examining the biological process of oxygen (O_2) transport through hemoglobin, which acts as a carrier in this facilitated transport process. Both the carrier and the solute-carrier complex are very similar in molar volume, justifying this approximation for these facilitated transport systems [38, 42].

For FSC FTM systems, Belaissaoui et al. [5], investigated the CO_2 facilitated transport effect through hybrid Poly (Allyl Amine) membrane. In their work, both reaction rate constant (k_f and k_r), are considered as concentration independent and the equilibrium rate constant can be expressed as:

$$K_{eq} = {k_f}/{k_r} \tag{2.34}$$

In this case, excess of carrier is assumed and its concentration is considered as constant [43, 44]. By substituting Eq. (2.34) in the differential equations from Eqs. (2.18a)–(2.18c), for FSC FTMs the one dimensional steady equations can be expressed in x-direction as:

$$D_A \frac{d^2 C_A}{dx^2} = k_f \left(C_A C_C - \frac{1}{K_{eq}} C_{AC} \right) \tag{2.35a}$$

$$D_{AC} \frac{d^2 C_A}{dx^2} = -k_f \left(C_A C_C - \frac{1}{K_{eq}} C_{AC} \right) \tag{2.35b}$$

$$C_C = \frac{C_T}{1 + K_{eq} C_{AO}} \tag{2.35c}$$

Here, the same boundary conditions were considered as mentioned in Eqs. (2.19a)–(2.19d). These differential equations from Eqs. (2.35a)–(2.35c) can be expressed in a dimensionless form as:

$$\frac{d^2 C_A^*}{dx^*} = \frac{\alpha_m K}{\varepsilon} \left(C_A^* C_C^* - \frac{1}{K} C_{AC}^* \right) \tag{2.36a}$$

$$\frac{d^2 C_{AC}^*}{dx^*} = -\frac{K}{\varepsilon} \left(C_A^* C_C^* - \frac{1}{K} C_{AC}^* \right) \tag{2.36b}$$

$$C_C^* = \frac{1}{1 + K} \tag{2.36c}$$

These differential equations are solved with the boundary conditions as mentioned below:

$$x^* = 0, C_A^* = 1, \frac{dC_{AC}^*}{dx^*} = 0 \tag{2.37a}$$

$$x^* = 1, C_A^* = \frac{C_A^L}{C_A^0}, \frac{dC_{AC}^*}{dx^*} = 0 \tag{2.37b}$$

where, $x^* = x/L$ and $C_A^* = \frac{C_A}{C_A^0}$. α_m is the mobility ratio between the carrier and solute, given by $D_{AC}C_T/D_A C_A^0$. ε and K represent, Inverse Damkohler number, and dimensionless reaction equilibrium constant, and they are respectively,

$$\varepsilon = \frac{D_{AC}}{k_r e^2} \tag{2.38}$$

$$K = \frac{k_f C_A^0}{k_r} \tag{2.39}$$

where, e indicates the selective layer thickness of the membrane. Further, C_A^0, the solute concentration at the feed side of the membrane is given by:

$$C_A^0 = k_{D,A} p'_A \tag{2.40}$$

where, p'_A the feed partial pressure of the solute A (e.g. CO_2). To solve the above differential equations of FSC FTM systems, Smith and Quinn [44] developed an analytical solution based on two main assumptions: the excess of carrier and the partial pressure of permeant component (e.g., CO_2 gas molecules) should be zero at downstream. In this analytical model, the facilitation factor (F) is given by:

$$F = \frac{1 + \frac{\alpha_m K}{1+K}}{1 + \frac{\alpha_m K}{1+K}\left(\frac{\tanh\lambda}{\lambda}\right)} \tag{2.41}$$

In this Eq. (2.41), $\left(\frac{\tanh\lambda}{\lambda}\right)$ represents the measure of facilitation effect [38], where λ is:

$$\lambda = \frac{1}{2}\sqrt{\frac{1 + (\alpha_m + 1)K}{\varepsilon(1 + K)}} \tag{2.42}$$

The maximum of "F" can be obtained as λ approaches infinity, indicating the reaction equilibrium. Whereas, when λ approaches zero, "F" approaches unity, showing reaction kinetics limitations. Furthermore, at the reaction equilibrium Eq. (2.41) can be rewritten as:

$$F = 1 + \frac{\alpha_m K}{1 + K} \tag{2.43}$$

By assuming an approximation of chemical equilibrium across the FTM matrix and modifying Eq. (2.30), the total flux of solute A (e.g. CO_2) gas molecules can be computed by the following equation [45]:

$$J_A = D_A \frac{k_{D,A}}{e}\left(p'_A - p''_A\right) + D_{AC} K_{eq} C_T \frac{k_{D,A}}{e}\left[\frac{p'_A}{1 + K_{eq} k_{D,A} p'_A} - \frac{p''_A}{1 + K_{eq} k_{D,A} p''_A}\right] \tag{2.44}$$

where, p'_A and p''_A are the feed and product partial pressure of the solute A, respectively. The facilitated factor (F) is given by:

$$F = 1 + \frac{D_{AC} K_{eq} C_T}{D_A\left(p'_A - p''_A\right)}\left[\frac{p'_A}{1 + K_{eq} k_{D,A} p'_A} - \frac{p''_A}{1 + K_{eq} k_{D,A} p''_A}\right] \tag{2.45}$$

At $p''_A = 0$ and $D_{AC} = D_A$, Eq. (2.45) reduces to Eq. (2.43).

Here, the factor "F" defines the physiochemical properties of the FTM system. Therefore, these models can be used to explain the real transport behaviour of the FTM.

2.5 Conclusion and Future Recommendations

The FTM, in particular, demonstrated high gas separation performance due to the active transport mechanism, similar to the biological natural membrane. However, understanding the reversible reactions that take place during the facilitated transport process requires deep theoretical analysis. Therefore, in this chapter, different types of facilitated transport reaction mechanisms, their reversible reaction chemistries, and gas transport models with the approximate analytical solutions for both mobile and fixed-site carrier systems are thoroughly discussed. In the first section, given the nature of carriers, two general types of facilitated transport reaction mechanisms are discussed, which are co-transport and counter-transport reaction mechanisms, along with three main mechanisms of the facilitated transport process of both MC and FSC-based FTMs, which include: vehicle mechanism, component hopping on fixed-site and mobile carriers, are comprehensively explained. To support these mechanisms, four different facilitated transport reaction (FTR) chemistry theories are briefly described in the second section, which includes the facilitated proton transport reaction, the nucleophilic addition reaction, the Pi complexation reaction, and the electro-chemical reaction. In the last section, the gas transport model based on the continuity equation has been discussed in detail, along with its approximate analytical solutions for both MC and FSC based FTM, which revealed that the theoretical analysis of transport mechanisms with their reaction chemistries and the mathematical modelling method are essential for understanding FTM-based CO_2 separation process and predicting operating conditions in order to optimize it for high CO_2 separation performance. Some of the research areas that need to be considered for the further FTM-based gas separation process are mentioned below:

- For enhanced carrier reactivity, facilitated transport revisable reaction chemistries needs to be further explored and diversified.
- Fundamentals of inter layer diffusion process in composite based FTMs need to be further evaluated.
- The complex mechanism of facilitated transport membranes need to be further investigated to further improve the perm-selectivity of the membrane.
- A more innovative approach to the problem of carrier stability in FTM is required.
- For a deeper theoretical understanding of FTM process, advanced simulation needs to be develop.

References

1. Luis P, Van Gerven T, Van der Bruggen B (2012) Recent developments in membrane-based technologies for CO_2 capture. Prog Energy Combust Sci 38(3):419–448
2. Karim SS, Farrukh S, Hussain A, Noor T, Younas M (2021) A comprehensive overview of dual-layer composite membrane for air (O_2/N_2) separation. Polym Polym Compos 29(9):S1630–S1640
3. Robeson LM (2008) The upper bound revisited. J Membr Sci 320(1–2):390–400
4. Freeman BD (1999) Basis of permeability/selectivity tradeoff relations in polymeric gas separation membranes. Macromolecules 32(2):375–380
5. Belaissaoui B, Lasseuguette E, Janakiram S, Deng L, Ferrari M-C (2020) Analysis of CO_2 facilitation transport effect through a hybrid poly (allyl amine) membrane: pathways for further improvement. Membranes 10(12):367
6. He X, Lei L, Chu Y (2018) Facilitated transport membranes for CO_2 removal from natural gas. In: Current trends and future developments on (bio-) membranes. Elsevier, pp 261–288
7. Han Y, Ho WW (2020) Recent advances in polymeric facilitated transport membranes for carbon dioxide separation and hydrogen purification. J Polym Sci 58(18):2435–2449
8. Tong Z, Ho WW (2017) Facilitated transport membranes for CO_2 separation and capture. Sep Sci Technol 52(2):156–167
9. Cussler E, Aris R, Bhown A (1989) On the limits of facilitated diffusion. J Membr Sci 43(2–3):149–164
10. Noble RD (1992) Generalized microscopic mechanism of facilitated transport in fixed site carrier membranes. J Membr Sci 75(1–2):121–129
11. Schultz JS, Goddard JD, Suchdeo SR (1974) Facilitated transport via carrier-mediated diffusion in membranes: Part I. Mechanistic aspects, experimental systems and characteristic regimes. AIChE J 20(3):417–445
12. Goddard JD, Schultz JS, Suchdeo SR (1974) Facilitated transport via carrier—mediated diffusion in membranes: Part II. Math Aspects Anal AIChE J 20(4):625–645
13. Ferraz H, Duarte L, Di Luccio M, Alves T, Habert A, Borges C (2007) Recent achievements in facilitated transport membranes for separation processes. Braz J Chem Eng 24(1):101–118
14. Kilner JA, Skinner SJ, Irvine S, Edwards P (2012) Functional materials for sustainable energy applications. Elsevier

15. Kislik VS (2009) Liquid membranes: principles and applications in chemical separations and wastewater treatment. Elsevier
16. Eljaddi T, Laurent L, Miloudi H (2017) Review on mechanism of facilitated transport on liquid membranes. J Membr Sci Res 3(3):199–208
17. Hassoune H, Rhlalou T, Verchère J-F (2009) Mechanism of transport of sugars across a supported liquid membrane using methyl cholate as mobile carrier. Desalination 242(1–3):84–95
18. White KM, Smith BD, Duggan PJ, Sheahan SL, Tyndall EM (2001) Mechanism of facilitated saccharide transport through plasticized cellulose triacetate membranes. J Membr Sci 194(2):165–175
19. Li Y, Wang S, He G, Wu H, Pan F, Jiang Z (2015) Facilitated transport of small molecules and ions for energy-efficient membranes. Chem Soc Rev 44(1):103–118
20. Caldin EF, Gold V (2013) Proton-transfer reactions. Springer, US
21. Brown TL, LeMay HE, Bursten BE, Brunauer LS (1997) Chemistry: the central science. Prentice Hall Englewood Cliffs, NJ, vol 13
22. Yang Z, Coutinho D, Feng F, Ferraris JP, Balkus K Jr (2004) Novel inorganic/organic hybrid electrolyte membranes. Prepr Pap-Am Chem Soc Div Fuel Chem 49(2):599
23. Gros G, Moll W, Hoppe H, Gros H (1976) Proton transport by phosphate diffusion—a mechanism of facilitated CO_2 transfer. J Gen Physiol 67(6):773–790
24. Michael BS (2007) March's advanced organic chemistry: reactions, mechanisms, and structure. Wiley, Hoboken
25. Jerry M (1992) Advanced organic chemistry: reactions, mechanisms and structure. A Wiley-Interscience Publication
26. Han Y, Ho WW (2019) Recent developments on polymeric membranes for CO_2 capture from flue gas. J Polym Eng 10(11):365
27. Danckwerts P (1979) The reaction of CO_2 with ethanolamines. Chem Eng Sci 34(4):443–446
28. Luna-Triguero A, Sławek A, Sánchez-de-Armas R, Gutiérrez-Sevillano JJ, Ania CO, Parra JB, Vicent-Luna JM, Calero S (2020) π-Complexation for olefin/paraffin separation using aluminosilicates. Chem Eng J 380:122482
29. Wang S, Liu Y, Zhang M, Shi D, Li Y, Peng D, He G, Wu H, Chen J, Jiang Z (2016) Comparison of facilitated transport behavior and separation properties of membranes with imidazole groups and zinc ions as CO_2 carriers. J Membr Sci 505:44–52
30. Li F, Li Y, Chung T-S, Kawi S (2010) Facilitated transport by hybrid POSS®–Matrimid®–Zn^{2+} nanocomposite membranes for the separation of natural gas. J Membr Sci 356(1–2):14–21
31. Wang Y, Yang FH, Yang RT, Heinzel JM, Nickens AD (2006) Desulfurization of high-sulfur jet fuel by π-complexation with copper and palladium halide sorbents. Ind Eng Chem Res 45(22):7649–7655
32. Li Y, Chung TS (2008) Highly selective sulfonated polyethersulfone (SPES)-based membranes with transition metal counterions for hydrogen recovery and natural gas separation. J Membr Sci 308(1–2):128–135
33. Peng D, Liu Y, Wang S, Tian Z, Xin Q, Wu H, Chen J, Jiang Z (2016) Facilitated transport membranes by incorporating different divalent metal ions as CO_2 carriers. RSC Adv 6(69):65282–65290
34. Wei Y, Yang W, Caro J, Wang H (2013) Dense ceramic oxygen permeable membranes and catalytic membrane reactors. Chem Eng J 220:185–203
35. Zhang L, Xu N, Li X, Wang S, Huang K, Harris WH, Chiu WK (2012) High CO_2 permeation flux enabled by highly interconnected three-dimensional ionic channels in selective CO_2 separation membranes. Energy Environ Sci 5(8):8310–8317
36. Sherman SR, Gray JR, Brinkman KS, Huang K (2012) Combustion-assisted CO_2 capture using MECC membranes. J Membr Sci 401:323–332
37. Xu N, Li X, Franks MA, Zhao H, Huang K (2012) Silver-molten carbonate composite as a new high-flux membrane for electrochemical separation of CO_2 from flue gas. J Membr Sci 401:190–194

38. Rea R, De Angelis MG, Baschetti MG (2019) Models for facilitated transport membranes: a review. Membranes 9(2):26
39. Noble RD, Way JD, Powers LA (1986) Effect of external mass-transfer resistance on facilitated transport. Ind Eng Chem Fundam 25(3):450–452
40. Rafiq S, Deng L, Hägg MB (2016) Role of facilitated transport membranes and composite membranes for efficient CO_2 capture—a review. ChemBioEng Rev 3(2):68–85
41. Olander DR (1960) Simultaneous mass transfer and equilibrium chemical reaction. AIChE J 6(2):233–239
42. Teramoto M (1994) Approximate solution of facilitation factors in facilitated transport. Ind Eng Chem Res 33(9):2161–2167
43. Ansaloni L, Zhao Y, Jung BT, Ramasubramanian K, Baschetti MG, Ho WW (2015) Facilitated transport membranes containing amino-functionalized multi-walled carbon nanotubes for high-pressure CO_2 separations. J Membr Sci 490:18–28
44. Smith DR, Quinn JA (1979) The prediction of facilitation factors for reaction augmented membrane transport. AIChE J 25(1):197–200
45. Janakiram S, Espejo JLM, Yu X, Ansaloni L, Deng L (2020) Facilitated transport membranes containing graphene oxide-based nanoplatelets for CO_2 separation: effect of 2D filler properties. J Membr Sci 616:118626

Chapter 3
Fabrication, Characterization, and Design of Facilitated Transport Membranes (FTMs)

Zarrar Salahuddin, Syed Shujaat Karim, Kiran Mustafa, and Sarah Farrukh

Abstract Facilitated transport membranes (FTMs) are an emerging class of membrane materials which exploit the various mechanisms of gas permeation and separation to improve the performance of membranes for gas separation. In this chapter, we study the different fabrication techniques used to synthesize membranes as well as carrier particles. Special focus is laid on the advanced and emerging technologies. Characterization techniques used to assess the properties of both FTMs and carrier media are also studied in detail, and some advancement in characterization techniques have also been outlined throughout the course of time. The development in fabrication and testing techniques to measure the gas separation performance are of prime focus. Furthermore, special attention is also given to the current trends and from those trends, a roadmap of the future is drawn. As a whole, this chapter outlines the entire path of development of membrane technology and the role that FTMs play in furthering this role in time to come.

Keywords Facilitated transport · Composite membrane · Porous support · Carrier particles · Gas separation performance

Z. Salahuddin · S. S. Karim · S. Farrukh (✉)
School of Chemical and Materials Engineering (SCME), National University of Sciences and Technology (NUST), Sector H-12, Islamabad, Pakistan
e-mail: sarah.farrukh@scme.nust.edu.pk

Z. Salahuddin
e-mail: zarrar.phdscme@student.nust.edu.pk

S. S. Karim
e-mail: sshujaat.che6scme@student.nust.edu.pk

K. Mustafa
Department of Chemistry, The Women University Multan, Multan, Pakistan

S. Farrukh et al. (eds.), *Facilitated Transport Membranes (FTMs) for CO_2 Capture: Overview and Future Trends*, Green Energy and Technology,
https://doi.org/10.1007/978-3-031-21444-8_3

3.1 Introduction

Over recent decades, conventional techniques for various industrial processes have quickly been replaced by membrane technology. As far as processes involving gas treatments are concerned, there has been a surge in research output relating to gas separation membranes, which are serving as the way forward for industrial gas treatment processes. Efficiency of these membranes is measured by the positive trade-off between selectivity and permeability. A gas separation membrane is only considered efficient, if it provides high selectivity of separation for one gas over another, while also giving good permeability values. The effectiveness of such membranes is in turn affected by the material selected for their preparation as well as by the method of fabrication being used. Different fabrication methods can give rise to isotropic dense or anisotropic membranes, which have varying properties [1].

As described earlier, fabrication techniques have a very important role to play in deciding the effectiveness of a membrane for gas separation processes, so we will also discuss in detail, some of the most essential fabrication techniques used for gas separation membranes. Some of the techniques for gas separation membranes include solution casting [2], phase inversion [3], dip coating [4] and spin coating [5] among several others.

Initially, all membranes were fabricated as flat sheets either by pouring solution into petri dishes [6] or by using a casting knife or doctor blade [7]. However, with the advent of newer technology, more advanced membrane modules were synthesized as well. Membrane modules are also prepared in tubular geometry, most notably in the form of hollow fiber modules [8]. The question might arise as to why new membrane geometries needed to be fabricated. The reason for that is that in the industry, large quantities of gas needs to be treated, which in turn would require large membrane areas. In such circumstances, using flat sheet membranes would mean having to allocate very large physical area for the membrane modules, however when using hollow fiber modules, large membrane area can be packed into smaller volumes as hollow fiber membranes have large cross-sectional area per unit volume (specific area). Apart from that, hollow fiber membranes also allow for easier cleaning as there is more possibility of back-flushing. Also, there is less requirement for pre-treatment of feed gas mixtures and less requirement for membrane maintenance [9].

Also, in the initial stages of development of gas separation membranes, various fabrication techniques were only focused on the fabrication of membranes using pure polymeric materials. One such work carried out by Tang et al. demonstrates the use of cellulose acetate fabricated by simple solution casting using various solvent mixtures [10]. It was seen that this method produced isotropic dense membranes. Another work mentioned earlier produced anisotropic, asymmetric membranes and was carried out by Loeb and Sourirajan. In fact, this was the first instance of successful fabrication of asymmetric membranes [3]. However, it was soon realized that purely polymeric membranes cannot fulfil the gas separation requirements of the current and future industrial systems, therefore new materials and methods for fabrication of gas separation membranes needed to be developed [11].

In light of these emerging challenges, completely new classes of membranes started to emerge. One such class of gas separation membranes is called *mixed matrix membranes*. Mixed matrix membranes (MMMs) are made by the combination of inorganic material in the form of micro or nano-particles as discrete phase incorporated in a polymeric matrix, which acts as a continuous phase [12]. MMMs comprise the advantages of both polymeric matrices, relating to processability, robustness and scalability with the advantages of more enhanced and superior gas separation performance provided by inorganic fillers. Therefore, MMMs can overcome some of the problems encountered by pure polymeric membranes, and can be used to enhance both the permeability and selectivity. Some fillers used for the preparation of MMMs include metal–organic frameworks (MOFs), zeolites, carbons [13] as well as zeolitic imidazolate frameworks (ZIFs) [14].

Until now, all membranes that we have discussed, whether they are pure polymeric membranes or MMMs, are single-layer isotropic dense membranes. Such membranes have been observed to provide high selectivity, especially in the case of MMMs, but these single-layer isotropic, dense membranes do not provide suitably high permeability [15, 16]. Therefore, researchers were focused on tackling this newfound problem. This gave rise to another class of gas separation membranes called *composite membranes*. Composite membranes provide the main advantage of exceptionally higher flux as compared to isotropic, dense membranes [17]. Composite membranes are all such membranes which consist of at least two structural elements. They can be either single-layer or multi-layer composite membranes. Single-layer composite membranes consist of a single thin, selective layer on top of a porous support. In this type of membrane, the gas separation is carried out by the thin, selective layer, whereas the porous support only provides mechanical strength. In a multi-layer composite membrane, there can be multiple thin, selective layers and the bottom layer would be the porous support which would serve the same purpose here as it would in the single-layer composite membranes, Composite membranes can be prepared by coating the porous support with the dense, thin, selective layer through dip coating, interfacial polymerization or plasma polymerization among other methods [18].

Research is perpetually underway to find more and more advanced and efficient gas separation membranes. Of the more recent research endeavours being undertaken in this field, there is one aspect which has been explored extensively as it makes use of existing technologies in a unique manner. This exploration has led to the development of a new category of gas separation membranes called *facilitated transport membranes*. Facilitated transport membranes (FTMs) are much like mixed matrix membranes, except that they incorporate reactive fillers within polymer matrices, which react with active gases such as CO_2, H_2S or SO_2, within the gas mixture and thus provide appreciably higher permeability and selectivity to one component of the gas mixture [19].

To fully understand and appreciate the effectiveness in gas separation brought about the phenomenon of facilitated transport, we must analyze how facilitate transport differs from the mechanism of gas permeation through ordinary gas separation membranes including MMMs. In most gas separation membranes, the mechanism at play for gas permeation is the *solution-diffusion mechanism.* In this mechanism, the permeating gas dissolves into the polymer at the feed side, then diffuses through the polymeric membrane and desorbs at the permeate side [20]. Therefore, this mechanism has no reaction occurring between the permeating gas and the membrane materials (polymer or filler). On the other hand, facilitated transport is dependent on a chemical reaction occurring between an active carrier within the membrane and the gas of interest. Once, this reaction has taken place, the carrier takes the gas of interest across the cross-section of the membrane at a much faster rate while inhibiting the permeation of all other components of the gas mixture [21]. As facilitated transport membranes are an emerging type of gas separation membranes, therefore this chapter focuses on this field of gas separation membranes.

In all these different types of gas separation membranes, we have thus far studied the various types of membranes being used and their respective advantages and drawbacks. However, there is one other major aspect which influences the effectiveness of gas separation membranes and that is the material selection for the fabrication of the membranes. Certain criteria points are of utmost importance when selecting a material to fabricate gas separation membranes. Some of the main criteria points include [22]:

- The membrane material must have high selectivity for the required component of the gas mixture.
- In case of composite membranes, all the layers must have good adhesion and compatibility.
- It should have high permeability towards the required component of the gas mixture.
- It should be physically and chemically stable.
- It should have a long shelf life.
- It should have fouling resistance and plasticization resistance.
- It should have high mechanical strength.
- It should be cost-effective and should be easily made into a variety of membrane modules.

All these criteria points are critical for the proper, long term functioning of any membrane material, and so must be of prime focus even before the fabrication methods are chosen for the synthesis of membrane modules.

3.2 Fabrication Techniques for FTMs

As discussed earlier, the fabrication technique used for FTMs greatly influences the effectiveness of the performance of the membrane. So, it is imperative to study the various fabrication techniques being used for the fabrication of FTMs.

3.2.1 Solution Casting

Solution casting is one of the most commonly used method for the fabrication of FTMs on lab scale. Depending on whether the FTM is in the form of a composite membrane or an isotropic MMM, different equipment may be used to carry out solution casting as illustrated in Fig. 3.1. In the case of a composite membrane, firstly a homogeneous solution is prepared which contains requisite amounts of the polymer which will serve as the main matrix of the membrane and inorganic filler particles also evenly dispersed within the solution. This homogeneous solution is then evenly distributed onto the surface of an already prepared porous support by using an automatic or manual membrane caster, which in turn consists of a doctor blade [23]. In the case of an isotropic MMM, a homogeneous solution is first prepared containing the correct amount of polymer dissolved thoroughly in solvent, and also containing the required amount of filler properly dispersed within the solvent. The final solution is then cast onto a petri dish and either dried in a vacuum oven at elevated temperature (for high boiling solvent) or left to stand for long period of time in the atmosphere for the solvent to evaporate [24, 25].

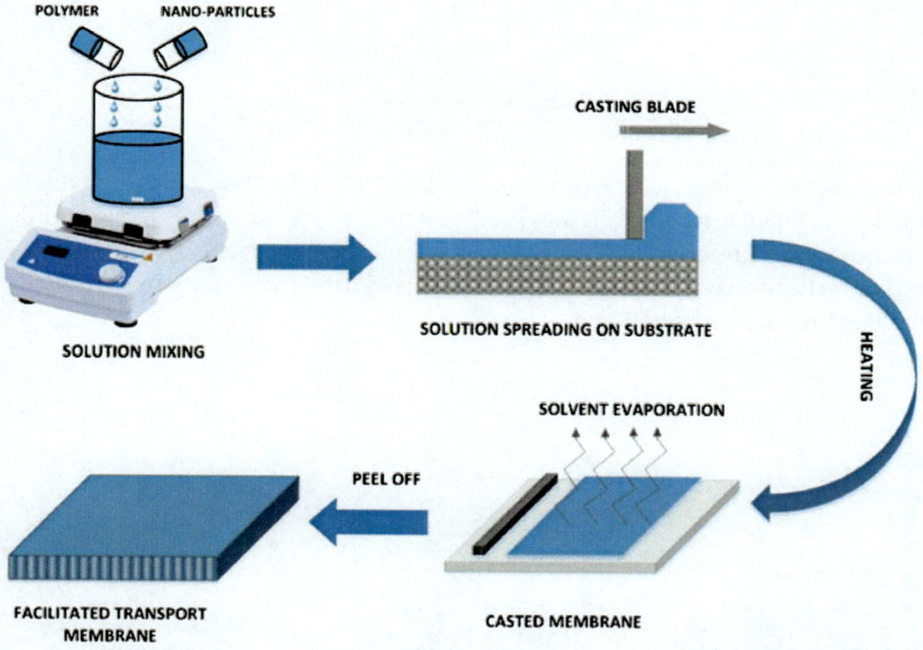

Fig. 3.1 Schematic diagram of solution mixing and casting method [22, 26]

3.2.2 Dip Coating

Dip coating is another commonly used and fairly simple technique used for the fabrication of composite FTMs as demonstrated in Fig. 3.2. In this technique, porous support is dipped into a solution containing polymer and filler which will ultimately become the thin, selective top layer of the composite membrane. After dipping the porous support into the solution for a few seconds, the final product is then lifted out of the solution at a controlled rate. The sample is then left to dry to form the final composite membrane [27]. The thickness of the thin, selective top layer depends on a number of factors described below [4]:

- Polymer solution concentration
- Dipping time
- Speed for the withdrawal of the porous substrate
- Evaporation rate and environmental impact.

Based on these parameters, Landau and Levich developed a predictive model for calculating the thickness of the coating of the selective layer on top of the porous support. The mathematical representation of the model is shown in Eq. (3.1) below [4]:

$$h = c * \frac{(\dot{\eta} \mathrm{U})}{\gamma^{1/6}(\rho g)^{1/2}} \tag{3.1}$$

Here, h is the coating thickness, c is a constant (0.944 for Newtonian liquids), η denotes the liquid viscosity, U is the speed of withdrawal of the porous support while γ is the surface tension of liquid against air and ρ is the liquid density. Through this method, selective layer having thickness less than 5 μm can be easily formed on the surface of porous support.

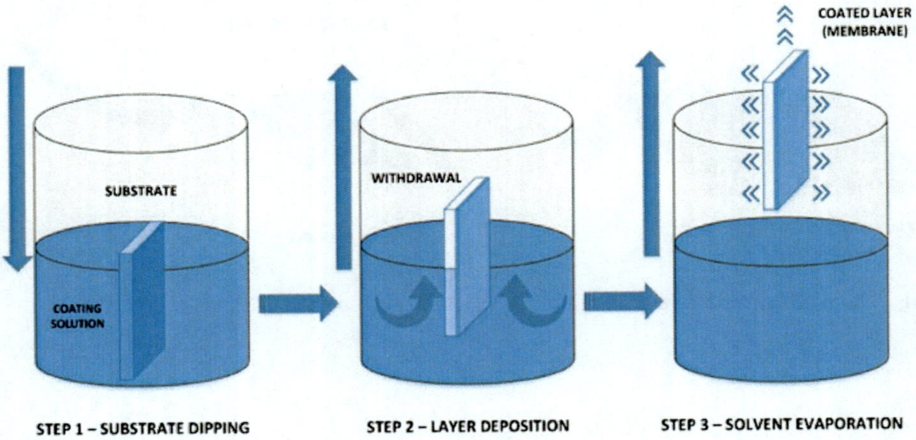

Fig. 3.2 Schematic diagram of dip coating preparation method [22, 28]

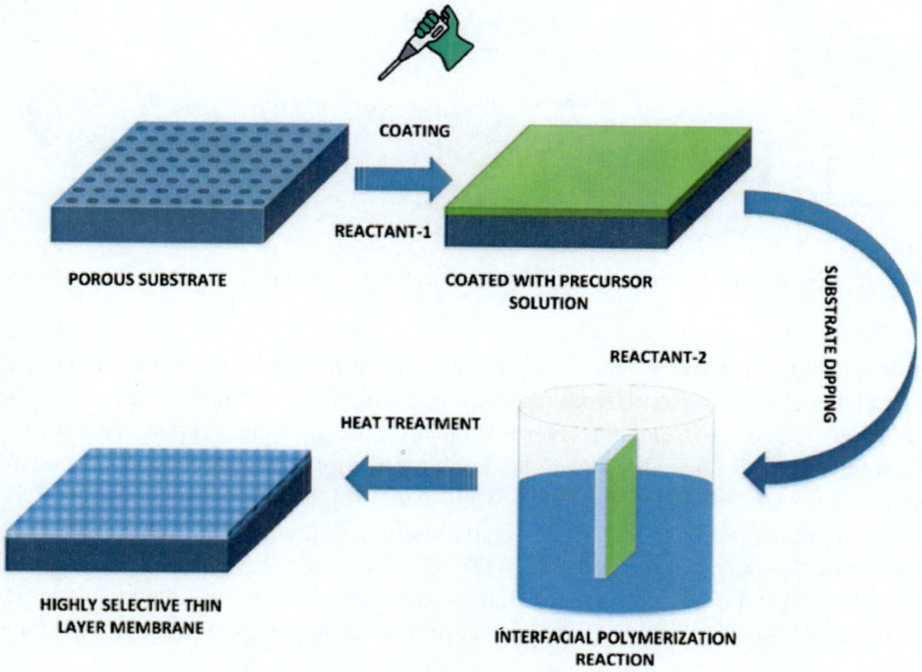

Fig. 3.3 Schematic diagram of interfacial polymerization method [22, 31]

3.2.3 Interfacial Polymerization

Interfacial polymerization is another method used for the synthesis and fabrication of composite FTMs as depicted in Fig. 3.3. This method was initially developed by Cadotte [29] and was perfected by North Star Research. In this fabrication technique, a polymeric porous support is dipped in a reactive coating solution. Thus, the reactive coating solution percolates into the pores of the porous support. Afterwards, the coated support layer is immersed in a reactant bath for some time. This immersion results in the formation of a highly cross-linked selective layer forming on the surface of the porous support. Ultra-thin selective layer can be deposited on the surface of the porous support using this method. Dense, selective layer with thickness of about 0.1 μm or less can be deposited using this technique. One disadvantage of this technique is that the selective layer thus formed being highly cross-linked is very rigid and cannot stand high tensile stresses [30].

3.2.4 Free Radical Polymerization

Free radical polymerization is a novel method, used to produce composite FTMs as shown in Fig. 3.4. In this method, free radical polymerization of a monomer is

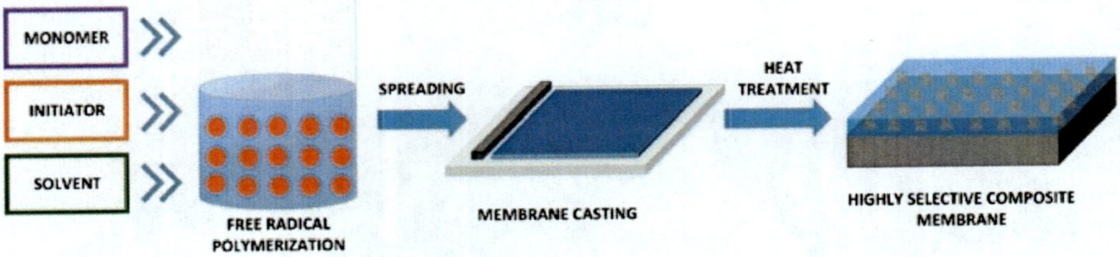

Fig. 3.4 Schematic diagram of free radical polymerization method [33, 34]

initiated using an initiator. Next, the polymer thus formed is blended with inorganic filler and another polymer to form the casting solution for the top, thin, dense, selective layer. This dense, selective layer is then cast onto a porous support. This forms the composite FTM by using free radical polymerization. This process is outlined in detail in a research work carried out by Wang et al. [32]. In their work, they used allylamine monomer and used α,α-azobisisobutyramide oxime as an initiator. After the allylamine has been polymerised, 20 wt% polyallylamine (PAAm) is blended with polyvinyl alcohol (PVA) to form the thin selective layer. This thin, selective layer is then coated onto polysulfone (PSf) porous support using an appropriate applicator.

3.2.5 Spin Coating

Spin coating is a method commonly used for the fabrication of composite FTMs in the industry as illustrated in Fig. 3.5. The equipment used for carrying out spin coating is called a spin coater. In a spin coater, thin coating is introduced onto a porous substrate by spreading a thin layer of the polymeric solution of the rotating substrate. The coating is spread onto the porous support through centrifugal forces, whereas the solvent evaporation greatly depends on the speed of spinning as well as the intrinsic properties of the solvent. Furthermore, solvent evaporation is accelerated by stronger air flow brought about when the spin coater is rotating at higher speeds [5].

3.2.6 Chemical Vapour Deposition

Chemical vapour deposition (CVD) is a method by which gaseous reactants react chemically at or in the vicinity of a heated porous substrate as shown in Fig. 3.6. This reaction causes the deposition of a thin, solid layer on the surface of the porous substrate. For the purpose of carrying out chemical vapour deposition, solid reagents are first vaporized and then reaction occurs near the surface of the substrate in the presence of high vacuum to produce a uniform coating. CVD can easily produce

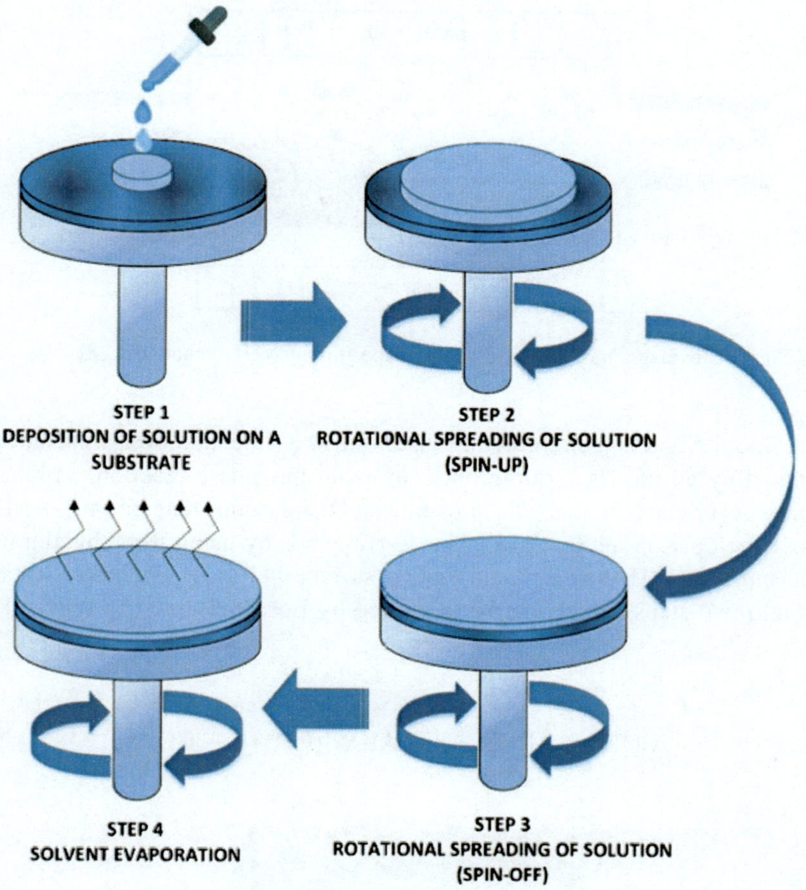

Fig. 3.5 Schematic diagram of spin-coating method [22, 35]

high purity materials with structural control at the nano meter scale. Thus, it can be used to produce high quality FTMs as well [36]. Other advantages of using CVD to produce FTMs include the fact that CVD can be used for materials which are naturally insoluble with one another, and it can be used for surface modification, for example narrowing the pores of a porous substrate by material deposition [37].

3.2.7 *Atomic Layer Deposition*

Atomic layer deposition (ALD) is a special class of chemical vapour deposition (CVD) which is characterized by the sequential use of self-terminating gas–solid reactions as illustrated in Fig. 3.7 [39]. This means that in ALD, the reactions occur between the gaseous reactants and the solid surface and no gas phase reactions should ideally take place. Another characteristic feature of ALD is that the gaseous reactants

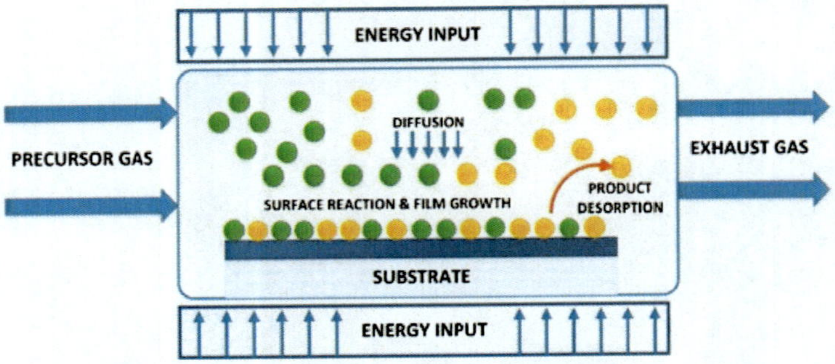

Fig. 3.6 Schematic diagram of chemical vapour deposition (CVD) method [22, 38]

are sent into the reaction chamber in pulses one at a time, and every reactant pulse is followed by an inert gas purge pulse to avoid gas phase reaction, as reactions must only occur between the solid substrate and the gaseous reagent [40]. ALD was initially developed in the late 1980s and early 1990s by using trimethylaluminium (TMA) and water (H_2O) on titania (TiO_2) surface [41, 42]. The overall reaction occurs into two steps, which can be described by Eqs. (3.2) and (3.3) given below [43]:

$$AlOH^* + Al(CH_3)_3 \rightarrow AlOAl(CH_3)_2{}^* + CH_4 \tag{3.2}$$

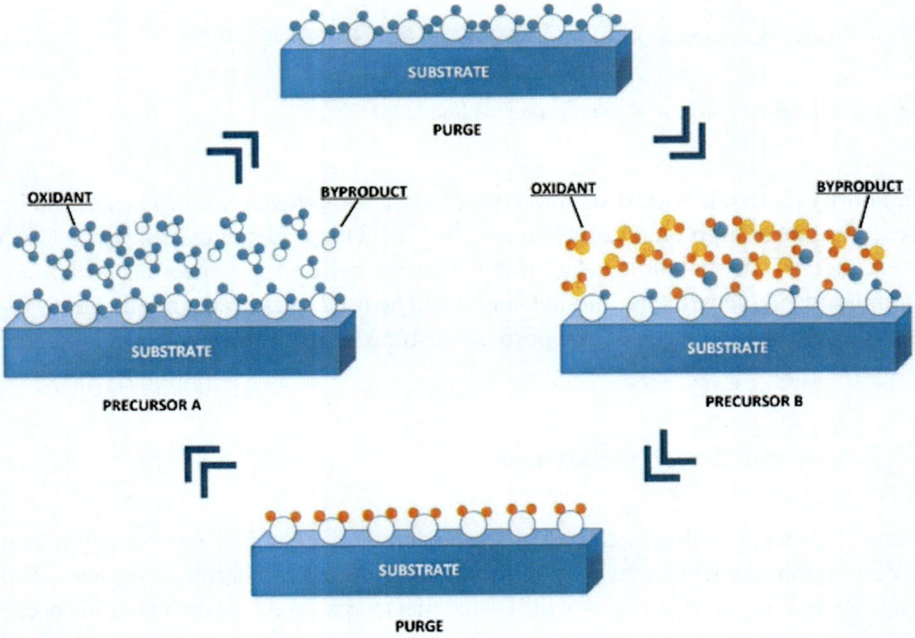

Fig. 3.7 Schematic diagram of atomic layer deposition (ALD) method [46, 47]

$$AlCH_3^* + H_2O \rightarrow AlOH^* + CH_4 \tag{3.3}$$

These alternating pulses of TMA and water vapour led to the formation of highly pure and well-defined coating of alumina (Al_2O_3). One major advantage of ALD is that this technique is conformal to the substrate surface, which means that it gives uniform thickness across the substrate surface and follows its cracks, dips and contours [44]. Another advantage is that it can be carried out at relatively low temperatures as compared to CVD [39]. ALD also encounters some drawbacks. Some of these drawbacks include the fact that it involves more complicated reactor designs as well as more complicated reaction chemistry [45]. ALD is nonetheless a very effective technique for depositing thin, selective layers on top of a composite porous support as it produces ultra-thin film deposits which have high purity and atomic level control.

3.2.8 Vacuum Method

As is evident from the name, this method of facilitated transport membrane fabrication makes use of reduced pressure conditions or vacuum. In one such work outlining the fabrication method carried out by Akhmetshina et al. [48], as depicted in Fig. 3.8, the supported ionic liquid membrane (SILM) was synthesized by first fixing the porous support membrane in a desiccator and the pressure within the desiccator was then reduced to 10^{-5} mbar. The support membrane was kept at this pressure within the desiccator for about 2 h, to remove air from its pores. Afterwards, ionic liquid was introduced into the desiccator and spread over the porous support, while maintaining the vacuum inside the desiccator. The SILM was then kept inside the desiccator for another two hours, and then after being removed, SILM was weighed and the ionic liquid (IL) uptake was measured as well as the thickness of the coated layer of IL. In this process, SILM was prepared under vacuum.

3.3 Fabrication Techniques for Carrier Medium

Carrier particles are a very vital part of all facilitated transport membranes, as the facilitated transport mechanism of gas permeation through a membrane would not be possible without them. There are two main classes of carrier particles, fixed site carriers and mobile carriers. Fixed site carriers are fixed within the polymeric matrix, while mobile carriers are free to move throughout the matrix [49]. Various methods are used for the synthesis and fabrication of carrier particles which are then incorporated into polymeric matrices for FTM preparation. Some of the most common methods will be discussed in this section.

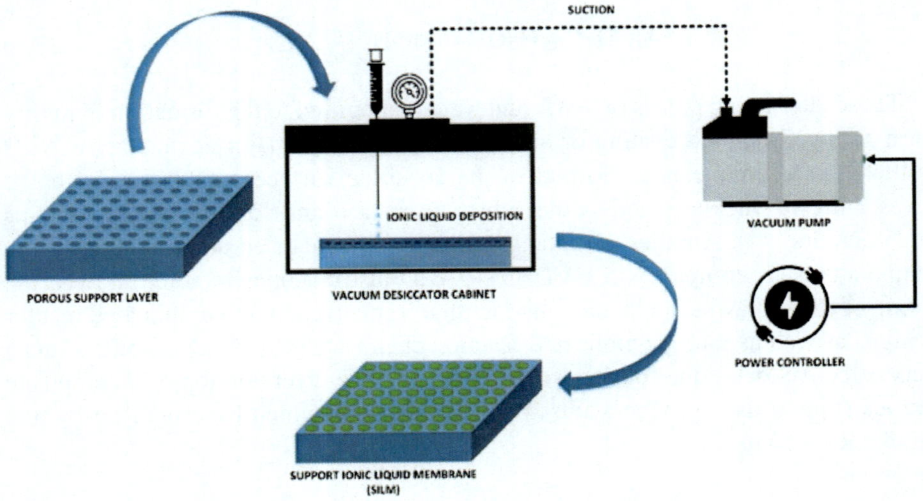

Fig. 3.8 Schematic diagram of vacuum fabrication method [48]

3.3.1 Modified Polymer Directed Chemical Synthesis

The biggest challenge associated with using MMMs for gas separation is the interfacial incompatibility encountered between polymer matrices and inorganic nanoparticles [50]. Various strategies have been used to overcome this huge challenge and unlock the full potential of MMMs. A very recent and advanced strategy which has only recently been developed is the polymer directed chemical synthesis [51]. Using this synthesis route, metal-induced microporous polymer (MMP), a new class of porous polymers, is synthesized. In the process of synthesizing MMPs, amine-bearing polymers are used along with small organic linkers and divalent metal ions. Initially, the small organic linkers coordinate to the metal ions and then the unit cell thus formed starts to self-assemble along the polymer chain. This forms a porous polymeric framework and as can be seen from its starting materials, this type of framework has the advantages of both inorganic fillers and polymeric materials, with greatly reduced interfacial incompatibility. MMPs can also be fabricated into nanoparticles, and then dispersed in polymeric matrices and this arrangement is expected to give stronger interfacial compatibility as well as better gas transport properties [52].

In one such work carried out by Yuan et al. [53], a new approach was adopted to increase CO_2 permeation even more by the use of nanoparticles containing high density of free amine groups. For this purpose, nanoparticles were made from high-valence metal-induced microporous polymer (HMMP). These nanoparticles were then dispersed in polyvinyl amine (PVAm) matrix to form facilitated transport MMMs. The resulting MMMs were then coated onto the surface of modified polysulfone (mPSf) porous support, thus forming a hybrid facilitated transport mixed matrix composite membrane. This novel technique of FTM fabrication has one major advantage that it eliminates or minimizes the interfacial incompatibility between polymer matrices and inorganic filler particles, thus improving the structural integrity of the FTM and its performance [54, 55].

3.3.2 *In-Situ Growth of Nanoparticles*

In-situ growth, as the name suggests, is a method by which nanoparticles are grown within the system in which they are to be incorporated. In one such work carried out by Zhang et al. [56], ZIF-8 was synthesized by in-situ growth on the surface of Zn–Al layered double hydroxide (LDH) nanosheets. Zn–Al LDH nanosheets were prepared following the procedure detailed in [57]. For the purpose of synthesizing ZIF-8@Zn–Al LDH, 0.25 g of 2-methyl imidazole (2-mIm) and 0.14 g of sodium formate were dissolved in 40 ml methanol. Next, 0.12 g of the as-synthesized LDH was added into this solution. After vigorous stirring at room temperature, i.e. 25 °C, for half an hour, the mixture was poured into a Teflon-lined stainless steel vessel. The vessel was sealed and placed in a convective oven at 100 °C for 2 h. The resulting product was then separated by centrifugation and washed three times with methanol. ZIF-8@LDH powder was obtained after drying under vacuum at 120 °C for 6 h.

3.3.3 *Rapid Room Temperature Synthesis*

This method of synthesizing nanoparticles combines precursor solutions at room temperature (25 °C), without the use of auxiliary stabilizing agents or activation using conventional heating, microwave or ultrasonic irradiation as has been reported in various literature sources [58–60]. Another remarkable characteristic of this method is that even in the absence of activation or stabilizing agents, it still gives well-shaped ZIF-8 nanocrystals in the form of powders or colloidal dispersions. In this method which was followed by Cravillon et al. [61], solutions of zinc nitrate hexahydrate ($Zn(NO_3)_2 \cdot 6H_2O$) and 2-methyl imidazole (2-mIm) in methanol are mixed at room temperature. In this work, special care is taken to add 2-mIm in excess as opposed to the recipes published in other literature sources. In this work, excellent results were obtained when $Zn(NO_3)_2 \cdot 6H_2O$, 2-mIm and methanol are added in a molar ratio of 1:8:700.

3.3.4 *Co-precipitation Method*

Co-precipitation method is often thought of as the solid-state analogue or the side branch of the sol–gel method. It is an efficient, soft chemical route for the synthesis of mixed oxide materials [62]. This method involves dissolving the starting materials in a common solvent, then adding a precipitating agent so as to form a homogeneous and single phase inorganic solid. Co-precipitation follows a two-step mechanism, in which the first step is the hydrolysis of the salt with water, followed by poly-condensation to form nanoparticles [63]. To understand the experimental procedure followed to synthesize nanoparticles using co-precipitation, we take the example of the synthesis of Zn–Al LDH as fabricated by Liu et al. [64]. In this method, zinc nitrate hexahydrate ($Zn(NO_3)_2.6H_2O$) and aluminium nitrate nonahydrate ($Al(NO_3)_3{\cdot}9H_2O$) are dissolved in deionized water at a Zn/Al molar ratio of 2:1. The total ionic concentration (Zn^{+2} + Al^{+3}) is kept at 0.15 M. After the two salts are thoroughly dissolved, urea is added into the solution in a three-neck flask such that the ratio of urea and nitrate ions (NO_3^-) is constant at 3:1. The final mixture was heated at 100 °C with constant, vigorous stirring for 24 h. The product was then separated by centrifugation and then washed three times with deionized water. This product is then dried at 60 °C under vacuum for 24 h.

3.3.5 *Hydrothermal Method*

Hydrothermal method is a chemical reaction in water in a sealed pressure vessel, which is in fact a type of reaction at both high temperature and pressure [65]. An example of the use of hydrothermal method for the synthesis of carrier particles is described in a research work carried out by Mubashir et al. [66]. In this work, NH_2-MIL-53(Al) crystals were synthesized by using aluminium chloride ($AlCl_3$), acid and Dimethylformamide (DMF) in a molar ratio of 1:1.4:184.5. The synthesis was carried out using hydrothermal heating method. This work in turn uses the synthesis procedure outlined in [67]. The solution mixture is prepared by mixing 22.5 ml DMF and 7.5 ml of water. Next, 0.7 g of $AlCl_3.6H_2O$ was dissolved in 15 ml of this solution and this was coded as solution A. Simultaneously, 0.5 g of amino terephthalic acid was dissolved in the remaining 15 ml of the solution mixture, and the resultant was coded as solution B. Next, solutions A and B were mixed and stirred at 200 rpm for 5–10 min. After stirring, the resultant solution mixture was heated in a pressure vessel at 150 °C for hydrothermal growth. Yellow powder of the product was obtained by separating through centrifugation at 4000 rpm for 15 min. Then, the samples were dried at 150 °C in an oven. The powdered product was then activated by removing 2-terephthalic acid by washing with DMF under continuous reflux for 24 h. The powder thus obtained was washed four times with acetone, and heated overnight at 30 °C. Thus, NH_2-MIL-53(Al) was successfully synthesized.

3.3.6 *Stöber Method*

Stöber method is based on a series of chemical reactions which permits the controlled growth of spherical silica particles of uniform size by means of hydrolysis of alkyl silicates and subsequent condensation of silicic acid in alcoholic solutions. Here, ammonia is used as a morphological catalyst and the size distribution of particles obtained in suspension range between 0.05 and 2 μm [68]. To fully demonstrate the working of this method, we take the example of nanoparticle synthesis by Stöber method as carried out by Zhang et al. [69]. In their research work, they added 14 ml tetraethyl orthosilicate (TEOS) in a solution containing 250 ml ethanol, 25 ml water and 6 ml of 25% aqueous ammonia solution. This solution was stirred for 24 h at 25 °C. The silica spheres thus formed were purified by repeated centrifugation and then desiccation at 50 °C until a constant weight is obtained. As can be seen from the aforementioned description of the Stöber method, it does not require high temperatures or pressures or extreme chemical reactions, therefore this could serve as a suitable soft chemical route for the synthesis of carrier particles for FTMs.

3.3.7 *Solvothermal Method*

Solvothermal synthesis route is a process taking place in a closed system that requires high temperature and pressure to cause a chemical reaction or decomposition of the precursor materials to form the desired product directly from a solution [70]. The main difference between hydrothermal synthesis and Solvothermal synthesis of nanoparticles is that hydrothermal synthesis refers to a chemical reaction taking place in an aqueous solution above the boiling point of water, whereas Solvothermal synthesis refers to a chemical reaction taking place in a non-aqueous solution at a relatively high temperature. Solvothermal synthesis is an important branch of inorganic synthesis and can be used to synthesize nanomaterials [71]. An example to properly understand the working of solvothermal synthesis method is the research work carried out by Thür et al. [72]. In this work, UiO-67 and its bipyridine derivative were prepared using standard Solvothermal method. For this purpose, 1.27 mmol linker (biphenyl-4,4′-dicaroxylate (BPDC), bipyridine-4,4′-dicarboxylate (BPyDC) or a combination of both) was dissolved in 100 ml DMF in a 250 ml Schott bottle and sonicated for 15 min. Next, 1.27 mmol zirconium chloride ($ZrCl_4$) was added, followed by 2.15 ml acetic acid and then 0.5 ml of 37% HCl. The mixture was then placed in an oven at 120 °C for 48 h. Thus, metal–organic framework (MOF) was formed as crystals. After MOF formation, the crystals were washed three times with DMF and three times with ethanol. After each washing step, the mixture was centrifuged for 30 min at 4000 rpm. After thorough washing, MOF formulations were dried overnight in an oven at 80 °C. In this way, MOF crystals which are commonly used as carrier media in FTMs can be easily synthesized. As the crystals fabricated by Thür et al. were then used to synthesize Matrimid based FTMs [72]. A major advantage of using solvothermal method for synthesizing nanoparticles is that there is a wide range of precursors which can be used as the choice of solvent is not restricted to water as is the case with hydrothermal synthesis route [73].

3.3.8 Schiff Base Condensation Reaction

Schiff Base Condensation Reaction is an advanced method to produce self-exfoliated nanosheets by condensation reaction between two organic linkers to form covalent organic nanosheets (CONs). This method is explained in a research work carried out by Wang et al. [74]. In this method, 1,3,5-Triformylphloroglucinol (Tp) is first prepared through Duff formylation reaction as reported in literature [75]. Next, Tp is reacted with 3,5-diamino-1,2,4-triazole (DT) in a sealed Pyrex tube to form TpDT CONs. The Pyrex tube is normally charged with 90 mg Tp, 45 mg DT, 3 ml dioxane, 1 ml dimethylacetamide (DMAc), 3 ml mesitylene and 1 ml of 6 M aqueous acetic acid. This tube is then sealed under vacuum after being degassed by three freeze–pump–thaw cycles. The tube is then kept at 120 °C for three days. The solid products thus obtained are filtered and washed with hot DMF, dioxane, methanol, acetone and tetrahydrofuran (THF). Afterwards, Soxhlet extraction was performed using DMF/methanol solvent. Dried TpDT CONs were finally obtained after extraction, by drying under vacuum at 120 °C for 12 h. Therefore, it can be seen that this condensation reaction method can be used to synthesize advanced materials which can then be used in producing FTMs for gas separation, as was done in the same work by Wang et al. [74]. In this same work, TpDT CONs are then used to produce FTMs with Pebax® serving as the polymer matrix.

3.3.9 Facile Modification Method

In this synthesis technique, carrier particles (MOFs, ZIFs) that are indigenously prepared are then modified using a modification agent and this changes the properties of the carrier particles to make them even more suited for use in FTMs. One such example is observed in a work carried out by Nordin et al. [76]. In this work, they initially synthesized ZIF-8 using the precursors $Zn(NO_3)_2$, 2-mIm and water in the 1:6:500 respectively. Firstly, 2 g zinc nitrate hexahydrate ($Zn(NO_3)_2.6H_2O$) in 12.11 g of deionized water. For ligand solution preparation, 3.312 g 2-mIm was dissolved in 48.45 g of deionized water. After dissolution of 2-mIm, 3 ml triethylamine (TEA) is added to the ligand solution, then slowly the metal salt solution is added to the ligand solution as well. On addition of the metal salt solution, the final solution starts to turn cloudy. This solution was then stirred vigorously for half an hour, then the reaction products are centrifuged. The final product was washed several times with deionized water and then the washed powder is dried in an oven at 60 °C for 12 h. The final powdered product was then ground to fine particles and dried again in an oven at 100 °C for 12 h [77]. For facile modification of ZIF-8, 1 g of the prepared ZIF-8 powder is dispersed in predefined volume of ammonium hydroxide solution, adding an additional 10 ml deionized water into the solution. The suspension is then sonicated for 60 min to break the interplanar bonds within the bulk material. Afterwards, the suspension is vigorously stirred for 24 h at specified temperatures. The sample was separated by centrifugation and washed three times with deionized water. The obtained powder is then dried overnight in an oven at 60 °C. Thus, modified ZIF-8 is obtained which is then used to prepare asymmetric MMMs with polysulfone (PSf) as the polymer matrix.

3.3.10 Co-condensation Method

Co-condensation method, also referred to as *one-pot route,* is a unique method for the functionalization of mesoporous silica. It involves the addition of an organic group, normally silane-containing reactive groups such as chlorides, ethoxy or methoxy groups, during the mesoporous synthesis and the entire functionalization is carried out in a unique way [78]. In one such research work highlighting the use of co-condensation method for nanofiller synthesis and functionalization, carried out by Han et al. [79], Santa Barbara Amorphous 15 (SBA-15), a mesoporous silica sieve having pore size distribution between 5 and 15 nm, was functionalized in an aqueous solution through direct co-condensation of 2,2′-bipyridine linked methallylsilane with tetraethyl orthosilicate (TEOS) under acidic conditions. Furthermore, this functionalized SBA-15 mesoporous silica was also impregnated with palladium (Pd) and rhodium (Rh). In another work carried out by Gottuso et al. [80], functionalized mesoporous silica nanoparticles (MSNs) are synthesized using three different organo-substituted silica precursors, namely hexadecyltrimethoxysilane, triethoxy-3-(2-imidazolin-1-yl)propylsilane and (3-mercaptopropyl)triethoxysilane. Molar ratios of hexadecyltrimethoxysilane, triethoxy-3-(2-imidazolin-1-yl)propylsilane and (3-mercaptopropyl)triethoxysilane with tetraethyl orthosilicate (TEOS) were kept at 1:4, 1:9 and 1:19 respectively. All condensation reactions take place in a water/oil emulsion. Here again, we see that mesoporous silica is functionalized using silicate-based functional groups to form nanoparticles for use in FTMs.

3.3.11 Sonochemical Method

Sonochemistry is one of the earliest techniques used to prepare nanomaterials [81]. Sonochemistry is the research area in which molecules undergo a chemical reaction because of the application of high-energy ultrasound radiation that is sound waves possessing a frequency between 20 and 10 MHz. The physical phenomenon which is responsible for the sonochemical method is acoustic cavitation. Acoustic cavitation is the formation, growth and collapse of bubbles in a liquid [82]. Moreover, the sonochemical method is a very beneficial technique in preparation of amorphous products, insertion of nanomaterials into mesoporous materials, deposition of nanoparticles on ceramic and polymeric surfaces, and morphology control of nanomaterials [83]. One of the earliest works using the sonochemical method for the synthesis of nanomaterials was carried out by Suslick et al. [84]. In this work, it was also concluded that acoustic cavitation brought about by the sonochemical method gives rise to extreme local heating in otherwise cold liquids. These microscopically harsh yet macroscopically mild conditions brought about by acoustic cavitation lead to the breakage of bonds in bulk compounds and produce nanomaterials. In their work, they also found out that high heating and cooling rates are needed for the formation of nanomaterials from bulk materials. In sonochemical method, it was found that the heating and cooling rates were as large as 2×10^9 K s^{-1}, which is why it is a very effective method for the formation of nanoparticles. Suslick et al. synthesized nearly pure, amorphous iron using iron pentacarbonyl ($Fe(CO)_5$), which is a volatile, organo-metallic compound.

After this successful synthesis, sonochemical method gained traction in the field of nanomaterial synthesis. This method is now also used for the synthesis 2D nanomaterials (nanosheets) from bulk layered materials such as MoS_2 [85], $NbSe_2$ [86] and WS_2 [87] among many others. Therefore, this method is a very efficient means of producing nanofillers for FTMs.

3.3.12 Etching and Soaking Method

The working of the etching and soaking method for the formation of nanomaterials can be understood by observing a research work carried out by He et al. [88]. In their work, the researchers synthesized hierarchical metal organic frameworks (H-MOFs). H-MOFs were initially fabricated because ordinary MOFs have micropores having diameters less than 2 nm and this limits the use in many applications. H-MOFs eliminate this problem because they have a system of micropores and mesopores. In their work, H-MOF nanostructures (nanosheets and nanocubes) were synthesized by a competitive coordination strategy which evolved through an etching process using a competitive ligand. Next, Pd nanoparticles are immobilized on H-MOF by a simple soaking method using Pd^{+2} precursors.

3.3.13 Ion Exchange Method

Ion exchange method is a highly advanced technique to produce nanoparticles. It is thought to be a very versatile technique as it can be used to prepare complex heterostructures as well as thermodynamically metastable phases such as core–shell arrangements. These structures cannot be prepared with conventional techniques of nanomaterial synthesis [89, 90]. One such work used ion exchange method to synthesize porous ionic polymer (PIP) which served as the filler particle in FTMs [91]. In this work, firstly porous organic polymer (POP) is synthesized by a method outlined in [92]. For synthesis of TPFC POP, 1.28 g of triptyene was dissolved in 10 ml dichloroethane in 250 ml three-necked flask. Next, 2.28 g of formaldehyde dimethyl acetal (FDA) was added to the mixture. Afterwards, 4.88 g anhydrous ferric chloride ($FeCl_3$) was added. The three-necked flask was then stirred for five hours at 45 °C in an oil bath and then stirred for 19 h at 80 °C. The mixture was then cooled to room temperature, filtered and washed three times with methanol. The solid product was placed in a Soxhlet apparatus and extracted with methanol for 24 h. After that the mixture was dried, this yielded a brown solid product which is identified as TPFC POP. TPFC POP was then chloromethylated and ionized to form PIP. For this purpose, 0.6 g TPFC and 3 g paraformaldehyde were added to 350 ml flask. Afterwards, 18 ml acetic acid, 60 ml concentrated hydrochloric acid and 9 ml phosphoric acid were added to this mixture. The flask was then sealed and held at 90 °C for three days. The cooled brown powder was then filtered and washed three times with water and methanol, and dried at 80 °C for 12 h. The brown product thus obtained was identified as PIP-Py-Cl. In this work, PIP-Py-Cl and other such compounds (denoted as PIP-Py-X) act as nanofillers and are used within Polymer of Intrinsic Microporosity-1 (PIM-1) matrix to form FTMs.

3.3.14 Modified Hummers Method

Over the years, many different methods have been used for the synthesis of graphene oxide (GO) from graphite. The first instance of the synthesis of GO was demonstrated by Brodie in 1859 [93]. In their work, they added a portion of potassium chlorate to slurry of graphite made in fuming nitric acid. This method was improved upon by Staudenmaier in 1898 [94]. In their procedure, they used a mixture of concentrated sulphuric acid and fuming nitric acid to form the slurry of graphite, and they performed gradual addition of chlorate to the reaction mixture. This small change in the procedure led to the formation of highly oxidized GO. This method was further improved and perfected when an alternative method was demonstrated by Hummers in 1958 [95]. In this alternate method, potassium permanganate ($KMnO_4$) was used with sodium nitrate ($NaNO_3$) in a slurry of graphite made in concentrated sulphuric acid. In all these methods to obtain graphene oxide, one thing has been established after extensive research that the most effective method for graphite exfoliation to obtain GO, a non-conductive hydrophilic carbon material, is by the use of strong oxidizing agents [96]. Since then, a modified Hummers method has been developed and is adopted by researchers for the synthesis of graphene oxide [97].

In one such work carried out by Ying et al. [98], 0.5619 g of graphite powder was added into a mixture of 9.9 ml concentrated sulphuric acid and 1.875 g phosphorus pentoxide (P_2O_5) and 1.875 g potassium persulfate at a temperature of 80 °C in a water bath with stirring for four and a half hours. The mixture was then dilute with 750 ml deionized water and solid product was collected by vacuum filtration on a polycarbonate membrane having a pore size of about 200 nm. Afterwards, the collected product was washed with deionized water to remove residual acid until the pH of the solids approached that of a neutral solution. The washed, neutralized solid was then dried at room temperature for 24 h. Thereafter, the dried solid was added to 90 ml of concentrated sulphuric acid in an ice-water bath with continuous stirring. After the graphite oxide was fully dispersed, 11.25 g of potassium permanganate was added to the mixture. This mixture was then transferred to a water bath at 35 °C for 2 h, until the reaction did not release any more heat. Subsequently, the mixture was diluted with 187.5 ml deionized water in an ice-water bath. Then, it was placed in a water bath at 35 °C for another two hours. Afterwards, 525 ml deionized water and 15 ml of 30% hydrogen peroxide were added under constant stirring. The contents of the reaction mixture showed a colour change to bright yellow. Once effervescence ceased in the reaction mixture, 30 ml of 37% hydrochloric acid was added. Thereafter, the solution was placed in ambient environment for 48 h. The sediment of this settled solution was then collected and centrifuged with 10% hydrochloric acid and deionized water one after another for five times each to remove the residual acid and metal ions. Solid product after washing, was dried in a blast air oven at 50 °C for 48 h. This is a classical example of the use of modified Hummers method for the synthesis of GO nanosheets and this protocol has also been outlined in [99]. The as-prepared GO was then used in this study to prepare facilitated transport supported ionic liquid membrane (FT-SILM). Therefore, this method warrants itself as a suitable approach for the synthesis and fabrication of carrier particles to be used in FTMs.

3.4 Testing and Characterization of FTMs and Carrier Particles

Testing and characterization of FTMs and carrier particles is a vital part of understanding and gauging the usefulness and effectiveness of the synthesis and fabrication techniques used for FTMs and carrier media. The testing and characterization methods used most commonly are summarized in this section.

3.4.1 Gas Permeation Testing

In case of FTMs, as gas separation is the main objective so it is vital to check the permeability of various gases through the membrane. For this purpose, gas permeability testing system is used. As an example of the use of such a system for gas permeation testing, we take the system used by Ayub et al. [24]. In this work, the gas permeability test system consists of a membrane cell which is arranged in dead-end filtration mode. This means that in the cell, feed gas is pumped from the top and the permeate leaves through the bottom. The membrane cell consists of two porous ceramic plates which are held in place by screws and the membrane to be tested is placed in between these two plates. Feed gas line is connected to a gas cylinder. The pressure gauge fitted at the cylinder is the point of pressure control, and as the system is pressure-controlled so feed and permeate pressure are the two important parameters. The permeate gas exits into a small tube from where it enters into bubble flow meter. The bubble flow meter is a straight cylinder of constant cross-section with a bulb fitted at one end. The bulb is filled with soap solution. When gas enters through the narrow tube into the cylinder, it causes some of the soap solution to rise with it in the form of a bubble. The volume that the bubble travels in time gives the volumetric flow rate Q of the gas with the active membrane area being denoted as A. The pressure drop across the membrane is denoted as ΔP and membrane thickness denoted as L. Using these parameters. The permeability P of any gas can be calculated using Eq. (3.4) below:

$$P = \frac{Q \times L}{\Delta P \times A} \tag{3.4}$$

To understand how effectively a membrane separates a mixture of gases, we must know the selectivity α_{AB} of the membrane for gas A over gas B. This can be calculated as in Eq. (3.5) shown below [15]:

$$\alpha_{AB} = \frac{P_A}{P_B} \tag{3.5}$$

In the case of FTMs, often gas permeation and separation performance is tested under humidified conditions as illustrated in Fig. 3.9. This is done because the carrier particles within the membrane matrix may be mobile and their mobility is often aided by the humidity present within a gas. To investigate this phenomenon, research work was carried out by Saeed et al. [100]. In their research work, gas permeation rig was set up so as to allow testing of gas permeation to be carried out under both dry and humidified conditions. For this purpose, a two-way valve was installed at the entry of the feed gas. For dry conditions, humidifier would be bypassed and for humid conditions, the feed gas would first pass through the humidifier. The permeability and selectivity of the membrane for the gases were measured and calculated as above, but humidification was an added step to check the effect of humidity on the separation performance of FTMs. It was found from this research that CO_2/CH_4 separation performance for biogas upgrading was maximum at maximum water uptake, which means that humidity has a positive impact on the separation performance of FTMs.

Another aspect that was investigated in this research work by Saeed et al. is the use of sweep gas. Sweep gas is used to maintain a minimum concentration of CO_2 on the permeate side so that driving force for gas separation can be maximized. Furthermore, sweep gas was also maximized to ensure maximum uptake of moisture by the membrane, as it was believed and understood by the researchers and later

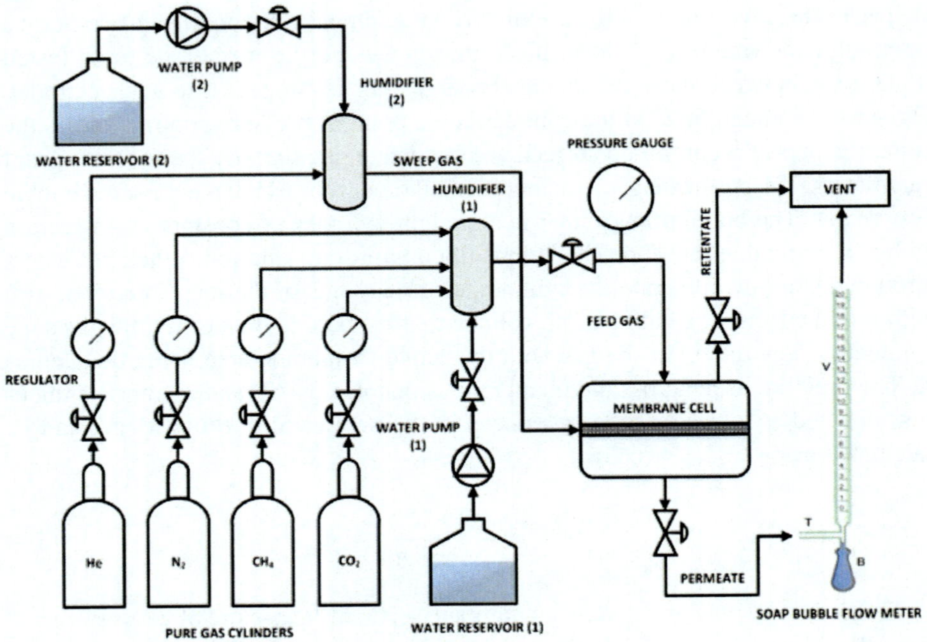

Fig. 3.9 Schematic diagram of pure gas permeation testing system for FTM [22, 100]

proven to be true that maximum water uptake yielded the best separation performance for CO_2 over CH_4.

The same approach as for pure gas permeation testing can be employed in the case of a mixed gas permeation testing system, as shown in Fig. 3.10. However, after permeation through FTM, the moisture present in the product gas is removed by using a separator. Then the permeate gas composition is determined using chromatography, in which the gas separation performances of FTM are recorded after reaching a steady state process [74].

3.4.2 *Fourier Transform Infra-Red (FT-IR) Spectroscopy*

Spectroscopy is the science that studies the interaction between matter and electromagnetic radiation [101]. Of all the portions of the electromagnetic spectrum, the energy of infra-red radiation corresponds to the amount of energy required to cause molecular vibrations. Moreover, infra-red radiation promotes transitions between vibrational energy levels, leading to spectra that allow for identification of functional groups and also which represent a very characteristic and sensitive fingerprint of a specific molecule. Infra-red radiation is arbitrarily divided into three sections, i.e. near-IR, mid-IR and far-IR radiation. The energy of mid-IR radiation corresponds to the energy required to cause fundamental molecular vibrations such as stretching,

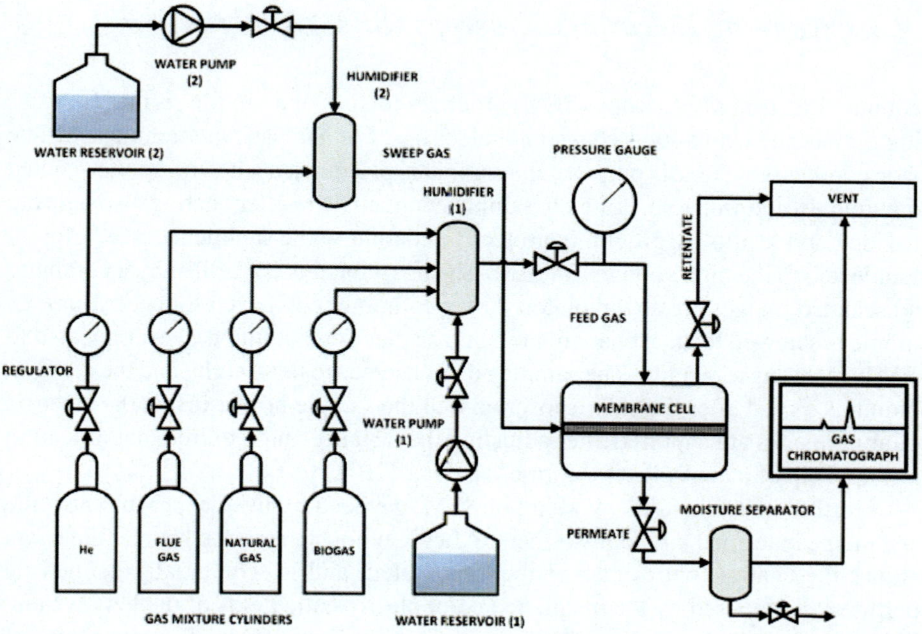

Fig. 3.10 Schematic diagram of mixed gas permeation testing system for FTM [22, 74, 100]

bending, wagging and scissoring. Fourier Transform Infra-Red (FT-IR) Spectroscopy operates in the mid-IR region of the electromagnetic spectrum which ranges between wave number values of 400–4000 cm^{-1} [102, 103].

FT-IR spectroscopy is a rapid, non-destructive and time-saving method that can detect a range of functional groups and is sensitive to changes in molecular structure [104]. The fundamental of infra-red spectroscopic technology is based on atomic vibrations of a molecule that only absorbs specific frequencies and energies of infra-red radiation. Molecules can only be detected and classified by FT-IR spectroscopy because different molecules have different infra-red spectra. FT-IR spectroscopy is carried out by equipment called FT-IR spectrometer. The FT-IR spectrometer essentially consists of an interferometer which measures the energy being transmitted to the sample. The infra-red radiation being emitted by the black body (source) is transmitted to the interferometer where the spectral encoding takes place. The resultant interferometer signal is transmitted to the sample where some signals are transmitted through while other bounce off the sample surface. Only specific energy wavelengths are absorbed by the sample. The beam eventually passes on to the detector, from where signals are transmitted to the processing computer which applies Fourier transformation on the signals [105]. Fourier Transform is carried out on the output signals from the IR detector because Fourier transform is used to interpolate functions and smooth signals, thus clearing out the high spatial frequency edges in pixelated images [106].

3.4.3 Scanning Electron Microscopy (SEM) Analysis

Scanning Electron Microscopy (SEM) produces images of a sample surface by scanning the surface with a focused beam of electrons. For analysis, the samples must be made conductive. For this purpose, the sample is placed on a stub and sputter-coated with gold. In addition to making the sample conductive, sputter coating also helps to provide good results by providing protective coating to the sample surface before it is subjected to the high voltages of SEM. The instrument is controlled by a computer and scanned images are also monitored by a computer connected to the instrument. Sample is screwed to an exchanging rod and an electron column is also connected to SEM by creating a vacuum. The sample position needs to be straight, and the electron column is closed after the stub is inserted into the sample holder inside the electron column. Images are captured after adjusting the distance and required magnification with the help of a controller for further analysis [107, 108].

To further understand the working of SEM, we need to dive deeper into how the final image is formed by the instrument. After leaving the electron source, the beam follows the central (optic) axis of the lens system and is sequentially defined by apertures and focused by the magnetic and/or electrostatic fields of the lens system. Within the final (objective) lens, a system of scan coils act to displace the beam off the optic axis, so that it can be focused to a location on the specimen. At any particular time, there is only one ray path through the scanning system and the beam reaches only one location on the specimen. The SEM image is a geometric construction created through computer control by focusing the beam to a sequence of discrete x–y locations on the specimen, and measuring the effect of the interaction of the beam at every location on the specimen. For a single gray-scale SEM image, this interaction could be the output from a single electron detector such as the Everhart–Thornley detector. It should be noted here that there is no "true image" formed from SEM, as is formed in a light optical microscope. In SEM, at each location sampled by the electron beam, each signal is measured by an appropriate detector and the analogue measurement is converted to an equivalent digital value (using an analogue-to-digital converter, ADC). The beam x–y location and the intensity of the signal of interest generate a digital stream of data packets. Various signals are available to be detected by a detector, which include secondary electrons (SE), backscattered electrons (BSE), absorbed current, generated X-rays, cathodoluminiscence among other types of signals [109].

3.4.4 X-ray Diffraction (XRD) Analysis

X-ray Diffraction (XRD) is a non-destructive technique that provides detailed information about the crystallographic structure, chemical composition and physical properties of materials [110]. X-rays are a form of electromagnetic radiation having wavelength ranging from 0.01 to 10 nm. To put it simply, X-rays are generated when electrons are accelerated under an applied potential difference and turned into electromagnetic radiation [111].

For the production of X-rays, a cathode filament machined into a cathode cup is activated, causing intense heating of the cathode filament [112]. The heating of the filament leads to the release of electrons in a process called *thermionic emission* [113]. The released electrons form into an electron cloud at the filament surface, however repulsive forces prevent the ejection of these electrons from the electron cloud [114]. Upon application of a high voltage to both the cathode and the anode, there is an acceleration of electrons ejected to an electrically positive anode [112]. Once the high kinetic energy electrons finally reach the anode target, the process of X-ray production is initiated. Tungsten is most often used as the anode target; however other materials may also be employed. Electrons come extremely close to the nucleus of the target atoms, causing deceleration and change in direction, converting the kinetic energy of electrons into electromagnetic radiation in a process called *breaking radiation* or *bremsstrahlung* [115]. The output is stream of X-ray energies.

Diffraction effects are observed when electromagnetic radiation impinges on periodic structures with geometrical variations on the length scale of the wavelength of the incident radiation. The interatomic distances in crystals and molecules amount to 0.15–0.4 nm. This corresponds to the wavelengths of X-rays, which have photon energies between 3 and 8 keV. There are three different types of interactions which occur between X-rays and matter. Firstly, electrons may be released from their bound atomic states in the process of photoionization. Since energy and momentum are transferred from X-ray radiation to the electrons, this type of interaction falls into the category of inelastic scattering. Second type of interaction is another inelastic scattering process known as Compton scattering. Compton scattering is an interaction between moderate energy X-rays and an outer orbital electron, in such a way that some of the energy of the X-rays is absorbed by the electron and the X-ray is deflected at an angle to its original path of incidence. The third type of interaction between X-rays and matter is known as Thomson scattering. This is an elastic type of scattering in which on interaction, the electron oscillates as a Hertz dipole with the frequency of the incoming radiation, and emits dipole radiation. The wavelength λ of the X-rays is conserved in Thomson scattering. It is the Thomson component in the scattering of X-rays, that is used in structural investigations in X-ray diffraction [116, 117]. The scattering occurring in X-ray diffraction can be studied using Bragg's Law, which is presented in Eq. (3.6) below [118]:

$$2d \sin \theta = n\lambda \tag{3.6}$$

X-ray diffraction analysis is carried out in an X-ray diffraction instrument. Although there are many variations of this instrument, all of these variations consist of the components which will be discussed here. Firstly, the most important component of this instrument is an X-ray source. Synchrotrons are most commonly used for the generation of X-rays in various applications including in X-ray diffraction instruments, because of several advantages including its extremely high brilliance.

Brilliance is defined as the number of photons emitted per unit target area over a unit angle of emission per unit energy (photons s^{-1} m rad^{-1} mm^{-2} per 0.1% of radiation bandwidth. This means that synchrotrons make much more efficient use of incident energy of electrons and cause far less heating as compared to conventional X-ray sources [119]. Once the X-ray has been produced, collimation needs to be performed. Proper collimation is important because it prevents unnecessary exposure outside the area of interest and it improves image quality by producing less scatter radiation from these areas [120]. The most commonly used collimator is a crystal monochromator. Crystal monochromators placed in the beam path of an X-ray tube allow for either the selection of a characteristic line from the anode or the selection of energy band from a continuous spectrum [121]. After passing through the crystal monochromator, a filtered X-ray beam is focused on the sample which is mounted on the sample support. The most important aspects to be considered when mounting the sample on the sample support is the sample area, sample surface and sample thickness. Sample area must not be too big for the sample support and the surface of the sample support must be flat with no curvature or roughness [122]. After passing through the sample, the scattered XRD beams are then sent to the two-dimensional (2D) detector. This is a new type of detector, which not only produces detection signals but also 2D image processing and 2D diffraction pattern manipulation and interpretation [123].

3.4.5 *Transmission Electron Microscopy (TEM) Analysis*

Transmission electron microscopy (TEM) is a high-resolution technique used to reveal structural details, size distribution and morphology of nanoparticles made up of lipids and proteins as well as hard nanoparticles composed of metallic nanoparticles, carbon and plastics. TEM is based on the interaction between high-energy electron beam and a thin sample. The amplitude and phase variations of the transmitted beam provide images of the nanoparticle. Based on sample preparation, TEM can be classified as negative-stain, freeze-fractured and cryogenic TEM [124]. As TEM studies the structure of samples by analysing the electrons that pass right through the sample (primary electrons) [125] as opposed to SEM, which studies the structure of samples by analysing the electrons that bounce off the sample (back-scattered electrons) or electrons ejected from the sample (secondary electrons) [126], so samples prepared for TEM analysis need to be very thin, having thickness of around 100–200 nm so that electrons can easily pass through, depending on the material density and accelerating voltage applied to the electrons [127]. Therefore, in the past, the biggest problem associated with material characterization using TEM has been the preparation of thin foil specimens. However, a novel technique for sample preparation, known as focused ion beam (FIB) technique, has recently been developed [128] and research is still underway to perfect this technique for preparing specimens for TEM analysis [129]. Despite this major challenge, however, TEM continues to be the most popular technique for characterizing nanomaterials, and that is mainly because of the fact that TEM can capture chemical information and produce images of nanomaterials at a spatial resolution equal to the level of atomic dimensions [130]. As far as the instrumentation used for TEM analysis is considered, it is the same as that used for SEM as described in Sect. 4.3 above. The only difference in both techniques is the types of electrons which are analyzed to study the structure which is also explained earlier in this section.

3.4.6 *Water Uptake Tests*

As was evident from the research work carried out by Saeed et al. [100], gas separation performance is improved by using humidified gas. The effect of humidity on the membrane properties (membrane swelling) can be analyzed by using moisture uptake tests. In their research work, Saeed et al. carried out moisture uptake tests by subjecting the polyvinyl amine (PVAm)/cellulose nanocrystals (CNC) FTMs to a humid environment at room temperature. At first, the membranes were placed in a closed chamber with relative humidity (RH) of 0%. Then, the degree of swelling was measured by subjecting the membrane samples to 87% RH. The membranes were placed in such a way that there was no direct contact of the membranes with the saturated salt solution. The degree of swelling of the membranes was measured with the increase in weight of membranes after every 24 h. The water uptake and degree of swelling tests were carried out for a period of ten days, and every reading until day 10 was taken after a period of 24 h. The degree of swelling (DoS) can be calculated using Eq. (3.7) given below [131]:

$$DoS(\%) = \frac{S - D}{D} * 100\% \tag{3.7}$$

Here, S is the mass of swelled membrane and D is the mass of dry membrane.

3.4.7 *Atomic Force Microscopy (AFM) Analysis*

In 1986, Gerd Binnig and Heinrich Rohrer shared the Nobel Prize in Physics for inventing the scanning tunnelling microscope and discovering that it can image individual surface atoms with unprecedented resolution [132]. The success of the scanning tunnelling microscope has led to the development of a number of other scanning probe microscopes, which work on the basis of mechanically scanning a sharp tip over a sample surface [133]. The atomic force microscope is one of the most successful of these new devices [134]. The working principle of an atomic force microscope (AFM) is that the force exerted by a sharp tip on a surface can be used to map the surface based on the forces existing between the atoms on the surface and their interactions with the tip of the cantilever of AFM. The same principle is the basis of the working of scanning tunnelling microscope, as whatever the force may be, whether it is mechanical, magnetic or electrostatic, all such microscopes consist of five major components [135]:

- A sharp tip mounted on a soft cantilever spring
- A way of sensing the deflection of the cantilever
- A feedback system to monitor and control the deflection (and hence the interaction force)
- A mechanical scanning system (usually piezoelectric) that moves the sample with respect to the tip in a raster pattern
- A display system that converts the measured data into an image.

Although, all force microscopes have similar basic components, their implementation varies considerably. For example, in an atomic force microscope, the cantilever is micro fabricated from silicon, silicon oxide or silicon nitride using photolithographic techniques [136]. Another critical component of AFM is the sensor. It is understood that the sensor should have sub-angstrom level sensitivity and should apply no force on the cantilever. For this purpose, electron tunnelling detection was initially used. However, it was found that this method had a major drawback. The effectiveness of electron tunnelling detection was greatly reduced if the surface was contaminated. So, in modern AFMs, sensors work on the basis of optical detection. Optical detection schemes are divided into two types, namely interferometry [137] and beam deflection [138]. In a particular beam-deflection AFM, light from a diode laser is reflected from a mirror-like cantilever surface. The direction of the reflected light beam is sensed by a position-sensitive photodetector. Although interferometer systems have taken up many forms over the years, the more recent ones are as simple to implement as beam-deflection systems [139]. One of the main advantages of using interferometry is that it does not need cantilevers with mirror-like reflecting surfaces, which is important in applications where magnetic and electrostatic imaging is being performed, as those systems would use fine wire cantilevers [140].

One of the most outstanding results obtained from the AFM was to achieve atomic level resolution of the surface. It was found that this can be achieved by having the cantilever tip in contact with the sample while scanning. In theory, this system seems to be implausible as the size of the cantilever tip is so small and it may get damaged fairly easily, however this can be achieved by keeping the area of contact between the cantilever tip and the surface of sample exceedingly small. By doing so, loading forces as small as 10^{-7} to 10^{-10} N can be achieved, thus enabling AFM to be used just like a stylus profilometer to study the topography of the surface atoms [141]. The earliest successful experiment of atomic resolution was obtained for layered materials like graphite, molybdenum sulfide and boron nitride [142]. Lastly, to ensure that external vibrations do not influence the movement of the cantilever tip, AFM equipment design should be made rigid and compact, especially on the path from the cantilever to the sample [135].

3.4.8 Porosity Tests

Porosity is defined as the ratio of volume of pores to the volume of bulk material, and is usually expressed as a percentage [143]. Methods to measure porosity can be divided into two main types, void-sensitive methods and matrix-sensitive methods. Void sensitive methods use a fluid to measure characteristic properties of the pores that are accessible from the surface of the sample, namely open pores and open porosity. Matrix-sensitive methods exploit the interaction of electromagnetic radiation with the sample, and typically provide a measure of the total porosity (open and closed pores) [144]. Void-sensitive methods include helium pycnometry [145], mercury porosimetry [146], nitrogen adsorption [147] and thermosporometry [148]. On the other hand, matrix sensitive methods include scanning electron microscopy (SEM) [149], atomic force microscopy [150], focused ion beam SEM [151], confocal scanning microscopy [152], nuclear magnetic resonance [153], X-ray computed tomography [154] and terahertz time domain spectroscopy (THz-TDS) [155]. In previous sections, we have studied the various types of matrix-sensitive methods to study porosity. In this section, we will look into some of the most commonly used void-sensitive methods to measure the porosity of porous solids. In this section, the porosity of porous membrane supports used for FTMs will also be measured using gravimetric analysis, as was done in a research work carried out by Singh et al. [156].

(a) **Mercury Porosimetry**

Mercury porosimetry is one of the most widely used methods for the determination of pore distribution in solids. It allows for the determination of both mesopores (having pore radius of 1–25 nm) and macro-pores (having pore radius greater than 25 nm). The pore size distribution is determined by the volume of mercury that enters the pores under pressure. The measurement is based on Eq. (3.8) as shown below [157]:

$$p = \frac{2\pi\sigma\cos\alpha}{r_p} \tag{3.8}$$

Here, p is the pressure under which the mercury is absorbed into the pores, σ is the surface tension of mercury (Hg), α is the contact angle between the solid surface and mercury liquid and r_p is the radius of cylindrical pores. If the applied pressure is between 0.1 and 200 MPa, then pore sizes in the range of 3.75–7500 nm can be easily determined from this method. However, as the pores are not perfectly cylindrical, so there is bound to be marked difference between the actual pore size and the pore size measured by mercury porosimetry. The actual pore size can be more accurately measured using electron microscopy.

(b) **Gas Adsorption Analysis**

The specific pore volume, pore size and pore size distribution of microporous materials are determined by gas adsorption measurements at relatively low pressures (low values of p/p_0 = ratio of pressure and saturation pressure). This method is based on the pressure dependence of capillary condensation on the diameter of the pores in which this condensation takes place. To accurately measure the pore size distribution, the desorption isotherm is also determined. This makes a clear distinction between true adsorption and capillary condensation [157]. Capillary condensation is described by the Kelvin equation (Eq. 3.9) represented below [158]:

$$\ln\frac{p}{p_0} = \frac{V}{RT} = \frac{2\sigma\cos\theta}{r_p} \tag{3.9}$$

Here, V is the molar volume, σ is surface tension of adsorbate, θ is the contact angle between the adsorbate and the solid and r_p is the radius of the cylindrical pores. Another important parameter in porous solids is the specific surface area (measured in m^2/g). A surface area measurement is usually the measure of the accessible surface. This surface area is determined by first collecting physio-sorption data, which is a relationship between the amount of gas adsorbed and the equilibrium pressure. Then, various methods can be used to derive the characteristics of the nanomaterial. The most commonly used method to derive the specific surface area of a nanomaterial is the Brunauer–Emmett–Teller (BET) method [159]. In this method, special care must be taken for sample preparation, and adsorbed gases, moisture and other impurities must be removed prior to testing. Although, this method has its shortcomings, but it is fast and inexpensive for calculating the specific surface area (SSA), denoted as SA_{BET}, as long as care is taken in sample preparation and implementation of calibrants [160]. In the BET method, the amount of gas (usually N_2) that gives monomolecular coverage is measured, thus the total surface area is calculated. This monomolecular adsorption can be explained by the Langmuir equation (Eq. 3.10) [161]:

$$\frac{V}{V_M} = \frac{K(\frac{p}{p_0})}{[1 + K\left(\frac{p}{p_0}\right)]} \tag{3.10}$$

Here, K is a constant, V is the total volume of adsorbed gas and V_M is the volume of the monolayer. However, the Langmuir equation is only valid for p/p_0 values below 0.1. Above this value, multi-layer adsorption becomes important and so a theoretical model was developed by Brunauer et al. [162], which yielded the famous BET equation. This is represented in Eq. (3.11) below:

$$\frac{V}{V_M} = \frac{cp}{[p_0 - p][1 + (c - 1)p/p_0]} \tag{3.11}$$

In addition to the surface area, the difference in the surface area measured by BET method and by mercury porosimetry can give the micropores fraction of porous materials like zeolites, MOFs, ZIFs and other fillers used in FTMs. There is one major disadvantage encountered with BET characterization, that it can only be used for dried samples. Therefore, agglomeration or aggregation may lead to smaller values of specific surface area and it may be necessary to carry out disaggregation or fractionation on post-dried samples before analysis, so that a more accurate value of specific surface area may be calculated [163].

In BET specific surface area (SSA) analysis, N_2 adsorption isotherms are prepared by first degassing the porous solids at a temperature of 120 °C (393 K) at a pressure of 10^{-5} torr (0.0013 Pa). N_2 gas adsorption was then carried out at this pressure and at a temperature of − 196 °C (77 K). Desorption cycle is carried out at room temperature and at ambient pressure. In this way, N_2 adsorption isotherms are obtained for determining the specific surface area of porous solids [164].

(c) **Gravimetric Porosity Tests**

Gravimetric analysis has long been used to determine the porosity of porous membranes [165, 166], and as a porous support plays a vital role in the fabrication of composite FTMs, so a study of the use of this method of porosity measurement is also very important. To understand how this measurement technique works, let us consider a research work carried out by Karim et al. [167]. For this method, membrane samples were dried at 50 °C under vacuum for 2 h to remove moisture present in the membrane pores. Then, it was weighed using analytical balance. This weight was denoted as the weight of fully dried sample, W_d. Thereafter, the membrane samples were placed in a displacement fluid for 24 h and then weighed again. This weight was marked as the weight of wet samples, W_w. Then, membrane porosity (ε) was calculated using Eq. (3.12) represented below [168]:

$$\varepsilon = \frac{(W_w - W_d)/\rho_w}{\frac{(W_w - W_d)}{\rho_w} + W_d/\rho_p} * 100\% \tag{3.12}$$

Here, ρ_w is the density of displacement fluid, which was 2-propanol in this research work and ρ_p is the density of polymer which makes up the polymeric membrane, which is polyvinyl alcohol (PVA) in this research work. It is known that 2-propanol has a density of 0.785 g/cm^3 whereas PVA has a density of 1.19 g/cm^3. Using these values and the weights of wet and dried samples, the porosity of the porous support membrane samples can be easily calculated.

3.4.9 Nuclear Magnetic Resonance (NMR) Analysis

Nuclear magnetic resonance, as a phenomenon, was discovered by Purcell et al. at the Massachusetts Institute of Technology (MIT) in 1946 [169]. Independently, Felix Bloch also discovered the use of nuclear magnetic resonance at Stanford University, which later led to the development of magnetic resonance imaging (MRI) [170]. Today, NMR has developed itself into a powerful and complex analytical tool that can be used to identify compounds in solution as well as in solid state. It can equally be used to gather structural information of molecular compounds with excellent degree of accuracy. NMR collects information concerning nuclei of certain atoms present in the sample when they are subjected to a very powerful, constant, static magnetic field. These nuclei are then subjected to a second oscillating magnetic field, which is actually emitted by electromagnetic radiation in the radio-frequency regime. Moreover, the second magnetic field is about 10,000 times weaker in intensity than the first constant magnetic field. The results from NMR analysis are represented in the form of NMR spectrum. The spectrum represents signals of resonance which result from the absorption by the sample of definite frequencies sent by the electromagnetic source [171].

All atomic nuclei and all fundamental particles are represented by a set of intrinsic parameters. Spin is one of those intrinsic parameters, which represents the behaviour of atoms in an environment where its presence is influenced by orientation and direction, such as in a magnetic field. Every nucleus with a non-zero quantum number, is influenced by a magnetic field, and this influence is expressed by its spin quantum number [172]. Therefore, such nuclei can behave as a small magnet with a magnetic moment. If this nuclear magnet is then submitted to a magnetic field of intensity B_0, then there will be an interaction between the nucleus having magnetic moment μ and the external magnetic field of intensity B_0. Thus, these two magnetic fields will have energy of interaction (also called a Hamiltonian), represented as in Eq. (3.13) below:

$$E = -\overrightarrow{\mu}.\overrightarrow{B_0} = -\mu B_0 \cos\theta \tag{3.13}$$

In a nuclide represented by ${}^{A}_{Z}X$, the nucleus will have a non-zero spin quantum number if the number of protons Z and the number of nucleons A are not both even number, for example ${}^{1}_{1}H$, ${}^{13}_{6}C$ and ${}^{14}_{7}N$. All these nuclei having a non-zero spin quantum number will give an NMR signal and can be studied using NMR spectroscopy [173].

In any macroscopic sample containing nuclei having non-zero spin quantum numbers, all nuclei are oriented in a random direction in the absence of a magnetic field. In this state, the nuclei are said to be in a *degenerate state*. When an external strong magnetic field is applied, it causes two groups of nuclei to form, those having high energy E_2 and those having low energy E_1. Thus, the application of the strong magnetic field breaks off the degeneracy of the sample of nuclei [174]. These two orientations are called *spin-up* and *spin-down* states. This division of nuclei into two groups based on orientation is called *Zeeman splitting* [175]. The difference in energy ΔE between the two spin states is calculated through the Maxwell–Boltzmann Distribution using Eq. (3.14) shown below [176, 177]:

$$\Delta E = E_2 - E_1 = \gamma \frac{h}{2\pi} B_0 \tag{3.14}$$

Here, h is the Planck's constant, B_0 is the magnetic field intensity (in Tesla, T) and γ is the gyromagnetic ratio (or the gyromagnetic constant). The gyromagnetic ratio is the ratio of the magnetic momentum of a particle to its angular momentum and it is expressed in radians per second per Tesla (rad s^{-1} T^{-1}) [178]. The energy difference between the two spin states ΔE is equal to the amount of energy absorbed by the nucleus to show an NMR signal. Moreover, ΔE is proportional to the magnetic field intensity, B_0. At a magnetic field intensity of 1.4 T, the energy absorbed by a proton (hydrogen nuclide) is equal to 3.95×10^{-26} J, which is hardly enough energy to be detected by an spectroscopic technique. Therefore, a strong magnetic field is needed for this technique.

Given a sample of nuclides and knowing the suitable magnetic field intensity that needs to be applied, we can find the frequency of radiation at which the low-energy nuclides transition into the higher energy state. This can be done by irradiating the sample of nuclides with electromagnetic radiation of variable frequency. This experiment can be continued until the nuclei start to absorb at a particular frequency of radiation, this frequency is called the *Larmor frequency.* The Larmor frequency can be calculated by the Larmor Equation, which is given as Eq. (3.15) below [179]:

$$f_L = \frac{\gamma B_0}{2\pi} \tag{3.15}$$

As the gyromagnetic ratio of a proton is 2.68×10^8 rad T^{-1} s^{-1}, so at a magnetic field intensity ranging from 1 to 4 T, the Larmor frequency ranges between 43 to 170 MHz. This frequency is in the radiofrequency regime of the electromagnetic spectrum. Joseph Larmor, an Irish physicist, was responsible for determining the Larmor frequency, which gives us the energy required for the excitation of nuclei from a low energy spin state to a higher energy spin state. He carried out his work long before the discovery of nuclear magnetic resonance by Bloch and Purcell. Therefore, he arrived at his results independently. In fact, his results laid the foundation for a thorough study of NMR, until the advent of quantum mechanics [180, 181].

Until now, the type of NMR that we have been studying is known as the continuous wave NMR (CW-NMR). In this type of NMR spectroscopy, the nuclei are continuously bombarded with radiofrequency radiation in a strong magnetic field as was discussed until now. However, one problem with this type of NMR is that the nucleus that is shrouded in the strong magnetic field is too weak to be measured on its own. So, to get a resonance signal from this nucleus, the strong magnetic field is superimposed with a weak magnetic field that is originating from a coil that is fed with alternating current radiofrequency. As the alternating current creates magnetic field in pulses, so the NMR signal obtained from this type of arrangement is also called *pulsed NMR* [182].

After the pulse has been absorbed by the nucleus, it needs some time to return to its equilibrium value. This is known as nuclear relaxation and the time required to achieve this relaxation is called relaxation time. It is done in two ways. One is the reconstitution of the population to its original state. This is known as longitudinal relaxation or spin–lattice relaxation and it takes a time constant T_1 to complete. The second type of relaxation is the relaxation due to loss of phase coherence. This is achieved in time constant T_2 and is known as transverse relaxation or spin–spin relaxation [171]. Spin–lattice relaxation is the irreversible evolution of a spin system towards thermal equilibrium with the orbital degrees of freedom of the medium in which the spins are embedded, called the lattice [183]. Spin–lattice relaxation or longitudinal relaxation requires a stimulus such as a magnetic field fluctuating at a frequency equal to the frequency of the transition [184]. Spin–spin relaxation or transverse relaxation refers to the exchange of magnetization in a direction perpendicular to the applied magnetic field [185]. In addition to loss of phase coherence, transverse relaxation is also due to the heteronuclear Overhauser effect which results from cross-relaxation between dipolar-coupled spins [186].

Now that we have thoroughly studied how magnetic resonance is achieved and how signals are generated. Now, we will look at the main aspects of sample preparation for solution-state NMR as solution-state NMR is the classical approach used for molecular structure determination, in which the test sample is being studied while in solution [187]. The most commonly used solvents for preparing samples for NMR include deuterated chloroform ($CDCl_3$), acetone-d_6, DMSO-d_6, acetonitrile-d_3, methanol-d_4 and heavy water (D_2O). Other more sustainable and green alternative solvents are also being investigated including 2-methyl-tetrahydrofuran (2-Me-THF), iso-propyl acetate and n-heptane [188]. Deuterated solvents are used to suppress the large solvent signal that would otherwise occur in the proton NMR spectrum, if ordinary solvents were used [189]. In addition to these components of the sample, an internal reference is also added into the sample, which would give the strongest signal and against which all other sample components would be compared. Tetramethylsilane (TMS) is the most commonly used internal reference for proton NMR spectroscopy. This is because it contains 12 protons which give a strong, sharp resonance line with a chemical shift at a low resonance frequency, which is lower than almost all other proton resonances. Thus, it does not interfere with other references. Chemical shift is a very important concept which forms the cornerstone for the chemical applications of NMR. Chemical shift is actually the concept that a nucleus does not experience the entire intensity of the magnetic field applied to the complete sample, instead it only experiences the magnetic field which reaches it after the shielding and screening of the electrons surrounding it. In this way, NMR studies and identifies each nucleus separately and not the sample as a whole. Every single nucleus, therefore, has its own chemical shift relative to its position in the sample as a whole [190].

3.5 Current Trends and Future Recommendations

Although, facilitated transport membranes (FTMs) have their many advantages which propelled their initial use, including the fact that they provide high selectivity in conjunction with high flux, because they incorporate a filler into the polymer matrix which reacts reversibly with our gas of interest (mainly CO_2) [19]. However, they have some major drawbacks which have limited their use in industrial applications. One major drawback is the instability of the carrier particles, as they are highly prone to degradation and evaporation [191]. On the other hand, there are emerging examples of FTMs being used in pilot-plant scale and field tests in various industries like cement industry, and this development is indeed encouraging [192, 193]. Moreover, it can be seen that recent research works are very focused on supported ionic liquid (SIL) FTMs [194, 195]. Looking at the current trends, it can be said that the research is based on the right track as far as FTMs are concerned. However, new and improved synthesis methods for carrier particles must be developed so that they do not disintegrate at the high temperature and pressure conditions encountered in the industry. Moreover, polymer plasticization and polymer degradation at those high temperatures must also be worked upon. Also, work needs to be done not only on gas separation, but also on the long-term durability of FTMs in the industrial environment. Tests need to be conducted to check the durability of FTMs and strategies need to be developed to improve the durability. In spite of all its advantages, in reality, membrane technology for carbon capture still does not possess any real economic advantage over conventional techniques. For this purpose, FTMs can play a vital role in providing better carbon capture at lower cost and lesser energy demand. Work is underway in this regard [196], but efforts need to be more concerted and coordinated to achieve the goal.

3.6 Conclusion

Various methods for the synthesis and characterization of carrier particles and facilitated transport membranes (FTMs) are studied and investigated. The procedures followed for the synthesis of various types of fillers including fixed site carriers, mobile carriers and hybrid carriers, have been outlined. The advantages and shortcomings of the methods of fabrication have also been mentioned in detail. In addition to that, the utilization of various characterization techniques for analysing the different properties of carrier particles and FTMs has also been studied. Several barriers to the development of FTM systems on industrial scale have been identified and recommendations to overcome those barriers have been proposed. As a whole, this work provides a thorough representation of the state-of-the-art developments in FTM systems and corresponding carrier particle fabrication.

References

1. Ismail AF, Khulbe KC, Matsuura T (2015) Membrane fabrication/manufacturing techniques. In: Gas separation membranes. Springer, Berlin, pp 193–220
2. Van der Bruggen B (2000) Membrane technology, pp 1–47
3. Loeb S, Sourirajan S (1962) Sea water demineralization by means of an osmotic membrane. ACS Publications
4. Puetz J, Aegerter MA (2004) Dip coating technique. In: Aegerter MA, Mennig M (eds) Sol-gel technologies for glass producers and users. Springer, Boston, pp 37–48
5. Larson RG, Rehg TJ (1997) Spin coating. In: Liquid film coating. Springer, Berlin, pp 709–734
6. Salahuddin Z, Farrukh S, Hussain A (2018) Optimization study of polyethylene glycol and solvent system for gas permeation membranes. Int J Polym Anal Charact 23(5):483–492

7. Dai Z, Ansaloni L, Deng L (2016) Recent advances in multi-layer composite polymeric membranes for CO_2 separation: a review. Green Ener Environ 1(2):102–128
8. Julian H, Wenten I (2012) Polysulfone membranes for CO_2/CH_4 separation: state of the art. IOSR J Eng 2(3):484–495
9. Turken T, Sengur-Tasdemir R, Ates-Genceli E, Tarabara VV, Koyuncu I (2019) Progress on reinforced braided hollow fiber membranes in separation technologies: a review. J Water Process Eng 32:100938
10. Tang M-W, King WM, Wensley CG (1989) Air dried cellulose acetate membranes. Google Patents
11. Aroon M, Ismail A, Matsuura T, Montazer-Rahmati M (2010) Performance studies of mixed matrix membranes for gas separation: a review. Separ Purif Technol 75(3):229–242
12. Basile A, Gugliuzza A, Iulianelli A, Morrone P (2011) Membrane technology for carbon dioxide (CO_2) capture in power plants. In: Basile A, Nunes SP (eds) Advanced membrane science and technology for sustainable energy and environmental applications. Woodhead Publishing, pp 113–159. https://doi.org/10.1533/9780857093790.2.113
13. He Z, Kumar Reddy KS, Karanikolos G, Wang K (2018) Chapter 5—CO_2/CH_4 separation (natural gas purification) by using mixed matrix membranes. In: Basile A, Favvas EP (eds) Current trends and future developments on (bio-) membranes. Elsevier, pp 155–181. https://doi.org/10.1016/B978-0-12-813645-4.00005-2
14. Prakash M, Sakhavand N, Shahsavari R (2013) H_2, N_2, and CH_4 gas adsorption in zeolitic imidazolate framework-95 and-100: ab initio based grand canonical Monte Carlo simulations. J Phys Chem C 117(46):24407–24416
15. Mehmood O, Farrukh S, Hussain A, Younas M, Salahuddin Z, Pervaiz E, Ayoub M (2021) Investigation of cellulose acetate/gamma-cyclodextrin MOF based mixed matrix membranes for CO_2/CH_4 gas separation. 11(2):313–330
16. Salahuddin Z, Farrukh S, Hussain A, Noor T, Kwapinski WJP (2021) Mixed and single gas permeation performance analysis of amino-modified ZIF based mixed matrix membrane. Polym Polym Compos 29(9):S707–S718
17. Qiblawey H (2022) Development of novel composite membranes in water/wastewater treatment. Membranes 12(3):260
18. Pinnau I (2000) Membrane separations | membrane preparation. In: Wilson ID (ed) Encyclopedia of separation science. Academic Press, Oxford, pp 1755–1764. https://doi.org/10.1016/B0-12-226770-2/05241-8
19. Doong SJ (2012) Membranes, adsorbent materials and solvent-based materials for syngas and hydrogen separation. In: Kilner JA, Skinner SJ, Irvine SJC, Edwards PP (eds) Functional materials for sustainable energy applications. Woodhead Publishing, pp 179–216. https://doi.org/10.1533/9780857096371.2.179
20. Kaldis SP, Pantoleontos GT, Koutsonikolas DE (2018) Chapter 12—Membrane technology in IGCC processes for precombustion CO_2 capture. In: Basile A, Favvas EP (eds) Current trends and future developments on (bio-) membranes. Elsevier, pp 329–357. https://doi.org/10.1016/B978-0-12-813645-4.00012-X
21. Basile A, Gugliuzza A, Iulianelli A, Morrone P (2011) Membrane technology for carbon dioxide (CO_2) capture in power plants. In: Advanced membrane science and technology for sustainable energy and environmental applications. Elsevier, pp 113–159
22. Karim SS, Farrukh S, Hussain A, Noor T, Younas M (2021) A comprehensive overview of dual-layer composite membrane for air (O_2/N_2) separation 29(9):S1630–S1640
23. Raza A, Yousaf MF, Farrukh S, Hussain A (2021) Thickness effect on permeance of CO_2/CH_4 gases in CA coated PVDF composite membranes. Trans Indian Ceram Soc 80(2):89–95
24. Ayub A, Farrukh S, Jan R, Azeem M, Salahuddin Z, Hussain AJAN (2021) Gas barrier properties evaluation for boron nitride nanosheets-polymer (polyethylene-terephthalate) composites. Appl Nanosci 11 (1):91–99
25. Hussain A, Farrukh S, Hussain A, Ayoub M (2019) Carbon capture from natural gas using multi-walled CNTs based mixed matrix membranes. Environ Technol 40(7):843–854

26. Xie K, Fu Q, Qiao GG, Webley PA (2019) Recent progress on fabrication methods of polymeric thin film gas separation membranes for CO_2 capture. J Membr Sci 572:38–60
27. Yousaf MF, Farrukh S, Hussain A, Hussain A (2021) Fabrication of CA/PEG/MWCNTs composite membranes for enhanced CO_2/CH_4 separation 23(4):448–462
28. Mohammadzadeh A, Zadeh SKN, Saidi MH, Sharifzadeh M (2020) Mechanical engineering of solid oxide fuel cell systems: geometric design, mechanical configuration, and thermal analysis. In: Design and operation of solid oxide fuel cells. Elsevier, pp 85–130
29. Cadotte J, King R, Majerle R, Petersen R (1981) Interfacial synthesis in the preparation of reverse osmosis membranes. J Macromol Sci Part A Chem 15(5):727–755
30. Castro RP, Baker RW, Wijmans JG (1991) Multilayer interfacial composite membrane. Google Patents
31. Cheng XQ, Zhang YL, Wang ZX, Guo ZH, Bai YP, Shao L (2014) Recent advances in polymeric solvent-resistant nanofiltration membranes. Adv Polym Technol 33(S1)
32. Cai Y, Wang Z, Yi C, Bai Y, Wang J, Wang S (2008) Gas transport property of polyallylamine–poly(vinyl alcohol)/polysulfone composite membranes. J Membr Sci 310(1):184–196. https://doi.org/10.1016/j.memsci.2007.10.052
33. Gurnani P, Perrier S (2020) Controlled radical polymerization in dispersed systems for biological applications. Prog Polym Sci 102:101209
34. Baruah C, Sarmah JK (2022) Guar gum-based hydrogel and hydrogel nanocomposites for biomedical applications. In: Micro-and nano-engineered gum-based biomaterials for drug delivery and biomedical applications. Elsevier, pp 473–492
35. Mustafa HAM, Jameel DA (2021) Modeling and the main stages of spin coating process: a review. J Appl Sci Technol Trends 2(03):91–95
36. Choy KL (2003) Chemical vapour deposition of coatings. Prog Mater Sci 48(2):57–170. https://doi.org/10.1016/S0079-6425(01)00009-3
37. Asatekin A, Gleason KK (2011) Polymeric nanopore membranes for hydrophobicity-based separations by conformal initiated chemical vapor deposition. Nano Lett 11(2):677–686
38. Sun L, Yuan G, Gao L, Yang J, Chhowalla M, Gharahcheshmeh MH, Gleason KK, Choi YS, Hong BH, Liu Z (2021) Chemical vapour deposition. Nat Rev Methods Primers 1(1):1–20
39. Tilli M et al (2015) Chapter 6—Thin films on silicon. In: Tilli M, Motooka T, Airaksinen V-M, Franssila S, Paulasto-Kröckel M, Lindroos V (eds) Handbook of silicon based MEMS materials and technologies, 2nd edn. William Andrew Publishing, Boston, pp 124–205. https://doi.org/10.1016/B978-0-323-29965-7.00006-3
40. Leskelä M, Niinistö J, Ritala M (2014) Atomic layer deposition. In: Hashmi S, Batalha GF, Van Tyne CJ, Yilbas B (eds) Comprehensive materials processing. Elsevier, Oxford, pp 101–123. https://doi.org/10.1016/B978-0-08-096532-1.00401-5
41. Higashi G, Fleming C (1963) Sequential surface chemical reaction limited growth of high quality aluminum oxide dielectrics. JAPL 55:1989
42. Soto C, Tysoe WT (1991) The reaction pathway for the growth of alumina on high surface area alumina and in ultrahigh vacuum by a reaction between trimethyl aluminum and water. J Vacuum Sci Technol A: Vacuum Surf Films 9(5):2686–2695
43. Dillon A, Ott A, Way J, George S (1995) Surface chemistry of Al_2O_3 deposition using $Al(CH_3)3$ and H_2O in a binary reaction sequence. Surf Sci 322(1–3):230–242
44. Bishop CA (2011) Atomic layer deposition. In: Bishop CA (ed) Vacuum deposition onto webs, films and foils, 2nd edn. William Andrew Publishing, Oxford, pp 331–336. https://doi.org/10.1016/B978-1-4377-7867-0.00019-2
45. Oviroh PO, Akbarzadeh R, Pan D, Coetzee RAM, Jen T-C (2019) New development of atomic layer deposition: processes, methods and applications. Sci Technol Adv Mater 20(1):465–496
46. Becker M, Sierka M (2019) Atomistic simulations of plasma-enhanced atomic layer deposition. Materials 12(16):2605
47. Barron AR (2013) Chemistry of electronic materials
48. Akhmetshina AI, Yanbikov NR, Atlaskin AA, Trubyanov MM, Mechergui A, Otvagina KV, Razov EN, Mochalova AE, Vorotyntsev IVJM (2019) Acidic gases separation from gas mixtures on the supported ionic liquid membranes providing the facilitated and solution-diffusion transport mechanisms. J Membr 9(1):9

49. Ohyanagi M, Nishide H, Suenaga K, Tsuchida E (1988) Effect of polymer matrix and metal species on facilitated oxygen transport in metalloporphyrin (oxygen carrier) fixed membranes. Macromolecules 21(6):1590–1594
50. Su NC, Sun DT, Beavers CM, Britt DK, Queen WL, Urban J (2016) Enhanced permeation arising from dual transport pathways in hybrid polymer–MOF membranes. Ener Environ Sci 9(3):922–931
51. Qiao Z, Zhao S, Sheng M, Wang J, Wang S, Wang Z, Zhong C, Guiver MD (2019) Metal-induced ordered microporous polymers for fabricating large-area gas separation membranes. Nat Mater 18(2):163–168
52. Wang B, Qiao Z, Xu J, Wang J, Liu X, Zhao S, Wang Z, Guiver MD (2020) Unobstructed ultrathin gas transport channels in composite membranes by interfacial self-assembly. Adv Mater 32(22):1907701
53. Yuan Y, Qiao Z, Xu J, Wang J, Zhao S, Cao X, Wang Z, Guiver MD (2021) Mixed matrix membranes for CO_2 separations by incorporating microporous polymer framework fillers with amine-rich nanochannels. J Membr Sci 620:118923
54. Xu R, Wang Z, Wang M, Qiao Z, Wang J (2019) High nanoparticles loadings mixed matrix membranes via chemical bridging-crosslinking for CO_2 separation. J Membr Sci 573:455–464
55. Japip S, Liao KS, Chung TS (2017) Molecularly tuned free volume of vapor cross-linked 6FDA-Durene/ZIF-71 MMMs for H_2/CO_2 separation at 150°C. J Adv Mater 29(4):1603833
56. Zhang N, Wu H, Li F, Dong S, Yang L, Ren Y, Wu Y, Wu X, Jiang Z, Cao X (2018) Heterostructured filler in mixed matrix membranes to coordinate physical and chemical selectivities for enhanced CO_2 separation. J Membr Sci 567:272–280
57. Liu P-F, Tao K, Li G-C, Wu M-K, Zhu S-R, Yi F-Y, Zhao W-N, Han L (2016) In situ growth of ZIF-8 nanocrystals on layered double hydroxide nanosheets for enhanced CO_2 capture. Dalton Trans 45(32):12632–12635
58. Huang L, Wang H, Chen J, Wang Z, Sun J, Zhao D, Yan Y (2003) Synthesis, morphology control, and properties of porous metal–organic coordination polymers. Microporous Mesoporous Mater 58(2):105–114
59. Ni Z, Masel RI (2006) Rapid production of metal-organic frameworks via microwave-assisted solvothermal synthesis. J Am Chem Soc 128(38):12394–12395
60. Qiu L-G, Li Z-Q, Wu Y, Wang W, Xu T, Jiang X (2008) Facile synthesis of nanocrystals of a microporous metal–organic framework by an ultrasonic method and selective sensing of organoamines. Chem Commun 31:3642–3644
61. Cravillon J, Münzer S, Lohmeier S-J, Feldhoff A, Huber K, Wiebcke M (2009) Rapid room-temperature synthesis and characterization of nanocrystals of a prototypical zeolitic imidazolate framework. Mater Sci 21(8):1410–1412
62. West AR (1999) Basic solid state chemistry. Wiley, New York
63. Babooram K (2008) Novel solution routes to ferroelectrics and relaxors. In: Ye Z-G (ed) Handbook of advanced dielectric, piezoelectric and ferroelectric materials. Woodhead Publishing, pp 852–883. https://doi.org/10.1533/9781845694005.7.852
64. Liu J, Song J, Xiao H, Zhang L, Qin Y, Liu D, Hou W, Du N (2014) Synthesis and thermal properties of ZnAl layered double hydroxide by urea hydrolysis. Powder Technol 253:41–45. https://doi.org/10.1016/j.powtec.2013.11.007
65. Xu Q-T, Li J-C, Xue H-G, Guo S-P (2018) Binary iron sulfides as anode materials for rechargeable batteries: crystal structures, syntheses, and electrochemical performance. J Power Sour 379:41–52. https://doi.org/10.1016/j.jpowsour.2018.01.022
66. Mubashir M, Yeong YF, Lau KK, Chew TL, Norwahyu JJS, Technology P (2018) Efficient CO_2/N_2 and CO_2/CH_4 separation using NH_2-MIL-53 (Al)/cellulose acetate (CA) mixed matrix membranes. Sep Purif Technol 199:140–151
67. Cheng X, Zhang A, Hou K, Liu M, Wang Y, Song C, Zhang G, Guo X (2013) Size- and morphology-controlled NH_2-MIL-53 (Al) prepared in DMF-water mixed solvents 42(37):13698–13705
68. Stöber W, Fink A, Bohn E (1968) Controlled growth of monodisperse silica spheres in the micron size range. J Colloid Interface Sci 26(1):62–69. https://doi.org/10.1016/0021-9797(68)90272-5

69. Zhang H, Guo R, Zhang J, Li X (2018) Facilitating CO_2 transport across mixed matrix membranes containing multifunctional nanocapsules. ACS Appl Mater Interfaces 10(49):43031–43039. https://doi.org/10.1021/acsami.8b15269
70. Nair GB, Dhoble SJ (2021) Phosphor-converted LEDs. In: Nair GB, Dhoble SJ (eds) The fundamentals and applications of light-emitting diodes. Woodhead Publishing, pp 87–126. https://doi.org/10.1016/B978-0-12-819605-2.00004-5
71. Feng SH, Li GH (2017) Chapter 4—Hydrothermal and solvothermal syntheses. In: Xu R, Xu Y (eds) Modern inorganic synthetic chemistry, 2nd edn. Elsevier, Amsterdam, pp 73–104. https://doi.org/10.1016/B978-0-444-63591-4.00004-5
72. Thür R, Van Velthoven N, Slootmaekers S, Didden J, Verbeke R, Smolders S, Dickmann M, Egger W, De Vos D, Vankelecom IFJ (2019) Bipyridine-based UiO-67 as novel filler in mixed-matrix membranes for CO_2-selective gas separation. J Membr Sci 576:78–87. https://doi.org/10.1016/j.memsci.2019.01.016
73. Zhong H, Mirkovic T, Scholes GD (2011) Nanocrystal synthesis. In: Andrews DL, Scholes GD, Wiederrecht GP (eds) Comprehensive nanoscience and technology. Academic Press, Amsterdam, pp 153–201. https://doi.org/10.1016/B978-0-12-374396-1.00051-9
74. Wang M, Quan K, Zheng X, Cao Y, Cui X, Xue M, Pan F (2020) Facilitated transport membranes by incorporating self-exfoliated covalent organic nanosheets for CO_2/CH_4 separation. Sep Purif Technol 237:116457. https://doi.org/10.1016/j.seppur.2019.116457
75. Chong JH, Sauer M, Patrick BO, MacLachlan MJ (2003) Highly stable keto-enamine salicylideneanilines. Org Lett 5(21):3823–3826. https://doi.org/10.1021/ol0352714
76. Md Nordin NAH, Racha SM, Matsuura T, Misdan N, Abdullah Sani NA, Ismail AF, Mustafa A (2015) Facile modification of ZIF-8 mixed matrix membrane for CO_2/CH_4 separation: synthesis and preparation. RSC Adv 5(54):43110–43120. https://doi.org/10.1039/C5RA02230D
77. Nordin NAHM, Ismail AF, Mustafa A, Goh PS, Rana D, Matsuura T (2014) Aqueous room temperature synthesis of zeolitic imidazole framework 8 (ZIF-8) with various concentrations of triethylamine. RSC Adv 4(63):33292–33300. https://doi.org/10.1039/C4RA03593C
78. Colilla M, Vallet-Regí M (2017) Ordered mesoporous silica materials. In: Ducheyne P (ed) Comprehensive biomaterials II. Elsevier, Oxford, pp 644–685. https://doi.org/10.1016/B978-0-12-803581-8.10231-0
79. Han YR, Kim JS, Park W-J, Lee C-H, Cheon J, Jun C-H (2021) Recyclable transition metal catalysis using bipyridine-functionalized SBA-15 by co-condensation of methallylsilane with TEOS. Chem Asian J 16(3):197–201. https://doi.org/10.1002/asia.202001152
80. Gottuso A, Armetta F, Cataldo A, Nardo VM, Parrino F, Saladino ML (2022) Functionalization of mesoporous silica nanoparticles through one-pot co-condensation in w/o emulsion. Microporous Mesoporous Mater 335:111833. https://doi.org/10.1016/j.micromeso.2022.111833
81. Gedanken A (2004) Using sonochemistry for the fabrication of nanomaterials. Ultrason Sonochem 11(2):47–55. https://doi.org/10.1016/j.ultsonch.2004.01.037
82. Suslick Kenneth S (1990) Sonochemistry. Science 247(4949):1439–1445. https://doi.org/10.1126/science.247.4949.1439
83. Qiao SZ, Liu J, Max Lu GQ (2017) Chapter 21—Synthetic chemistry of nanomaterials. In: Xu R, Xu Y (eds) Modern inorganic synthetic chemistry, 2nd edn. Elsevier, Amsterdam, pp 613–640. https://doi.org/10.1016/B978-0-444-63591-4.00021-5
84. Suslick KS, Choe S-B, Cichowlas AA, Grinstaff MW (1991) Sonochemical synthesis of amorphous iron. Nature 353(6343):414–416. https://doi.org/10.1038/353414a0
85. Khan R, Riaz A, Javed S, Jan R, Akram MA, Mujahid M (2018) Synthesis and characterization of MoS_2/TiO_2 nanocomposites for enhanced photocatalytic degradation of methylene blue under sunlight irradiation. Key Eng Mater 778:137–143. https://doi.org/10.4028/www.scientific.net/KEM.778.137
86. Khan R, Riaz A, Rabeel M, Javed S, Jan R, Akram MA (2019) TiO_2@$NbSe_2$ decorated nanocomposites for efficient visible-light photocatalysis. Appl Nanosci 9(8):1915–1924. https://doi.org/10.1007/s13204-019-01020-6

87. Han G-Q, Liu Y-R, Hu W-H, Dong B, Li X, Chai Y-M, Liu Y-Q, Liu C-G (2015) WS2 nanosheets based on liquid exfoliation as effective electrocatalysts for hydrogen evolution reaction. Mater Chem Phys 167:271–277. https://doi.org/10.1016/j.matchemphys.2015.10.043
88. He S, Chen Y, Zhang Z, Ni B, He W, Wang X (2016) Competitive coordination strategy for the synthesis of hierarchical-pore metal–organic framework nanostructures. Chem Sci 7(12):7101–7105. https://doi.org/10.1039/C6SC02272C
89. Cho G, Park Y, Hong Y-K, Ha D-H (2019) Ion exchange: an advanced synthetic method for complex nanoparticles. Nano Convergence 6(1):17. https://doi.org/10.1186/s40580-019-0187-0
90. Cho G, Park Y, Youngun H, Ha D-H (2019) Ion exchange: an advanced synthetic method for complex nanoparticles. Nano Convergence. https://doi.org/10.1186/s40580-019-0187-0
91. Wang C, Guo F, Li H, Xu J, Hu J, Liu H, Wang M (2020) A porous ionic polymer bionic carrier in a mixed matrix membrane for facilitating selective CO_2 permeability. J Membr Sci 598:117677. https://doi.org/10.1016/j.memsci.2019.117677
92. Wang S, Li X, Wu H, Tian Z, Xin Q, He G, Peng D, Chen S, Yin Y, Jiang Z, Guiver MD (2016) Advances in high permeability polymer-based membrane materials for CO_2 separations. Energy Environ Sci 9(6):1863–1890. https://doi.org/10.1039/C6EE00811A
93. Brodie BC (1859) On the atomic weight of graphite. Philos Trans R Soc Lond 149:249–259. https://doi.org/10.1098/rstl.1859.0013
94. Staudenmaier L (1898) Verfahren zur Darstellung der Graphitsäure. Ber Dtsch Chem Ges 31(2):1481–1487. https://doi.org/10.1002/cber.18980310237
95. Hummers WS, Offeman RE (1958) Preparation of graphitic oxide. J Am Chem Soc 80(6):1339–1339. https://doi.org/10.1021/ja01539a017
96. Higginbotham AL, Lomeda JR, Morgan AB, Tour JM (2009) Graphite oxide flame-retardant polymer nanocomposites. ACS Appl Mater Interfaces 1(10):2256–2261. https://doi.org/10.1021/am900419m
97. Kaur M, Kaur H, Kukkar D (2018) Synthesis and characterization of graphene oxide using modified Hummer's method. AIP Conf Proc 1953(1):030180. https://doi.org/10.1063/1.5032515
98. Ying W, Cai J, Zhou K, Chen D, Ying Y, Guo Y, Kong X, Xu Z, Peng X (2018) Ionic liquid selectively facilitates CO_2 transport through graphene oxide membrane. ACS Nano 12(6):5385–5393. https://doi.org/10.1021/acsnano.8b00367
99. Huang H, Mao Y, Ying Y, Liu Y, Sun L, Peng X (2013) Salt concentration, pH and pressure controlled separation of small molecules through lamellar graphene oxide membranes. Chem Commun 49(53):5963–5965. https://doi.org/10.1039/C3CC41953C
100. Saeed U, Jahan Z, Niazi MBK, Pervaiz E, Sher F (2020) Biogas upgrading with novel cellulose nano-crystals and polyvinyl amine nanocomposite membranes. Polym Testing 91:106867. https://doi.org/10.1016/j.polymertesting.2020.106867
101. Skoog DA, Holler FJ, Crouch SR (2017) Principles of instrumental analysis. 7 edn. Cengage learning, U.S
102. Workman JJ (1996) Interpretive spectroscopy for near infrared. Appl Spectrosc Rev 31(3):251–320. https://doi.org/10.1080/05704929608000571
103. Duxbury G (2000) Infrared Spectroscopy: From Free Radicals to the Infrared Sky. 1st edn. Wiley & Sons, London
104. Cocchi M, Foca G, Lucisano M, Marchetti A, Pagani MA, Tassi L, Ulrici A (2004) Classification of cereal flours by chemometric analysis of MIR spectra. J Agric Food Chem 52(5):1062–1067. https://doi.org/10.1021/jf034441o
105. Undavalli VK, Ling C, Khandelwal B (2021) Chapter 6—Impact of alternative fuels and properties on elastomer compatibility. In: Khandelwal B (ed) Aviation fuels. Academic Press, pp 113–132. https://doi.org/10.1016/B978-0-12-818314-4.00001-7
106. Bernath PF (2019) Fourier transform techniques. In: Worsfold P, Poole C, Townshend A, Miró M (eds) Encyclopedia of analytical science, 3rd edn. Academic Press, Oxford, pp 81–89. https://doi.org/10.1016/B978-0-12-409547-2.14518-4

107. Sun D, Siddiqui MOR, Iqbal K (2019) Specialty testing techniques for smart textiles. In: Smith WC (ed) Smart textile coatings and laminates, 2nd edn. Woodhead Publishing, pp 99–116. https://doi.org/10.1016/B978-0-08-102428-7.00004-3
108. Altawell N (2022) Machine olfaction device nanostructure coating. In: Altawell N (ed) Introduction to machine olfaction devices. Academic Press, pp 139–157. https://doi.org/10.1016/B978-0-12-822420-5.00008-8
109. Goldstein JI, Newbury DE, Michael JR, Ritchie NWM, Scott JHJ, Joy DC (2018) Image formation. In: Goldstein JI, Newbury DE, Michael JR, Ritchie NWM, Scott JHJ, Joy DC (eds) Scanning electron microscopy and X-ray microanalysis. Springer, New York, pp 93–110. https://doi.org/10.1007/978-1-4939-6676-9_6
110. Sima F, Ristoscu C, Duta L, Gallet O, Anselme K, Mihailescu IN (2016) Laser thin films deposition and characterization for biomedical applications. In: Vilar R (ed) Laser surface modification of biomaterials. Woodhead Publishing, pp 77–125. https://doi.org/10.1016/B978-0-08-100883-6.00003-4
111. Seibert JA (1997) The AAPM/RSNA physics tutorial for residents. X-ray Gener 17(6):1533–1557
112. McCollough CH (1997) The AAPM/RSNA physics tutorial for residents. X-ray Prod 17(4):967–984
113. Hall EH (1927) Photo-Electric Emission, Thermionic Emission and Peltier Effect: (From the Point of View of Dual Electric Conduction). Proc Nat Acad Sci 13(2):43–46
114. Seibert JA (2004) X-ray imaging physics for nuclear medicine technologists. Part 1: Basic principles of x-ray production. J Nucl Med Technol 32(3):139–147
115. Seibert JA, Boone JM (2005) X-ray imaging physics for nuclear medicine technologists. Part 2: X-ray interactions and image formation. J Nucl Med Technol 33(1):3–18
116. Verhey LJ, Petti PL (2010) 7 - Principles of Radiation Physics. In: Hoppe RT, Phillips TL, Roach M (eds) Leibel and Phillips Textbook of Radiation Oncology (Third Edition). W.B. Saunders, Elsevier, Philadelphia, pp 95–119. https://doi.org/10.1016/B978-1-4160-5897-7.00007-X
117. Birkholz M (2006) Thin film analysis by X-ray scattering. 2006 edn. John Wiley & Sons 2006. https://doi.org/10.1107/S0021889806034698
118. Dudley M, Huang XR (2001) X-ray Topography. In: Buschow KHJ, Cahn RW, Flemings MC et al. (eds) Encyclopedia of Materials: Science and Technology. Elsevier, Oxford, pp 9813–9825. https://doi.org/10.1016/B0-08-043152-6/01780-0
119. Adams FC (2019) X-Ray Absorption and Diffraction | Overview. In: Worsfold P, Poole C, Townshend A, Miró M (eds) Encyclopedia of Analytical Science (Third Edition). Academic Press - Elsevier, Oxford, pp 391–403. https://doi.org/10.1016/B978-0-12-409547-2.14263-5
120. Bomer J, Wiersma-Deijl L, Holscher HC (2013) Electronic collimation and radiation protection in paediatric digital radiography: revival of the silver lining. Insights Imaging 4(5):723–727. https://doi.org/10.1007/s13244-013-0281-5
121. Streli C, Wobrauschek P, Kregsamer P (2017) X-Ray Fluorescence Spectroscopy, Applications. In: Lindon JC, Tranter GE, Koppenaal DW (eds) Encyclopedia of Spectroscopy and Spectrometry (Third Edition). Academic Press - Elsevier, Oxford, pp 707–715. https://doi.org/10.1016/B978-0-12-803224-4.00315-0
122. Bish DL, Reynolds RC (2018) Sample Preparation For X-ray Diffraction. In: David LB, Jeffrey EP (eds) Modern Powder Diffraction. De Gruyter, pp 73–100. https://doi.org/10.1515/9781501509018-007
123. He BB (2003) Introduction to two-dimensional X-ray diffraction. Powder Diffr 18(2):71–85. https://doi.org/10.1154/1.1577355
124. Parupudi A, Mulagapati SHR, Subramony JA (2022) Chapter 1 - Nanoparticle technologies: Recent state of the art and emerging opportunities. In: Kesharwani P, Singh KK (eds) Nanoparticle Therapeutics. Academic Press - Elsevier, pp 3–46. https://doi.org/10.1016/B978-0-12-820757-4.00009-0
125. Song K (2017) 4 - Interphase characterization in rubber nanocomposites. In: Thomas S, Maria HJ (eds) Progress in Rubber Nanocomposites. Woodhead Publishing Series in Composites

Science and Engineering - Elsevier, pp 115–152. https://doi.org/10.1016/B978-0-08-100409-8.00004-8
126. McMullan D (1995) Scanning electron microscopy 1928–1965. Scanning: The Journal of Scanning Microscopies (Wiley) 17(3):175–185. https://doi.org/10.1002/sca.4950170309
127. Tomoda BT, Yassue-Cordeiro PH, Ernesto JV, Lopes PS, Péres LO, da Silva CF, de Moraes MA (2020) Chapter 3—Characterization of biopolymer membranes and films: Physicochemical, mechanical, barrier, and biological properties. In: de Moraes MA, da Silva CF, Vieira RS (eds) Biopolymer membranes and films. Elsevier, pp 67–95. https://doi.org/10.1016/B978-0-12-818134-8.00003-1
128. Saka H (2003) Chapter 14—Transmission electron microscopy. In: Yasuda E-i, Inagaki M, Kaneko K, Endo M, Oya A, Tanabe Y (eds) Carbon alloys. Elsevier Science, Oxford, pp 223–238. https://doi.org/10.1016/B978-008044163-4/50014-0
129. Park K-h (1990) Cross-sectional TEM specimen preparation of semiconductor devices by focused ion beam etching. MRS Online Proceedings Library (OPL) 199:271–280. https://doi.org/10.1557/PROC-199-271
130. Kumar PS, Pavithra KG, Naushad M (2019) Chapter 4—Characterization techniques for nanomaterials. In: Thomas S, Sakho EHM, Kalarikkal N, Oluwafemi SO, Wu J (eds) Nanomaterials for solar cell applications. Elsevier, pp 97–124. https://doi.org/10.1016/B978-0-12-813337-8.00004-7
131. Romero Nieto D, Lindbråthen A, Hägg M-B (2017) Effect of water interactions on polyvinylamine at different pHs for membrane gas separation. ACS Omega 2(11):8388–8400. https://doi.org/10.1021/acsomega.7b01307
132. Binnig G, Rohrer H, Gerber C, Weibel E (1982) Surface studies by scanning tunneling microscopy. Phys Rev Lett 49(1):57. https://doi.org/10.1103/PhysRevLett.49.57
133. Wickramasinghe HK (1989) Scanned-probe microscopes Sci Am 261(4):98–105
134. Binnig G, Quate CF, Gerber C (1986) Atomic force microscope. Phys Rev Lett 56(9):930. https://doi.org/10.1103/PhysRevLett.56.930
135. Rugar D, Hansma P (1990) Atomic force microscopy. Phys Today 43(10):23–30. https://doi.org/10.1063/1.881238
136. Albrecht TR, Quate CF (1988) Atomic resolution with the atomic force microscope on conductors and nonconductors. J Vacuum Sci Technol A 6(2):271–274. https://doi.org/10.1116/1.575441
137. Thompson DO, Chimenti DE (2012) Preface: review of progress in quantitative nondestructive evaluation. AIP Conf Proc 1430(1):3–5. https://doi.org/10.1063/1.4716208
138. Meyer G, Amer NM (1988) Novel optical approach to atomic force microscopy. Appl Phys Lett 53(12):1045–1047. https://doi.org/10.1063/1.100061
139. Rugar D, Mamin H, Erlandsson R, Stern J, Terris B (1988) Force microscope using a fiber-optic displacement sensor. Rev Sci Instrum 59(11):2337–2340. https://doi.org/10.1063/1.1139958
140. Radmacher M, Cleveland J, Hansma PJS (1995) Improvement of thermally induced bending of cantilevers used for atomic force microscopy. Scanning 17(2):117–121
141. Binnig G, Gerber C, Stoll E, Albrecht T, Quate C (1987) Atomic resolution with atomic force microscope. Europhys Lett 3(12):1281
142. Heinzelmann H, Meyer E, Brodbeck D, Overney G, Güntherodt HJ (1992) Atomic-scale contrast mechanism in atomic force microscopy. Zeitschrift für Physik B Condensed Matter 88(3):321–326. https://doi.org/10.1007/BF01470920
143. White WB (2012) Hydrogeology of Karst Aquifers. In: White WB, Culver DC (eds) Encyclopedia of caves, 2nd edn. Academic Press, Amsterdam, pp 383–391. https://doi.org/10.1016/B978-0-12-383832-2.00054-2
144. Naftaly M, Tikhomirov I, Hou P, Markl D (2020) Measuring open porosity of porous materials using THz-TDS and an index-matching medium. Sensors 20(11). https://doi.org/10.3390/s20113120
145. AlKhatib HS, Hamed S, Mohammad MK, Bustanji Y, AlKhalidi B, Aiedeh KM, Najjar S (2010) Effects of thermal curing conditions on drug release from polyvinyl acetate–polyvinyl pyrrolidone matrices 11(1):253–266

146. Ridgway C, Bawuah P, Markl D, Zeitler JA, Ketolainen J, Peiponen K-E, Gane P (2017) On the role of API in determining porosity, pore structure and bulk modulus of the skeletal material in pharmaceutical tablets formed with MCC as sole excipient 526(1–2):321–331
147. Ferrero C, Jiménez-Castellanos M (2002) The influence of carbohydrate nature and drying methods on the compaction properties and pore structure of new methyl methacrylate copolymers 248(1–2):157–171
148. Luukkonen P, Maloney T, Rantanen J, Paulapuro H, Yliruusi J (2001) Microcrystalline cellulose-water interaction—a novel approach using thermoporosimetry 18(11):1562–1569
149. Boissier C, Feidt F, Nordstierna L (2012) Study of pharmaceutical coatings by means of NMR cryoporometry and SEM image analysis 101(7):2512–2522
150. Tuntikulwattana S, Mitrevej A, Kerdcharoen T, Williams DB, Sinchaipanid N (2010) Development and optimization of micro/nanoporous osmotic pump tablets 11(2):924–935
151. Heng D, Tang P, Cairney JM, Chan H-K, Cutler DJ, Salama R, Yun J (2007) Focused-ion-beam milling: a novel approach to probing the interior of particles used for inhalation aerosols 24(9):1608–1617
152. Barman S, Bolin D (2018) A three-dimensional statistical model for imaged microstructures of porous polymer films 269(3):247–258
153. Strange JH, Webber JBW, Schmidt S (1996) Pore size distribution mapping 14(7–8):803–805
154. Markl D, Zeitler JA, Rasch C, Michaelsen MH, Müllertz A, Rantanen J, Rades T, Bøtker J (2017) Analysis of 3D prints by X-ray computed microtomography and terahertz pulsed imaging 34(5):1037–1052
155. Markl D, Bawuah P, Ridgway C, van den Ban S, Goodwin DJ, Ketolainen J, Gane P, Peiponen K-E, Zeitler JA (2018) Fast and non-destructive pore structure analysis using terahertz time-domain spectroscopy 537(1–2):102–110
156. Singh G, Bulasara VKJD, Treatment W (2015) Preparation of low-cost microfiltration membranes from fly ash 53(5):1204–1212
157. Hagen J (2015) Industrial catalysis: a practical approach. Wiley
158. Camuffo D (2014) Chapter 5—Physics of drop formation and micropore condensation. In: Camuffo D (ed) Microclimate for cultural heritage, 2nd edn. Elsevier, Boston, pp 165–201. https://doi.org/10.1016/B978-0-444-63296-8.00006-8
159. Brunauer S, Emmett PH, Teller E (1938) Adsorption of gases in multimolecular layers. J Am Chem Soc 60(2):309–319. https://doi.org/10.1021/ja01269a023
160. Holbrook RD, Galyean AA, Gorham JM, Herzing A, Pettibone J (2015) Overview of nanomaterial characterization and metrology. In: Frontiers of nanoscience. Elsevier, vol 8, pp 47–87
161. Kecili R, Hussain CM (2018) Chapter 4—Mechanism of adsorption on nanomaterials. In: Hussain CM (ed) Nanomaterials in chromatography. Elsevier, pp 89–115. https://doi.org/10.1016/B978-0-12-812792-6.00004-2
162. Richardson JT (2013) Principles of catalyst development. Springer
163. Klobes P, Munro RG (2006) Porosity and specific surface area measurements for solid materials. NIST Recommended Practice Guide. Special Publication (NIST SP), U.S. GOVERNMENT PRINTING OFFICE WASHINGTON
164. Bagreev A, Bashkova S, Reznik B, Zibat V, Bandosz TJ (2002) Heterogeneity of sewage sludge derived materials as a factor governing their performance as adsorbents of acidic gases. In: Rodriguez-Reinoso F, McEnaney B, Rouquerol J, Unger K (eds) Studies in surface science and catalysis. Elsevier, vol 144, pp 217–224. https://doi.org/10.1016/S0167-2991(02)80137-9
165. Wang N, Burugapalli K, Song W, Halls J, Moussy F, Zheng Y, Ma Y, Wu Z, Li K (2013) Tailored fibro-porous structure of electrospun polyurethane membranes, their size-dependent properties and trans-membrane glucose diffusion. J Membr Sci 427:207–217. https://doi.org/10.1016/j.memsci.2012.09.052
166. Lebourg M, Antón JS, Ribelles JG (2008) Porous membranes of PLLA–PCL blend for tissue engineering applications. Eur Polymer J 44(7):2207–2218. https://doi.org/10.1016/j.eurpolymj.2008.04.033

167. Karim SS, Hussain A, Farrukh S, Younas M (2020) Effects of Coagulation Residence Time on the Morphology and Properties of Poly (vinyl) Alcohol (PVA) Asymmetric Membrane via NIPS Method for O_2/N_2 Separation. J Polym Environ 28(11):2810–2822. https://doi.org/10.1007/s10924-020-01816-6
168. Lu X, Li X (2009) Preparation of polyvinylidene fluoride membrane via a thermally induced phase separation using a mixed diluent. J Appl Polym Sci 114(2):1213–1219. https://doi.org/10.1002/app.30184
169. Purcell EM, Torrey HC, Pound RV (1946) Resonance absorption by nuclear magnetic moments in a solid. Phys Rev Lett 69(1–2):37. https://doi.org/10.1103/PhysRev.69.37
170. Rabi I, The discovery of NMR
171. Rouessac F, Rouessac A (2013) Chemical analysis: modern instrumentation methods and techniques. Wiley
172. Bovey FA (1989) Structure of chains by solution NMR spectroscopy. In: Allen G, Bevington JC (eds) Comprehensive polymer science and supplements. Pergamon, Amsterdam, pp 339–375. https://doi.org/10.1016/B978-0-08-096701-1.00017-3
173. Lee JH, Okuno Y, Cavagnero S (2014) Sensitivity enhancement in solution NMR: emerging ideas and new frontiers. J Magn Reson 241:18–31. https://doi.org/10.1016/j.jmr.2014.01.005
174. Odelius M, Laaksonen A (1999) Chapter 8—Combined MD simulation—NMR relaxation studies of molecular motion and intermolecular interactions. In: Balbuena PB, Seminario JM (eds) Theoretical and computational chemistry. Elsevier, vol 7, pp 281–324. https://doi.org/10.1016/S1380-7323(99)80041-7
175. Kondratyev VN (2018) Zeeman splitting in structure and composition of ultramagnetized spherical nuclei. Phys Lett B 782:167–169. https://doi.org/10.1016/j.physletb.2018.05.004
176. Bertini I, Luchinat C, Parigi G, Ravera E (2017) Chapter 1—Introduction. In: Bertini I, Luchinat C, Parigi G, Ravera E (eds) NMR of paramagnetic molecules, 2nd edn. Elsevier, Boston, pp 1–24. https://doi.org/10.1016/B978-0-444-63436-8.00001-6
177. Camuffo D (2014) Chapter 9—Consequences of the Maxwell–Boltzmann distribution. In: Camuffo D (ed) Microclimate for cultural heritage, 2nd edn. Elsevier, Boston, pp 347–366. https://doi.org/10.1016/B978-0-444-63296-8.00010-X
178. Schneider G, Mooser A, Bohman M, Schön N, Harrington J, Higuchi T, Nagahama H, Sellner S, Smorra C, Blaum K (2017) Double-trap measurement of the proton magnetic moment at 0.3 parts per billion precision. Science 358(6366):1081–1084
179. Davidovits P (2019) Chapter 17—Nuclear physics. In: Davidovits P (ed) Physics in biology and medicine, 5th edn. Academic Press, pp 279–291. https://doi.org/10.1016/B978-0-12-813716-1.00017-3
180. Warwick A (1993) Frequency, theorem and formula: remembering Joseph Larmor in electromagnetic theory. Notes Rec R Soc Lond 47(1):49–60
181. Slichter CP (2013) Principles of magnetic resonance, vol 1. Springer Science & Business Media
182. Packer KJ, Rees C (1972) Pulsed NMR studies of restricted diffusion. I. Droplet size distributions in emulsions. J Colloid Interface Sci 40(2):206–218
183. Goldman M (2001) Formal theory of spin–lattice relaxation. J Magn Reson 149(2):160–187
184. Claridge TDW (2009) Chapter 8—Correlations through space: the nuclear overhauser effect. In: Claridge TDW (ed) Tetrahedron organic chemistry series. Elsevier, vol 27, pp 247–302. https://doi.org/10.1016/S1460-1567(08)10008-3
185. Webb GA (2017) Theory of parameters in NMR spectroscopy. In: Lindon JC, Tranter GE, Koppenaal DW (eds) Encyclopedia of spectroscopy and spectrometry, 3rd edn. Academic Press, Oxford, pp 535–542. https://doi.org/10.1016/B978-0-12-803224-4.00243-0
186. Prestel A, Bugge K, Staby L, Hendus-Altenburger R, Kragelund BB (2018) Chapter 8—Characterization of dynamic IDP complexes by NMR spectroscopy. In: Rhoades E (ed) Methods in enzymology. Academic Press, vol 611, pp 193–226. https://doi.org/10.1016/bs.mie.2018.08.026
187. Chou JJ, Sounier R (2012) Solution nuclear magnetic resonance spectroscopy. In: Methods in Molecular Biology, vol 955. MIMB, vol 1940–6029. Humana Press, Totowa, NJ - Springer Protocols. https://doi.org/10.1007/978-1-62703-176-9_26

188. Babij NR, McCusker EO, Whiteker GT, Canturk B, Choy N, Creemer LC, Amicis CVD, Hewlett NM, Johnson PL, Knobelsdorf JA, Li F, Lorsbach BA, Nugent BM, Ryan SJ, Smith MR, Yang Q (2016) NMR chemical shifts of trace impurities: industrially preferred solvents used in process and green chemistry. Org Process Res Dev 20(3):661–667. https://doi.org/10.1021/acs.oprd.5b00417
189. Hanson JE (2013) NMR spectroscopy in nondeuterated solvents (No-D NMR): applications in the undergraduate organic laboratory. In: NMR spectroscopy in the undergraduate curriculum. In: ACS symposium series. American Chemical Society, vol 1128, pp 69–81. https://doi.org/10.1021/bk-2013-1128.ch005
190. Becker ED (2000) Chapter 4—Chemical shifts. In: Becker ED (ed) High resolution NMR, 3rd edn. Academic Press, San Diego, pp 83–117. https://doi.org/10.1016/B978-012084662-7/50048-X
191. Bara JE, Camper DE, Gin DL, Noble RD (2010) Room-temperature ionic liquids and composite materials: platform technologies for CO_2 Capture. Acc Chem Res 43(1):152–159. https://doi.org/10.1021/ar9001747
192. Janakiram S, Santinelli F, Costi R, Lindbråthen A, Nardelli GM, Milkowski K, Ansaloni L, Deng L (2021) Field trial of hollow fiber modules of hybrid facilitated transport membranes for flue gas CO_2 capture in cement industry. Chem Eng J 413:127405
193. Dai Z, Fabio S, Marino NG, Riccardo C, Deng L (2019) Field test of a pre-pilot scale hollow fiber facilitated transport membrane for CO_2 capture. Int J Greenhouse Gas Control 86:191–200
194. Huang G, Isfahani AP, Muchtar A, Sakurai K, Shrestha BB, Qin D, Yamaguchi D, Sivaniah E, Ghalei B (2018) Pebax/ionic liquid modified graphene oxide mixed matrix membranes for enhanced CO_2 capture. J Membr Sci 565:370–379. https://doi.org/10.1016/j.memsci.2018.08.026
195. Zhang X, Xiong W, Tu Z, Peng L, Wu Y, Hu X (2019) Supported ionic liquid membranes with dual-site interaction mechanism for efficient separation of CO_2. ACS Sustainable Chem Eng 7(12):10792–10799. https://doi.org/10.1021/acssuschemeng.9b01604
196. Ferrari M-C, Amelio A, Nardelli GM, Costi R (2021) Assessment on the Application of Facilitated Transport Membranes in Cement Plants for CO_2 Capture. Energies 14(16):4772. https://doi.org/10.3390/en14164772

Chapter 4
Facilitated Transport Membranes (FTMs) for Natural Gas Purification (CO_2/CH_4)

Syed Shujaat Karim, Sher Ahmad, and Sarah Farrukh

Abstract In recent years, facilitated transport membranes (FTMs) have emerged as one of the most remarkable and promising membrane-based technology for natural gas (NG) purification applications in order to produce high-calorific value fuel. By integrating different types of carriers such as fixed-site carrier (FSC) and mobile carriers (M.C), fillers (e.g. MOFs, ZIF, and MWCNTs etc.), and other hybrid materials within the membrane matrix, the FTMs demonstrate a rapid CO_2 flux across the membrane while maintaining high CO_2/CH_4 selectivity and high plasticization resistance. Due to a combination of two major mechanisms: solution diffusion and carrier mediated transport, these membranes overcome the permeability-selectivity trade-off limit and typically surpasses Robeson's upper bound line for dense polymeric membranes. This chapter mainly provides deep insights into the FTM-based NG purification process and its recent advancement for CO_2 separation, which include different membrane structural details, fabrication methods of both membrane and carrier, various operating conditions and their effects on CO_2/CH_4 separation performance of the membrane, as well as the FTMs that exhibited the simultaneous separation capability of both acidic gases (e.g. CO_2 and H_2S) from NG fuel were thoroughly discussed. In addition, the Robeson's plot of different FTMs, challenges, and future prospects of FTM development for the NG purification application were also described in detail.

Keywords Facilitated transport membranes · Natural gas purification · CO_2/CH_4 separation · Natural gas upgradation · Mixed matrix membranes · Composite membranes

---

S. S. Karim · S. Ahmad · S. Farrukh (✉)
School of Chemical and Materials Engineering (SCME), National University of Sciences and Technology (NUST), Sector H-12, Islamabad, Pakistan
e-mail: sarah.farrukh@scme.nust.edu.pk

S. S. Karim
e-mail: sshujaat.che6scme@student.nust.edu.pk

S. Ahmad
e-mail: sher.ahmad@scme.nust.edu.pk

S. Farrukh et al. (eds.), *Facilitated Transport Membranes (FTMs) for CO_2 Capture: Overview and Future Trends*, Green Energy and Technology,
https://doi.org/10.1007/978-3-031-21444-8_4

Nomenclature

2-D	Two dimensional
α	Selectivity
Ar	Argon
A-ZIF-8	Amine modified ZIF-8
bpydc	2,2′-bipyridine-5,5′-dicarboxylic acid
bmim[ace]	1-butyl-3-methylimidazolium acetate
bmim[BF_4]	1-butyl-3-methylimidazolium tetrafluoroborate
C_2H_6	Ethane
C_3H_8	Propane
C_4H_{10}	Butane
CA	Cellulose acetate
CH_4	Methane
CMS	Carbon molecular sieve
CNF	Carbon nanofiber
CO_2	Carbon dioxide
COFs	Covalent organic frameworks
DEA	Diethanolamine
DT	3,5,-diamino-1,2,4-triazole
FSC	Fixed-site carrier
FTMs	Facilitated transport membrane
GO	Graphene oxide
GPU	Gas permeance unit
H_2O	Water molecule
H_2S	Hydrogen sulfide
He	Helium
IRMOF-3	Amine functionalized isoreticular metal–organic framework-3
K.D	Kinetic diameter
LDHs	Layered double hydroxides
M.C	Mobile carrier
MFFK-1	Tetrafluoroethylene-vinylidene fluoride copolymers
MIL	Matériaux de l′Institut Lavoisier
MMMs	Mixed matrix membranes
mPSF	Modified polysulfone
N_2	Nitrogen
NCs	Nano-capsules
Ne	Neon
NG	Natural gas
O_2	Oxygen
P	Pressure (bar)
PA	Polyamide
PASS	Sodium polyacrylate
PDMS	Polydimethylsiloxane

Pebax	Polyether block amide
PES	Polyether sulfone
PI	Polyimide
PP	Polypropylene
PPO_{dm}	poly (2,6-dimethyl-1,4-phenylene oxide)
PSF	Polysulfone
PVA	Polyvinyl alcohol
PVAm	Polyvinyl amine
SDS	Sodium dodecyl sulphate
T	Temperature (°C)
Tp	1,3,5-Triformylphloroglucinol
TpDT	1,3,5-Triformylphloroglucinol, 3,5,-diamino-1,2,4-triazole
Vol.	Volume, v
W.C	Wet condition
Wt.	Weight
Xe	Xenon
ZIF	Zeolitic imidazolate framework

4.1 Introduction

Natural gas (NG) is the major source of fossil fuel after coal and oil. Because of its abundant deposits and generally clean-burning properties, NG has become a more essential source of energy for different production industries, power generation and transportation sectors. Aside from these sectors, NG also often used for different domestics purposes, such as cooking and heating [1]. According to a global projection of NG consumption, it is expected to rise from 95 trillion cubic feet in 2003 to 182 trillion cubic feet in 2030 [2]. As shown in Table 4.1, NG fuel is largely comprised of methane (CH_4) gas, with a few additional hydrocarbons. However, it also contains various impurities, such as water vapour (H_2O), hydrogen sulfide (H_2S), carbon dioxide (CO_2), nitrogen (N_2) gas, and trace amounts of noble gases, which lower the fuel's calorific value, cause atmospheric pollution, corrosion in the process unit, affect human health and also increase the volume of NG that has to be transported [3].

Therefore, the removal of impurities, particularly acidic gases (e.g. CO_2 and H_2S) from NG has become essential requirement in the past decades. This separation of acidic gases from NG fuel is also known as natural gas sweetening process. A wide-range of different well-established conventional technologies including, pressure swing adsorption, the amine absorption process, and cryogenic distillation are currently used for the purification of NG. However, these conventional technologies are quite expensive, require a large area and high energy for the separation process, and are environmentally unfriendly. To address this problem, the world market has

Table 4.1 Typical composition of natural gas and its physical properties

Feed gas components	Molecular weight (g/mol)	Kinetic diameter (Å)	Composition range (mole %)	Ref.
Methane (CH_4)	16.00	3.80	70.00–90.00	[6, 7]
Ethane (C_2H_6)	30.70	4.44	0.00–20.00	[6, 8]
Propane (C_3H_8)	44.00	4.30	0.00–20.00	[6, 9]
Butane (C_4H_{10})	58.12	4.69	0.00–20.00	[6, 10]
Carbon dioxide (CO_2)	44.00	3.30	0.00–8.00	[5, 6]
Oxygen (O_2)	32.00	3.46	0.00–0.20	[5, 6]
Nitrogen (N_2)	28.00	3.64	0.00–5.00	[5, 6]
Hydrogen sulfide (H_2S)	34.00	3.60	0.00–5.00	[6, 9]
Argon (Ar)	40.00	3.40	Trace amount	[5, 6]
Helium (He)	4.00	2.60	Trace amount	[5, 6]
Neon (Ne)	20.00	2.75	Trace amount	[5, 6]
Xenon (Xe)	131.00	3.96	Trace amount	[5, 6]

shifted its focus towards different alternative solutions, in which membrane technology has emerged as one of the best alternative technology for NG purification due to its remarkable characteristics. Some of the benefits of membrane technology over the existing conventional technologies for NG purification processing include, the ease of scale-up, energy efficient, smaller foot print, clean production and can be combined with other technology to form a hybrid system. The gas separation process using this technology can be carried out continuously with greater flexibility of adjusting process variables and does not require complex instrumentation to set-up [4, 5].

Membrane technology was first introduced since 1980s into the NG processing industry and until now, this technology for CO_2 capture from NG fuel, accounts for around 5% of the current market [4]. In which majority of the membranes for NG processing are based on polymeric membranes because they are relatively inexpensive as compared to the other types of membrane, easy to fabricate into different configurations of compact module (e.g. thin flat-sheet, tubular, and hollow fibers), light-weight, and have the flexibility of surface and structural modification [1, 11]. In general the dense polymeric membranes for gas separation follows the solution diffusion (S-D) mechanism, where the gas species of the mixture are separated due to the difference of diffusivity and solubility in the membrane. The major constraint of these membranes is the trade-off between the permeability and selectivity as defined by the Robeson's upper bound lines graph, which limits its development in the commercial market for NG processing. However, in case of facilitated transport based polymeric membranes as illustrated in the Fig. 4.1, the transport of CO_2 molecules occurred due to the combination of two main mechanism processes: (i) pure physical S-D mechanism and (ii) by carrier mediation process, in which the CO_2 molecule from

NG mixture formed a reversible bond with the carrier present in the membrane to formed a CO_2-carrier "complex", which then diffuses across the membrane due to the concentration gradient at a higher rate. At permeate side, the CO_2-carrier complex broke down due the lower pressure and released the CO_2 molecule, while regenerating the carrier, which then again reacts with the CO_2 molecule at the feed side. These carriers can be mobile carriers (M.C) that are confined in the membrane matrix and move freely across the membrane, or they can be fixed-site carriers (FSC) as functional groups that are covalently bonded to the polymer backbones. When M.C reversibly bonded with the CO_2 molecules, can freely moves and diffuses across the membrane, while in the case of FSC, the CO_2 molecules are transported by the hopping mechanism. Whereas, both carrier could also be used in a combination for higher performance of NG purification [11, 12].

In recent years, many researchers have reported various types of FTMs for NG purification, including pristine polymer-based FTMs, polymer blend-based FTMs, mixed matrix-based FTMs, and composite FTMs. These membranes are capable of efficiently separating CO_2 gas from the NG fuel. Whereas, a new type high-performance FTMs were also developed by some researchers that can simultaneously remove both acidic gases (CO_2 and H_2S) from the NG fuel. Some of the recently

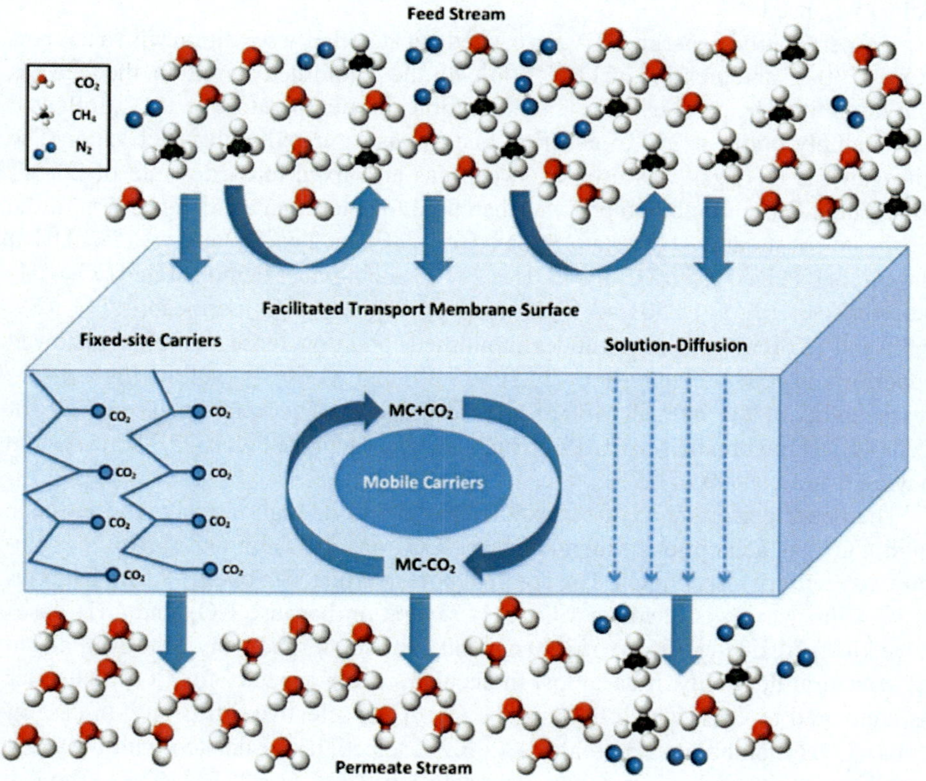

Fig. 4.1 Illustration of facilitated transport membrane (FTM) gas separation mechanisms for NG Purification

reported data of FTMs for NG purification applications from 2018 to present are discussed in the following Sect. 4.2.

4.2 Recent Advancements of FTMs for Natural Gas Purification (CO_2/CH_4)

In 2018, Ning Zhang and coworkers [13], explored ZIF-8@LDH hetero-structured as a filler in the polymeric membranes for enhancing the performance of CO_2/CH_4 separation. They used rapid room temperature synthesis method for the preparation of ZIF-8 and co-precipitation technique to make ZnAl-CO_3 LDH. While through the in-situ growth of ZIF-8 nano-particles on LDH, the hetero-structured filler ZIF-8@LDH was produced. Then they embedded single filler (ZIF-8 or LDH) at a concentration of 2 wt% and hetero-structured filler (ZIF-8@LDH) at a concentration of 2–5 wt% separately in the Pebax-1657 membrane matrix by using solution mixing and casting method for the fabrication of flat sheet FSC-based MMMs at a thickness of 60 ± 10 μm. The gas separation performance of fabricated membranes were then evaluated in pilot scale, where mixed gas stream of CO_2/CH_4 having composition of (30/70 vol%) was used.

The permeation experiments were carried out at both dry condition with a temperature of 30 °C and pressure of 1 bar. While for the humidified condition, the feed gas was humidified by passing it through a humidifier at a temperature of 35 °C, followed by an empty bottle at 25 °C, ensuring that it was saturated with water vapor. The nitrogen gas (N_2) gas was used as sweep gas at a fixed mass flowrate of 20 mL (STP)/min. Gas chromatography was then used to determine the composition of the permeate gas stream. At dry state, Pebax-1657/ZIF-8 at 2 wt.%, Pebax-1657/LDH at 2 wt.%, and Pebax-1657/ZIF-8@LDH at 2 wt.% membranes demonstrated CO_2/CH_4 selectivity of 17.31, 18.50 and 19.20, respectively, with CO_2 permeability of 5.89, 5.33, and 122 Barrer. Whereas under humidified condition, these membranes showed superior CO_2/CH_4 selectivity of 24.10, 28.20, and 31.60, as well as higher CO_2 permeability of 949, 619, and 1307 Barrer respectively. The water uptake capacity of Pebax/LDH membrane was 10.3%, which is higher than the Pebax/ZIF-8 membrane having water content of 6.2%.

The better performance of Pebax/ZIF-8 is due to its high porosity and intrinsic high CO_2 gas adsorption, which enhances CO_2 gas diffusion and CO_2/CH_4 solubility selectivity respectively. The nominal pore aperture size of ZIF-8 nanoparticles is 0.34 nm, which is ideal for CO_2/CH_4 separation because CO_2 and CH_4 gases have kinetic diameters (K.D) of 3.30 nm and 3.80 nm, respectively. However, due to its structural flexibility, it has a low molecular sieving capacity for CO_2 molecules as compared to CH_4, which limiting the CO_2/CH_4 selective diffusivity. In case of Pebax/LDH membranes, the higher CO_2/CH_4 selectivity associated with the facilitated transport mechanism of the carbonate ion (CO_3^{-2}) in LDH. The carbonate ion (CO_3^{-2}) acts as carrier for the facilitated transport of CO_2 gas molecules. Under humidified condition the reversible reaction between H_2O and CO_3^{-2} formed two (2) moles of bicarbonate ion (HCO_3^-). Which showed that, LDH has the ability to provide a facilitated transport channel for the CO_2 gas permeation and more tortuous path for the CH_4 gas molecules.

As compared to the single filler (ZIF-8 and LDH) based Pebax membranes, hetero-structured filler ZIF-8@LDH based Pebax membranes showed superior CO_2/CH_4 separation performance with enhanced CO_2 gas permeability due to three main characteristics: (a) synergistic effect of both ZIF-8 and LDH filler. (b) higher water uptake of 12.4% under humidified condition leads to increase in membrane swelling, which is favorable for the facilitated transport of CO_2 gas molecules. (c) increased in membrane fractional free volume due to the ZIF-8@LDH hetero-structured filler also increases in the gas diffusivity coefficient. In general, the enhanced solution diffusion mechanism is responsible for the increased physical selectivity, while the enhanced facilitate transport mechanism is responsible for the increased chemical selectivity of CO_2/CH_4 gas mixture.

Mubashir et al. [14], developed FSC-based MMMs containing amine-functionalized NH_2-MIL-53(Al) as a filler particles that were integrated at a concentration of 10–20 wt% into a low-cost cellulose acetate (CA) polymer membrane matrix. In this study, the NH_2-MIL-53(Al) filler particles were synthesized through hydrothermal heating method and NH_2-MIL-53(Al)/CA MMMs were fabricated by using solution mixing and casting method. The fabricated membranes were tested for the gas permeation by using both single pure gas and mixed gas having the composition of CO_2/CH_4 at 50/50 vol% at a pressure of 3 bar and temperature of 25 °C. In the pure gas testing system, MMM at optimized filler's loading of 15 wt%, exhibited the highest CO_2/CH_4 selectivity of 28.70 with CO_2 gas permeability of 52.60 Barrer. While in the mixed gas testing system, this membrane showed the maximum CO_2/CH_4 selectivity of 18.40 with CO_2 gas permeability of 34.80 Barrer. In comparison to pure gas testing, the CO_2/CH_4 selectivity and permeability of MMMs were shown to be reduced in case of mixed gas separation process. The optimized MMM's decrement in gas separation performance in case of mixed gas separation could be attributed to the competitive effects of one component of the gas mixture over another for sorption and transport penetration across the membrane. As a result of the lower CO_2 permeability, the CO_2/CH_4 selectivity also decreased. According to their studies, another benefit of integrating NH_2-MIL-53(Al) as a filler in the CA polymeric matrix is that it increases the plasticization resistance up to pressure of 30 bar, which is 200% more than the pristine CA based-polymeric membrane. As the feed pressure is increased, the CO_2 gas molecules adopt the mechanism of facilitated transport by interacting with the amine groups in the filler. This also means that increasing the feed pressure causes more CO_2 gas molecules to engage with the amine carrier until all of the amine carriers of the membrane have been interacted.

Zhang et al. [15], prepared multi-functional nano-capsules (NCs) by classical Stöber process. Then FSC MMMs were fabricated by the integrating NCs into the Pebax® MH 1657 membrane at a filler loading of (0–30) wt% through solution mixing and casting method for the application of CO_2/CH_4 separation. The thickness of these MMMs was controlled in the range of 75–95 μm. In these MMMs, NCs as filler provides good interfacial compatibility with the Pebax based membrane matrix and also reduces its crystallinity, which are the properties required for the high perm-selective MMMs. At a filler loading of 20 wt% NCs based MMMs showed the highest CO_2/CH_4 selectivity of 46 and CO_2 gas permeability of 1431 Barrer. While 30 wt% NCs based MMMs exhibited maximum CO_2 gas permeability of 1993 Barrer and CO_2/CH_4 selectivity of 39. The NCs also enhanced the water content uptake capabilities of the membrane matrix, which helps in the dissociation of acidic groups. Water molecules serve as a carrier for CO_2 gas molecules, forming bicarbonate ion (HCO_3^-) through a reversible reaction, allowing CO_2 gas molecules to be transported at a faster pace than other gas components like CH_4 molecules. NCs, on the other hand, have an abundance of carboxylic acid groups (–COOH) on the interior and transverse shell surfaces, which showed a strong affinity towards CO_2 molecules. These characteristics indicated that the NCs based MMMs showed higher CO_2/CH_4 perm-selectivity due to the combination of rapid facilitated transport process and selective solution diffusion mechanism.

Awad and Aljundi [16], developed a FSC FTM for the CO_2/CH_4 separation. In this composite FSC FTM, Polyamide (PA) thin selective layer was formed on the top surface of polysulfone (PSF) asymmetric membrane through the interfacial polymerization reaction of piperazine (PIP) and isophthaloyl chloride (IPC). While the PSF support layer was synthesized by the dry/wet phase inversion method with a controlled thickness of 200 μm. The resultant PA/PSF fabricated membrane showed a CO_2/CH_4 selectivity of 12.07, which was 178% higher than the pristine PSF asymmetric membrane, while CO_2 gas permeance was reduced from 2.411 to 2.124 GPU at a temperature of 25 °C and pressure of 5 bar. The improved CO_2 separation performance in the case of this composite membrane was due to the amide and amine functional groups present in the PA dense selective layer, which were activated in the presence of moisture and rendered the facilitated transport mechanism of CO_2 molecules in combination with the solution diffusion mechanism. However, the mass transfer resistance was also increased for the gas transport due to the coating of a PA layer on the top surface of the PSF asymmetric membrane, resulting in a slight reduction in CO_2 permeance.

Murugiah et al. [17], investigated the synergistic effect of binary nano-fillers mixture embedded in the glassy polymer matrix for the preparation of high performance FSC FTMs for CO_2/CH_4 separation. In these MMMs, the carbon nanofiber (CNF) and graphene oxide (GO) were incorporated at different filler loading percent (5-GO, 2-CNF/3-GO, 3-CNF/2-GO, 5-CNF) in the glassy poly (2,6-dimethyl-1,4-phenylene oxide)—(PPO_{dm}) based matrix using dry-phase inversion method. The resultant membranes were tested for gas separation performance at a fixed pressure of 3.5 bar and at room temperature. According to their studies, the GO based MMM membrane showed CO_2/CH_4 higher selectivity with low CO_2 permeability. Which indicated that the GO nanosheets, generated a rigidified polymer-filler interface and formed more tortuous path for the gas transport, which passes smaller gas molecules (CO_2) and impeded larger molecules (CH_4). In addition, GO also contain the functional group of carboxyl (–COOH) and hydroxyl (–OH), which have strong affinity towards the polar gas molecules and provide the facilitated transport for the CO_2 molecules. While the CNF based MMMs exhibited the high CO_2 permeability but low CO_2/CH_4 selectivity. The embedding of CNF in polymer matrix disrupted the polymer chain packing and increased its free volume, resulted in the improved gas permeation properties. Whereas, the addition of binary filler (GO/CNF) in PPO_{dm}— matrix, the MMMs exhibited superior perm-selective properties as compared to the single-filler based MMMs. At 2 wt.% of CNF and 3 wt.% of GO, the membrane showed the optimum gas separation performance having CO_2/CH_4 selectivity of 59.70 with CO_2 permeability of 62.32 Barrer. This result also indicated that the higher perm-selectivity of CO_2 gas molecules was obtained due to the combination of both facilitated transport mechanism and solution diffusion mechanism. While, CH_4 gas molecules permeated only through solution diffusion mechanism.

In 2019, Akhmetshina et al. [18], fabricated different support liquid membranes (SILMs) by a vacuum method for the NG purification. These SILMs composed of tetrafluoroethylene-vinylidene fluoride copolymers (MFFK-1) based support layer having thickness of 100 μm containing immobilized 1-butyl-3-methylimidazolium tetrafluoroborate-bmim[BF_4] and 1-butyl-3-methylimidazolium acetate-bmim[ace] ionic liquids, separately. The fabricated membranes were then tested for three different gas permeation systems: (a) single pure gas system, (b) binary gas system $CH_4/CO_2 = 85/15$ (v/v).% and (c) ternary gas system $CH_4/CO_2/H_2S = 76.5/18.2/5.3$ (v/v/v).%, at a transmembrane pressure of 1 bar and at temperature of 25 °C. In this study, the single gas permeation system tests result revealed that the bmim[ace]-MFFK-1 showed superior CO_2/CH_4 selectivity of 15.00 with a CO_2 permeability of 92.00 Barrer than the bmim[BF_4]-MFFK-1 having CO_2/CH_4 selectivity of 7.00 with a CO_2 permeability of 84.00 Barrer. In case of mixed gas bmim[BF_4]-MFFK-1 showed an increased in CO_2/CH_4 selectivity of 8.70 with reduced CO_2 permeability of 68.90 Barrer. While bmim[ace]-MFFK-1 exhibited a sharp increase in CO_2/CH_4 selectivity 96.90 and doubled CO_2 permeability of 186.10 Barrer. As explained by the Zhang et al. [19] that the higher gas separation performance can be achieved by the FTMs, when the active penetrants are lower in concentration in gas mixture. Whereas, the bmim[BF_4]-MFFK-1 showed better permeability to H_2S (142.00 Barrer) and CO_2 (70.00 Barrer) compared to CH_4 (8.60 Barrer) in the separation of the ternary gas mixture. While bmim[ace]-MFFK-1 showed closed permeability value of both acidic gases (110.40 and 100.30 Barrer, respectively) and have low value of CH_4 gas permeability, which also indicates that these SILMs based FTMs have the capability and promising potential for the removal of both acidic gases (CO_2 and H_2S) from the raw NG fuel.

Thür et al. [20], developed different novel MMMs using bipyridine-based UiO-67 MOFs as nano-fillers in the Polyimide (PI) matrix (Matrimid 5218). A solvothermal method was used to synthesize these bipyridine-based UiO-67 MOFs and they were incorporated with the 2,2′-bipyridine-5,5′-dicarboxylic acid (bpydc) linker at a variable molar percentage up to 100%. These MOFs were embedded in a 7 wt.% PI-based membrane using a solution mixing and casting process at a filler loading of 10 wt.%. The MMMs were then evaluated in a 50/50 vol% mixed gas permeation system at a feed pressure of 5 bar and a temperature of 30 °C. Their study revealed that, at an optimal bipyridine concentration of 33%, the MOFs enhanced both the membrane's selectivity and permeability by 100% and 63%, respectively, as compared to a pristine matrimid-5218 membrane. This increment in gas separation performance was attributed to the facilitated transport mechanism due to the interaction between the bipyridine site (carriers) present inside the MOFs and CO_2 molecules. However, the MMM developed interfacial defects at a higher nano-filler concentration of 30 wt%, but it still outperformed the pristine Matrimid membrane with a 350% increase in CO_2 gas permeability. They also indicated that, because of these nano-fillers' efficient performance, the bpydc-UiO-67-MOFs MMMs-based FTMs membranes could be a viable alternative for standard functionalized MOFs-based MMMs to obtain better performance in CO_2/CH_4 separation.

Covalent organic frameworks (COFs) have recently emerged as one of the most promising materials due to their attractive properties such as an ordered microporous structure, compatibility with the polymer interface, and thermal and chemical stability. In 2020, Wang's et al. [21], synthesized a 2-D self-exfoliated 1,3,5-Triformylphloroglucinol, 3,5,-diamino-1,2,4-triazole (TpDT) covalent organic nanosheets (CONs) using Schiff base condensation reaction. These TpDT-CONs were added as a filler at different loading of 0–2.0 wt% in the 4 wt% Pebax® MH 1657 based solution to form a TpDT-CONs based MMMs solutions. Then PVDF hollow fibers as support substrates were used to fabricate the TpDT-CONs-Pebax/PVDF composite membranes using dip-coating method. Afterwards, the resultant composite membranes were tested for the gas permeation properties using a single gas (pure) and a mixed gas (CH_4/CO_2 = 70/30 vol.%) testing systems at constant pressure of 1 bar and temperature of 25 °C under dry and humidified conditions. Both feed gas and the sweep gas—nitrogen (N_2) gas were saturated with water and applied to the shell side and the inner-porous side of the hollow fiber, respectively, in a humidified condition. In single gas testing system, TpDT-CONs-Pebax/PVDF MMMs showed an enhanced CO_2/CH_4 separation performance as compared to the pristine Pebax/PVDF composite membrane. According to their findings, increasing the TpDT-CONs concentration up to 1.0 wt% resulted in an increase in both gas permeance and selectivity. However, when the concentration was increased to 1.5 and 2.0 wt%, the gas separation performance was reduced. The MMMs having 1.0 wt% of TpDT-CONs exhibited a maximum CO_2 gas permeance and CO_2/CH_4 selectivity of 15.9 (GPU) and 31, respectively, under dry conditions. In humidified circumstances, the CO_2 gas permeance of the Pebax/PVDF composite membrane increased by 41%.

Similarly, MMMs (TpDT-CONs) at 1.0 wt% showed a 787% increase in CO_2 gas permeance, which had a value of 141.1 (GPU), with the improvement in CO_2/CH_4 selectivity up to 38 as compared to the dry condition. This substantial improvement in gas separation performance was enabled due to the facilitated transport process in combination solution diffusion mechanism and swelling of the hydrophilic polymeric matrix promoted by the water content in the humidified state. The CO_2/CH_4 selectivity of the Pebax/PVDF composite membrane was slightly reduced in mixed gas testing, which had a value of 21.4 with a CO_2 permeance of 61.5 (GPU). The TpDT-CONs based MMMs exhibited similar CO_2/CH_4 separation performance as showed in pure gas system. This enhancement in gas separation performance of composite membrane attributed due to the amine groups that is present abundantly in TpDT-CONs, functions as a carriers for CO_2 molecules to be transported more efficiently via reversible nucleophilic addition reactions. Furthermore, embedding TpDT-CONs in Pebax membrane matrix disturbed the packing of polymer chains, which increased the chain mobility and also offers a structured tunnel for the rapid transport of CO_2 molecules to the carriers for the facilitated transport process. Therefore we can say that the introduction of facilitated transport carriers via CONs in the membranes can provide an effective approach to design highly perm-selective membranes, which can be used as a viable alternative for the NG (CO_2/CH_4) purification.

Nab et al. [22], investigated the synergistic effects of diethanolamine (DEA) and carbon molecular sieve (CMS) as binary fillers embedded in the composite membranes for the high pressure CO_2/CH_4 separation application. These composite membranes were prepared using the dry–wet phase inversion method by casting an MMMs-based selective layer on top of the polypropylene (PP) porous support layer (thickness = 203.00 μm). The selective layers had a thickness in the range of 34.99–126.90 μm and were composed of a 20 wt% polyether sulfone (PES) polymer containing 10 wt% of CMS and a variable concentration of DEA ranging from (0 to 15) wt%. These fabricated asymmetric composite membranes were characterized for CO_2/CH_4 separation performance in a single (pure) gas testing system under a high feed pressure of 10–30 bar and at a temperature of 25 °C. The analysis of the gas permeation experiments revealed that the pristine PES/PP composite membrane showed the maximum CO_2/CH_4 selectivity of 8.15 with CO_2 permeance of 142.21 (GPU). Whereas, the new DEA-CMS/PES/PP composite membranes exhibited higher CO_2/CH_4 perm-selectivity as compared to the pristine PES composite membrane and have an enhanced mechanical strength to withstand the elevated pressure of 30 bar. The maximum CO_2 permeance across these composite membranes was attained at a lower feed gas pressure of 10 bar, and the minimum permeance value was obtained as the pressure increased to 30 bar. The CO_2/CH_4 selectivity, on the other hand, has shown the opposite trend. At a CMS and DEA concentration of 10 and 15 wt%, the composite membrane exhibited the most optimum gas separation performance having the values of CO_2/CH_4 selectivity of 16.04 and CO_2 permeance of 300.84 (GPU) at a pressure of 30 bar. In these composite membranes, the DEA acted as a carrier in the facilitated transport process, which has a strong affinity towards CO_2 molecules and a negligible affinity towards CH_4 gas. Another reason could be that these ionic species have occupied sites or vacant spaces that could be utilized to transfer CO_2 molecules. The CMS also contributed in the selective transport of CO_2 molecules as these fillers have the pore sizes smaller than the kinetic diameter (K.D) of CH_4 gas (0.38 nm) and greater than the K.D of CO_2 molecules (0.33 nm) providing the molecular sieving effects in the CO_2/CH_4 separation process, which allows the passage for CO_2 molecules and hindered the CH_4 to permeate across the membrane.

In 2021, Zhang et al. [23], developed an efficient high-performance FTM based polymeric multi-layer composite membrane for CO_2/CH_4 separation from Oilfield associated gas. In this composite membrane a gutter layer of crosslinked Poly-dimethylsiloxane—PDMS (1.5 wt%) was coated by wet-dry coating method with the thickness of 300 μm on the PSF ultrafiltration membrane, which avoids the infiltration of coating layer polymeric solution in the internal pore of the support membrane, in result preventing the excessive drop in gas permeance. After that, a selective layer of PPPS solutions at concentration between 0.05 and 0.20 wt% were mainly composed of polyvinyl amine (PVAm) was coated at a thickness of 0.250–0.333 μm on the PDMS/PSF support layer, with the addition of 20 mol% of polyvinyl alcohol (PVA), 1 mol% of sodium polyacrylate (PASS), and 10 mol% of sodium dodecyl sulphate (SDS) as additives to enhanced the CO_2/CH_4 separation process and compatibility between inter-layers of the composite membrane. The gas separation performance of the resultant composite membrane was evaluated by using a mixed gas system ($CH_4/CO_2 = 55/45$ vol%) under humidified condition (40–100% relative humidity) at a variable temperature in the range of 25–80 °C and feed pressure of 1.5–18 bar. This study revealed that, by increasing the PPPS concentration from 0.05 to 0.20 wt% thickened the selective layer, resulting in a decrement of CO_2 permeance while improving CO_2/CH_4 selectivity due to increased gas resistance across the membrane. Whereas, increment in the relative humidity from 40 to 100%, promoted the facilitated transport process with more CO_2 adsorption on the membrane matrix surface, which increased both CO_2/CH_4 selectivity as well as the CO_2 permeance of the membrane. The increased in temperature from 25 to 80 °C, also enhanced the CO_2 permeance, while increment in feed pressure from 1.5 to 18 bar caused decrement in CO_2/CH_4 gas separation performance with lowered CO_2 permeance. At 0.15 wt% PPPS attached on the 1.5 wt%—PDMS/PSF layers, the composited membrane exhibited the optimum gas separation performance having the CO_2/CH_4 separation value of 62 with CO_2 permeance of 3.451×10^{-7} mol m^{-2} s^{-1} Pa^{-1} (1029.85 GPU) at a feed pressure of 1.5 bar and temperature of 25 °C under humidified condition in the mixed gas system.

Sasikumar et al. [24] investigated the MOFs' high tunable properties, which allows different modifications for high efficient for the gas separation applications. For this purpose, a SiO_2/ZIF-8 nano-filler was prepared using a facile one-step method and then subjected to the post synthetic amine modification with ammonium hydroxide

(NH_4OH) solution forming a Amine (A)-ZIF-8@SiO_2 nano-filler. Then each nano-fillers at a concentration of 0.5 wt% were blended with 25 wt% of PSF to form different dope solutions, which were then used for the fabrication of FSC FTMs based mixed matrix hollow fibers with a skin layer thickness of 95.00–103.10 nm via dry–wet phase inversion through spinning system. Whereas, a Pebax-1657 of 3 wt% was coated on the surface of the PSF-hollow fibers using dip-coating method to eliminate or reduced the defects present on the outer-surface of the membranes. Then gas separation performance of fabricated MMMs hollow fibers was tested using pure single gas permeation system at fixed pressure of 4 bar and at a temperature of 30 °C. Which showed that by incorporating the nano-fillers in the hollow fiber, resulted in the increment of CO_2/CH_4 separation performance. In comparison to the pristine PSF hollow fiber membrane, the CO_2 permeability of PSF@ZIF-8, PSF@ZIF-8/SiO_2, and PSF@A-ZIF-8-SiO_2 Mixed matrix based composite membranes increased by 8.5, 20.8, and 39%, accordingly. In the case of PSF@ZIF-8 composite, the electrostatic field of ZIF-8 with the quadrupole moment of CO_2 and molecular sieve effects could be the reason for the increment in CO_2/CH_4 separation performance. As the ZIF-8 possessed the pore aperture size of 0.34 nm, allows the smaller molecules like CO_2 with K.D of 0.33 nm, while restrict the passage for larger molecules of CH_4 molecules with a K.D value of 0.38 nm. The ZIF-8/SiO_2 nano-fillers increased the CO_2 permeability but decreased selectivity of the membrane. On the other hand, by using a modified nano-filler, A-ZIF-8-SiO_2 provide enhancement in both CO_2 permeability and CO_2/CH_4 selectivity as compared to the pristine PSF membrane. At lower loading 0.5 wt% of A-ZIF-8-SiO_2 nano-filler, the PSF-composite membrane exhibited the maximum CO_2 permeance of 41.15 GPU and optimum CO_2/CH_4 selectivity of 27.78. In this case, the significant increment of gas separation performance was due to the presence of amine functional group attached in the membrane surface, which has strong affinity towards the CO_2 gas and act as fixed site-carrier in facilitated transport of CO_2 molecules in combination with solution diffusion mechanism.

Similarly, Cai et al. [25], fabricated a composite membranes containing IRMOF-3 modified MOFs for the application of high-performance CO_2/CH_4 separation. The IRMOF-3 was prepared through optimizing synthesis method. The modified support layers (mPSF) were fabricated by coating, cross-linked PDMS on the PSF ultrafiltration membranes. The PVAm-MOFs solutions were prepared by blending IRMOF-3 with 2 wt% PVAm solution at filler loading in the range of 20–40 wt%. The doped solutions were then coated on the mPSF to prepared different concentration of IRMOF-3-PVAm/PSF composite membranes at a thickness between 0.409 and 0.416 μm. They used mixed gas (CH_4/CO_2 = 90/10 vol%) testing system to evaluate the CO_2/CH_4 separation performance of the composite membranes at feed gas pressure range from 10 to 30 bar and at a fixed temperature of 25 °C. The gas permeation experiment results revealed that IRMOF-3-PVAm/mPSF composite membrane exhibited the characteristics of a facilitated transport mechanism. The IRMOF-3 nano-fillers have the pore aperture of 9.58 Å and contained primary amine group with more active hydrogen atoms, which allows more gas molecules to penetrate and showed higher affinity towards the CO_2 molecules resulting higher CO_2/CH_4 separation as compared to the pristine composite membrane. At a filler loading of 25 wt%, IRMOF-3 composite MMM showed a maximum CO_2/CH_4 selectivity of 61 with CO_2 permeance of 499 (GPU) at 10 bar pressure and 25 °C. Whereas, by increasing the feed gas pressure, CO_2 permeance across the membrane decreases as the amine carriers get saturated. While a continual decrement in CH_4 permeance is attributed to the selective adsorption of CO_2 molecules by amine carrier in the nano-channel of MOFs. Another reason for this trend could be the compaction of the membrane at higher pressures, which decreased the free volume present in the membrane, resulted in the lower mobility of the penetrating CH_4 molecules.

4.3 Trade-Off Robeson Plot of FTMs for NG Purification—CO_2/CH_4 Separation

From the data in Table 4.2, the following Fig. 4.2 is plotted, which depicts the FTMs Robeson upper bound plot and summarizes the recently reported literature of different FTMs for CO_2/CH_4 separation (NG purification). The two empirical lines (1991 and 2008 Robeson upper bound limit) represent the polymeric membranes' trade-off limit [26]. In terms of both CO_2 permeability and CO_2/CH_4 selectivity, polymeric-based FTMs displayed remarkable CO_2/CH_4 separation performance for NG purification. In contrast to the pristine polymer-based dense membranes utilized for CO_2 separation using only solution diffusion mechanism from NG fuel, the majority of FTMs exceeded the 2008 upper bound limit based on the most recently study data. Among different polymeric-based FTMs for CO_2/CH_4 separation, SILM (MFFK-1-bimm[ace]), MMM based FTM (Pebax/NCs), thin composite based FTM (PPPS/PDMS/PSF) and hybrid filler based FTMs (Pebax/TpDT-CONs, Pebax/ZIF-8@LDH and PES/DEA-CMS) showed significantly improved CO_2/CH_4 separation performance as compared to the other FTMs, which signified that the recent FTM technology has the potential to be utilized for efficient and high CO_2 separation performance for NG purification.

Table 4.2 Recent advancement data of polymeric based FTMs for CO_2/CH_4 separation (NG purification)

Materials	Carriers/fillers	Fabrication method	FTMs types	Module/thickness	Operating conditions	$P(CO_2)$	$P(CH_4)$	$\alpha_{\frac{(CO2)}{(CH4)}}$	Ref.
Pebax® MH 1657 (5 wt%)	ZIF-8 (2 wt%)	Rapid room temperature synthesis [a] Solution mixing and casting method [b]	FSC (MMMs)	Flat-sheet (60 ± 10 μm)	T = 30 °C P = 1 bar Mixed gas CH_4/CO_2 (70/30 vol%)	102.00 (Barrer)	5.89 (Barrer)	17.31	[13]
Pebax® MH 1657 (5 wt%)	ZIF-8 (2 wt%)	Rapid room temperature synthesis [a] Solution mixing and casting method [b]	FSC (MMMs)	Flat-sheet (60 ± 10 μm)	T = 35 °C P = 1 bar (W.C: 6.2%) Mixed gas CH_4/CO_2 (70/30 vol%)	949.00 (Barrer)	39.37 (Barrer)	24.10	[13]
Pebax® MH 1657 (5 wt%)	LDH (2 wt%)	Co-precipitation [a] Solution mixing and casting method [b]	FSC (MMMs)	Flat-sheet (60 ± 10 μm)	T = 30 °C P = 1 bar Mixed gas CH_4/CO_2 (70/30 vol%)	98.60 (Barrer)	5.33 (Barrer)	18.50	[13]
Pebax® MH 1657 (5 wt%)	LDH (2 wt%)	Co-precipitation [a] Solution mixing and casting method [b]	FSC (MMMs)	Flat-sheet (60 ± 10 μm)	T = 35 °C P = 1 bar (W.C: 10.3%) Mixed gas CH_4/CO_2 (70/30 vol%)	619.00 (Barrer)	21.95 (Barrer)	28.20	[13]

(continued)

Table 4.2 (continued)

Materials	Carriers/fillers	Fabrication method	FTMs types	Module/thickness	Operating conditions	$P(CO_2)$	$P(CH_4)$	$\alpha \frac{(CO2)}{(CH4)}$	Ref.
Pebax® MH 1657 (5 wt%)	ZIF-8@LDH (2 wt%)	In-situ growth [a] Solution mixing and casting method [b]	FSC (MMMs)	Flat-sheet (60 ± 10 μm)	T = 30 °C P = 1 bar Mixed gas CH_4/CO_2 (70/30 vol%)	122.00 (Barrer)	6.35 (Barrer)	19.20	[13]
Pebax® MH 1657 (5 wt%)	ZIF-8@LDH (2 wt%)	In-situ growth [a] Solution mixing and casting method [b]	FSC (MMMs)	Flat-sheet (60 ± 10 μm)	T = 35 °C P = 1 bar (W.C: 12.4%) Mixed gas CH_4/CO_2 (70/30 vol%)	1307.00 (Barrer)	41.36 (Barrer)	31.60	[13]
CA (10 wt%)	NH_2-MIL-53(Al) (15 wt%)	Hydrothermal heating [a] Solution mixing and casting method [b]	FSC (MMMs)	Flat-sheet	T = 25 °C P = 3 bar Pure gas	52.60 (Barrer)	1.83 (Barrer)	28.70	[14]
CA (10 wt%)	NH_2-MIL-53(Al) (15 wt%)	Hydrothermal heating [a] Solution mixing and casting method [b]	FSC (MMMs)	Flat-sheet	T = 25 °C P = 3 bar Mixed gas CH_4/CO_2 (50/50 vol%)	34.80 (Barrer)	1.89 (Barrer)	18.40	[14]
Pebax® MH 1657 (4 wt%)	NCs (20 wt%)	Stöber method [a] Solution mixing and casting method [b]	FSC (MMMs)	Flat-sheet (75–95 μm)	T = 25 °C P = 2 bar Humidified Pure gas	1431.00 (Barrer)	31.10 (Barrer)	46.00	[15]

(continued)

Table 4.2 (continued)

Materials	Carriers/fillers	Fabrication method	FTMs types	Module/thickness	Operating conditions	$P(CO_2)$	$P(CH_4)$	$\alpha \frac{(CO2)}{(CH4)}$	Ref.
Pebax® MH 1657 (4 wt%)	NCs (30 wt%)	Stöber method [a] Solution mixing and casting method [b]	FSC (MMMs)	Flat-sheet (75–95 μm)	T = 25 °C P = 2 bar Humidified Pure gas	1993.00 (Barrer)	51.10 (Barrer)	39.00	[15]
PA/PSF	– NHCO	Interfacial polymerization (selective layer) [b] Phase Inversion (support layer) [b]	FSC (composite)	Flat-sheet (200 μm support layer)	T = 25 °C P = 5 bar W.C (saturated)	2.411 (GPU) 482.20 (Barrer)	0.556 (GPU) 111.36 (Barrer)	4.33	[16]
PPO_{dm}	Binary nano-fillers CNF (2 wt%) - GO (3 wt%)	Dry-phase inversion method [b]	FSC (MMMs)	Flat-sheet (150 μm)	T = 25 °C P = 3.5 bar Pure gas	62.32 (Barrer)	1.043 (Barrer)	59.70	[17]
MFFK-1	bmim[BF_4]	Vacuum method [b]	MC (SILMs)	Flat-sheet (100 μm)	T = 25 °C P = 1 bar Pure gas	84.00 (Barrer)	12.00 (Barrer)	7.00	[18]
MFFK-1	bmim[ace]	Vacuum method [b]	MC (SILMs)	Flat-sheet (100 μm)	T = 25 °C P = 1 bar Pure gas	92.00 (Barrer)	6.00 (Barrer)	15.00	[18]

(continued)

Table 4.2 (continued)

Materials	Carriers/fillers	Fabrication method	FTMs types	Module/thickness	Operating conditions	$P(CO_2)$	$P(CH_4)$	$\alpha \frac{(CO2)}{(CH4)}$	Ref.
MFFK-1	bmim[BF_4]	Vacuum method [b]	MC (SILMs)	Flat-sheet (100 μm)	T = 25 °C P = 1 bar Binary gas mixture CH_4/CO_2 (85/15 vol%)	68.90 (Barrer)	7.90 (Barrer)	8.70	[18]
MFFK-1	bmim[ace]	Vacuum method [b]	MC (SILMs)	Flat-sheet (100 μm)	T = 25 °C P = 1 bar Binary gas mixture CH_4/CO_2 (85/15 vol%)	186.10 (Barrer)	2.00 (Barrer)	96.90	[18]
MFFK-1	bmim[BF_4]	Vacuum method [b]	MC (SILMs)	Flat-sheet (100 μm)	T = 25 °C P = 1 bar Ternary gas mixture $CH_4/CO_2/H_2S$ (76.5/18.2/5.3 vol%)	CO_2:70.00 (Barrer) H_2S:142.00 (Barrer)	8.60 (Barrer)	CO_2/CH_4 (8.00) H_2S/CH_4 (16.40)	[18]
MFFK-1	bmim[ace]	Vacuum method [b]	MC (SILMs)	Flat-sheet (100 μm)	T = 25 °C P = 1 bar Ternary gas mixture $CH_4/CO_2/H_2S$ (76.5/18.2/5.3 vol%)	CO_2:100.30 (Barrer) H_2S:110.40 (Barrer)	1.03 (Barrer)	CO_2/CH_4 (96.50) H_2S/CH_4 (106.20)	[18]

(continued)

Table 4.2 (continued)

Materials	Carriers/fillers	Fabrication method	FTMs types	Module/thickness	Operating conditions	$P(CO_2)$	$P(CH_4)$	$\alpha \frac{(CO2)}{(CH4)}$	Ref.
Matrimid-5218 (7 wt%)	UiO-67–33 (10 wt%)	Solvothermal method [a] Solution mixing and casting method [b]	FSC (MMMs)	Flat-sheet	T = 30 °C P = 5 bar Mixed gas CH_4/CO_2 (50/50 vol%)	27.00 (Barrer)	0.36 (Barrer)	75.00	[20]
(4 wt%) Pebax® MH 1657 / PVDF	TpDT-CONs (0.5 wt%)	Schiff base condensation reaction [a] Solution mixing and casting method [b]	FSC (composite)	Hollow Fiber (Skin layer: 10 μm)	T = 25 °C P = 1 bar Pure gas	14.10 (GPU) 141.00 (Barrer)	0.48 (GPU) 4.86 (Barrer)	29.00	[21]
(4 wt%) Pebax® MH 1657 / PVDF	TpDT-CONs (1.0 wt%)	Schiff base condensation reaction [a] Solution mixing and casting method [b]	FSC (composite)	Hollow Fiber (Skin layer: 10 μm)	T = 25 °C P = 1 bar Pure gas	15.90 (GPU) 159.00 (Barrer)	0.51 (GPU) 5.12 (Barrer)	31.00	[21]
(4 wt%) Pebax® MH 1657 / PVDF	TpDT-CONs (1.5 wt%)	Schiff base condensation reaction [a] Solution mixing and casting method [b]	FSC (composite)	Hollow Fiber (Skin layer: 10 μm)	T = 25 °C P = 1 bar Pure gas	13.90 (GPU) 139.00 (Barrer)	0.46 (GPU) 4.63 (Barrer)	30.00	[21]

(continued)

Table 4.2 (continued)

Materials	Carriers/fillers	Fabrication method	FTMs types	Module/thickness	Operating conditions	$P(CO_2)$	$P(CH_4)$	$\alpha\ \frac{(CO2)}{(CH4)}$	Ref.
(4 wt%) Pebax® MH 1657 / PVDF	TpDT-CONs (2.0 wt%)	Schiff base condensation reaction [a] Solution mixing and casting method [b]	FSC (composite)	Hollow Fiber (Skin layer: 10 μm)	T = 25 °C P = 1 bar Pure gas	11.80 (GPU) 118.00 (Barrer)	0.56 (GPU) 5.61 (Barrer)	21.00	[21]
(4 wt%) Pebax® MH 1657 / PVDF	TpDT-CONs (0.5 wt%)	Schiff base condensation reaction [a] Solution mixing and casting method [b]	FSC (composite)	Hollow Fiber (Skin layer: 10 μm)	T = 25 °C P = 1 bar Humidified Pure gas	91.80 (GPU) 918.00 (Barrer)	3.06 (GPU) 30.60 (Barrer)	30.00	[21]
(4 wt%) Pebax® MH 1657 / PVDF	TpDT-CONs (1.0 wt%)	Schiff base condensation reaction [a] Solution mixing and casting method [b]	FSC (composite)	Hollow Fiber (Skin layer: 10 μm)	T = 25 °C P = 1 bar Humidified Pure gas	141.10 (GPU) 1411.00 (Barrer)	3.71 (GPU) 37.13 (Barrer)	38.00	[21]
(4 wt%) Pebax® MH 1657 / PVDF	TpDT-CONs (1.5 wt%)	Schiff base condensation reaction [a] Solution mixing and casting method [b]	FSC (composite)	Hollow Fiber (Skin layer: 10 μm)	T = 25 °C P = 1 bar Humidified Pure gas	118.70 (GPU) 1187.00 (Barrer)	4.23 (GPU) 42.39 (Barrer)	28.00	[21]

(continued)

Table 4.2 (continued)

Materials	Carriers/fillers	Fabrication method	FTMs types	Module/thickness	Operating conditions	$P(CO_2)$	$P(CH_4)$	$\alpha \frac{(CO2)}{(CH4)}$	Ref.
(4 wt%) Pebax® MH 1657 / PVDF	TpDT-CONs (2.0 wt%)	Schiff base condensation reaction [a] Solution mixing and casting method [b]	FSC (composite)	Hollow Fiber (Skin layer: 10 μm)	T = 25 °C P = 1 bar Humidified Pure gas	75.80 (GPU) 758.00 (Barrer)	3.44 (GPU) 34.45 (Barrer)	22.00	[21]
(4 wt%) Pebax® MH 1657 / PVDF	TpDT-CONs (0.5 wt%)	Schiff base condensation reaction [a] Solution mixing and casting method [b]	FSC (composite)	Hollow Fiber (Skin layer: 10 μm)	T = 25 °C P = 1 bar Humidified Mixed gas CH_4/CO_2 (70/30 vol%)	92.58 (GPU) 925.80 (Barrer)	3.19 (GPU) 31.94 (Barrer)	28.98	[21]
(4 wt%) Pebax® MH 1657 / PVDF	TpDT-CONs (1.0 wt%)	Schiff base condensation reaction [a] Solution mixing and casting method [b]	FSC (composite)	Hollow Fiber (Skin layer: 10 μm)	T = 25 °C P = 1 bar Humidified Mixed gas CH_4/CO_2 (70/30 vol%)	143.00 (GPU) 1430.00 (Barrer)	3.97 (GPU) 39.72 (Barrer)	36.00	[21]
(4 wt%) Pebax® MH 1657 / PVDF	TpDT-CONs (1.5 wt%)	Schiff base condensation reaction [a] Solution mixing and casting method [b]	FSC (composite)	Hollow Fiber (Skin layer: 10 μm)	T = 25 °C P = 1 bar Humidified Mixed gas CH_4/CO_2 (70/30 vol%)	118.90 (GPU) 1189.00 (Barrer)	4.44 (GPU) 44.48 (Barrer)	26.73	[21]

(continued)

Table 4.2 (continued)

Materials	Carriers/fillers	Fabrication method	FTMs types	Module/thickness	Operating conditions	$P(CO_2)$	$P(CH_4)$	$\alpha \frac{(CO2)}{(CH4)}$	Ref.
(4 wt%) Pebax® MH 1657 / PVDF	TpDT-CONs (2.0 wt%)	Schiff base condensation reaction [a] Solution mixing and casting method [b]	FSC (composite)	Hollow Fiber (Skin layer: 10 μm)	T = 25 °C P = 1 bar Humidified Mixed gas CH_4/CO_2 (70/30 vol%)	75.80 (GPU) 758.00 (Barrer)	3.82 (GPU) 38.28 (Barrer)	19.80	[21]
(20 wt%) PES/PP	DEA (5.0 wt%) – CMS (10.0 wt%)	Solution mixing [a] Dry/wet phase inversion method [b]	FSC (composite)	Flat-sheet (Skin layer: 34.99 μm)	T = 25 °C P = 10–30 bar Pure gas	375.45–268.18 (GPU) 13,136.99–9383.61 (Barrer)	40.12–18.19 (GPU) 1403.52–636.60 (Barrer)	9.36–14.74	[22]
(20 wt%) PES/PP	DEA (10.0 wt%) – CMS (10.0 wt%)	Solution mixing [a] Dry/wet phase inversion method [b]	FSC (composite)	Flat-sheet (Skin layer: 126.90 μm)	T = 25 °C P = 10–30 bar Pure gas	411.67–289.70 (GPU) 52,240.92–36,762.93 (Barrer)	41.53–18.63 (GPU) 5271.53–2364.17 (Barrer)	9.91–15.55	[22]
(20 wt%) PES/PP	DEA (15.0 wt%) – CMS (10.0 wt%)	Solution mixing [a] Dry/wet phase inversion method [b]	FSC (composite)	Flat-sheet (Skin layer: 93.86 μm)	T = 25 °C P = 10–30 bar Pure gas	469.31–300.84 (GPU) 64,530.12–41,365.50 (Barrer)	42.92–18.76 (GPU) 5903.94–2578.89 (Barrer)	10.93–16.04	[22]

(continued)

Table 4.2 (continued)

Materials	Carriers/fillers	Fabrication method	FTMs types	Module/thickness	Operating conditions	$P(CO_2)$	$P(CH_4)$	$\alpha \frac{(CO2)}{(CH4)}$	Ref.
(0.15 wt.%) PPPS/ (1.5 wt.%) PDMS/PSF	R-NH_2	Solution mixing and wet-dry coating method [a, b]	FSC (composite)	Flat-sheet (Skin layer: 0.295 μm)	T = 25 °C P = 1.5 bar Humidified Mixed gas CH_4/CO_2 (55/45 vol%)	1029.85 (GPU) 303.80 (Barrer)	16.61 (GPU) 4.90 (Barrer)	62.00	[23]
(3 wt%) Pebax-1657/ (25 wt%) PSF	A-ZIF-8@SiO_2 (0.5 wt%)	Facile one-step method [a] Solution mixing and casting method [b]	FSC (composite)	Hollow Fiber (Skin layer: 0.103 μm)	T = 30 °C P = 4 bar Pure gas	41.15 (GPU) 4.23 (Barrer)	1.85 (GPU) 0.19 (Barrer)	22.25	[24]
(2 wt%) PVAm/ mPSF	IRMOF-3 (25 wt%)	Optimizing synthesis method [a] Solution mixing and casting method [b]	FSC (composite)	Flat-sheet (Skin layer: 0.409–0.416 μm)	T = 25 °C P = 10–30 bar Mixed gas CH_4/CO_2 (90/10 vol%)	499.00–271.00 (GPU) 204.09–112.73 (Barrer)	7.79–5.31 (GPU) 3.34–2.16 (Barrer)	61.00–52.00	[25]

[a] Fabrication method for carrier
[b] Fabrication method for polymer and polymeric based FTMs

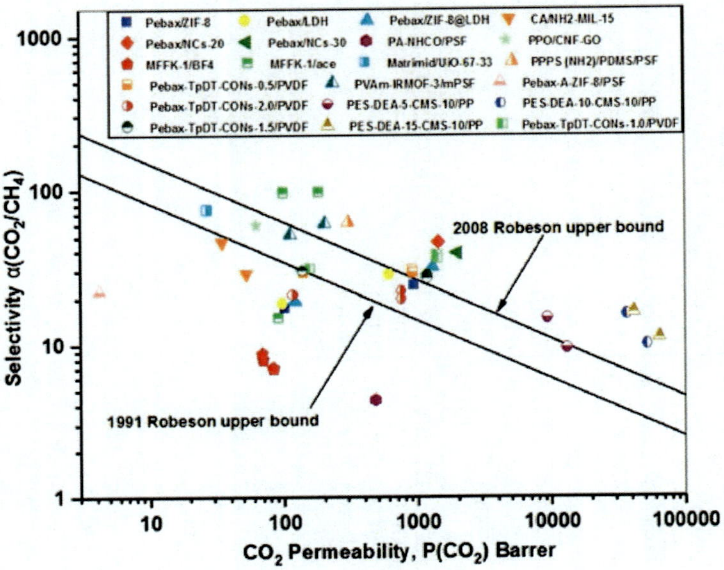

Fig. 4.2 Robeson's upper bound graph illustrating the NG purification (CO_2/CH_4) selectivity of different FTMs versus its CO_2 permeability from Table 4.2 data

4.4 Conclusion and Future Recommendations

In this chapter, the recent advancements in polymeric-based FTMs for the application of NG purification have been discussed in term of its CO_2 transport mechanism, Carrier mediation chemistry, fabrication method of both membranes and carriers, and CO_2/CH_4 separation performance in term of both its permeability and selectivity. These FTMs provide good CO_2 permeability, selectivity and high resistance to plasticization. Due to the coupled transport mechanism (solution diffusion and carrier mediation) of FTMs, both mobile carrier (M.C) and fixed-site carriers (FSC) based FTMs overcome the permeability-selectivity trade-off limit and surpasses the Robeson upper bound lines. The SILMs, hybrid filler based FTMs, and thin layer composite based FTMs are among the most promising membrane technology for CO_2/CH_4 separation. In addition, a particular type of SILMs such as MFFK-1—bmim [ace] and MFFK-1—bmim [BF_4] also have the capability of simultaneous removal of both the acidic gases CO_2 and H_2S from the NG fuel, which showed that these membranes also have the potential to be used for the removal of multiple impurities from raw NG stream. The raw NG also contain water (H_2O) molecules, for its dehydration a separate unit is also required for conventional technology processes. Whereas, in case of FTMs, using this humidified NG fuel stream is beneficial to promote rapid facilitated transport mechanism of CO_2 molecules and high CO_2/CH_4 separation performance. Another benefit of FTM such as recently reported MMMs-based FTMs and composite-based FTMs have superior structural resilience, making them suitable for NG purification application in harsh environments, where pristine-based polymeric membranes are unstable. Furthermore, although there is a significant development in polymeric-based FTMs for the application of NG purification, it still requires further research work, simulation and modelling, and comprehensive field tests for the customization of materials in order to resolve the existing technical gap and overcome the different issues associated with the implementation of this technology at commercial scale. Moreover, further improvement in the fabrication methods is also needed for the defect-free and low-cost production of these membranes on a large scale. In addition, the use of different fillers and surface modification methods needs to be considered for enhancing the membrane's physical strength and shelf life for long-term NG fuel processing and to maintain its integrity in harsh conditions.

References

1. He Z, Reddy KSK, Karanikolos G, Wang K (2018) CO_2/CH_4 separation (natural gas purification) by using mixed matrix membranes. Curr Trends Future Dev (Bio-) Membr 155–181
2. Hussain A, Nasir H, Ahsan M (2014) Process design analyses of CO_2 capture from natural gas by polymer membrane. J Chem Soc Pak 36(3)
3. Tanco MAL, Medrano JA, Gallucci F, Tanaka DAP (2018) Membrane optimization and process condition investigation for enhancing the CO_2 separation from natural gas. In: Current trends and future developments on (bio-) membranes. Elsevier, pp 469–509
4. Baker RW, Lokhandwala K (2008) Natural gas processing with membranes: an overview. Ind Eng Chem Res 47(7):2109–2121
5. Karim SS, Farrukh S, Hussain A, Noor T, Younas M (2021) A comprehensive overview of dual-layer composite membrane for air (O_2/N_2) separation. Polym Polym Compos 09673911211045852
6. Siažik J, Malcho M (2017) Accumulation of primary energy into natural gas hydrates. Procedia Eng 192:782–787
7. Ismail AF, Khulbe KC, Matsuura T (2015) Gas separation membranes. Switz Springer 10:978–973
8. Bao Z, Wang J, Zhang Z, Xing H, Yang Q, Yang Y, Wu H, Krishna R, Zhou W, Chen B (2018) Molecular Sieving of ethane from ethylene through the molecular cross-section size differentiation in gallate-based metal-organic frameworks. Angew Chem 130(49):16252–16257
9. Matteucci S, Yampolskii Y, Freeman BD, Pinnau I (2006) Transport of gases and vapors in glassy and rubbery polymers. Mater Sci Membr Gas Vap Sep 1:1–2
10. Agbaje TA, Vega LF, Khaleel M, Wang K, Karanikolos GN (2021) Membranes and adsorbents in separation of C4 hydrocarbons: a review and the definition of the current upper bounds. Sep Purif Technol 119530
11. Tong Z, Ho WW (2017) Facilitated transport membranes for CO_2 separation and capture. Sep Sci Technol 52(2):156–167
12. Han Y, Ho WW (2020) Recent advances in polymeric facilitated transport membranes for carbon dioxide separation and hydrogen purification. J Polym Sci 58(18):2435–2449
13. Zhang N, Wu H, Li F, Dong S, Yang L, Ren Y, Wu Y, Wu X, Jiang Z, Cao X (2018) Heterostructured filler in mixed matrix membranes to coordinate physical and chemical selectivities for enhanced CO_2 separation. J Membr Sci 567:272–280

14. Mubashir M, Yeong YF, Lau KK, Chew TL, Norwahyu J (2018) Efficient CO_2/N_2 and CO_2/CH_4 separation using NH_2-MIL-53 (Al)/cellulose acetate (CA) mixed matrix membranes. Sep Purif Technol 199:140–151
15. Zhang H, Guo R, Zhang J, Li X (2018) Facilitating CO_2 transport across mixed matrix membranes containing multifunctional nanocapsules. ACS Appl Mater Interfaces 10(49):43031–43039
16. Awad A, Aljundi IH (2018) Interfacial polymerization of facilitated transport polyamide membrane prepared from PIP and IPC for gas separation applications. Korean J Chem Eng 35(8):1700–1709
17. Murugiah P, Oh P, Lau K (2018) Concatenation of carbonaceous nanofillers for mixed matrix membrane development. In: IOP conference series: materials science and engineering, vol 1. IOP Publishing, p 012008
18. Akhmetshina AI, Yanbikov NR, Atlaskin AA, Trubyanov MM, Mechergui A, Otvagina KV, Razov EN, Mochalova AE, Vorotyntsev IV (2019) Acidic gases separation from gas mixtures on the supported ionic liquid membranes providing the facilitated and solution-diffusion transport mechanisms. Membranes 9(1):9
19. Zhang X, Tu Z, Li H, Huang K, Hu X, Wu Y, MacFarlane DR (2017) Selective separation of H_2S and CO_2 from CH_4 by supported ionic liquid membranes. J Membr Sci 543:282–287
20. Thür R, Van Velthoven N, Slootmaekers S, Didden J, Verbeke R, Smolders S, Dickmann M, Egger W, De Vos D, Vankelecom IF (2019) Bipyridine-based UiO-67 as novel filler in mixed-matrix membranes for CO_2-selective gas separation. J Membr Sci 576:78–87
21. Wang M, Quan K, Zheng X, Cao Y, Cui X, Xue M, Pan F (2020) Facilitated transport membranes by incorporating self-exfoliated covalent organic nanosheets for CO_2/CH_4 separation. Sep Purif Technol 237:116457
22. Nab F, Mukhtar H, Nasir R, Mohshim DFB, Arasu N, Man Z, Mannan HA (2020) Composite amine mixed matrix membranes for high-pressure CO_2–CH_4 separation: synthesis, characterization and performance evaluation. R Soc Open Sci 7(9):200795
23. Zhang C, Sheng M, Hu Y, Yuan Y, Kang Y, Sun X, Wang T, Li Q, Zhao X, Wang Z (2021) Efficient facilitated transport polymer membrane for CO_2/CH_4 separation from oilfield associated gas. Membranes 11(2):118
24. Sasikumar B, Bisht S, Arthanareeswaran G, Ismail A, Othman M (2021) Performance of polysulfone hollow fiber membranes encompassing ZIF-8, SiO_2/ZIF-8, and amine-modified SiO_2/ZIF-8 nanofillers for CO_2/CH_4 and CO_2/N_2 gas separation. Sep Purif Technol 264:118471
25. Cai X, Yuan Y, Sheng M, Wang J, Wang Z (2021) High-performance CO_2/CH_4 separation membrane fabrication with PVAm modified by the MOFs containing amine groups. J Nat Gas Sci Eng 89:103874
26. Robeson LM (2008) The upper bound revisited. J Membr Sci 320(1–2):390–400

Chapter 5
Facilitated Transport Membranes (FTMs) for Biogas Purification (CO_2/CH_4)

Syed Shujaat Karim, Ovaid Mehmood, Sarah Farrukh, and Muhammad Ayoub

Abstract Biogas is a sustainable energy source produced mostly from municipal wastewater treatment plants, landfills, and agricultural sources, mainly manure and energy crops, through the anaerobic digestion process. The major components of biogas are methane (CH_4) and carbon dioxide (CO_2), with small proportions of hydrogen sulfide (H_2S), ammonia (NH_3), hydrogen (H_2), nitrogen (N_2), carbon monoxide (CO), and oxygen (O_2). However, impurities in the form of CO_2 gas should be removed prior to commercial use of biogas to harness the maximum energy and increase the calorific value. Several purification processes have been developed, including chemical and physical absorption, pressure swing adsorption (PSA), cryogenic separation, and membrane technologies. Whereas substantial progress has been made in the field of FTMs for CO_2 separations over the years, which makes FTMs one of the most promising technology for biogas upgradation. Therefore, this chapter is mainly focused on the thorough review of different types of FTMs and their recent progress for the biogas purification application, which includes the details of different membrane structural details, fabrication method for both FTMs and carriers, and various parameters that influence the CO_2/CH_4 separation performance. In addition, the permeability-selectivity trade-off Robeson's upper bound graph that demonstrate the CO_2/CH_4 performance of different FTMs for biogas purification is also discussed in this literature.

Keywords Facilitated transport membranes · Biogas purification · CO_2/CH_4 separation · Biogas upgradation · Mixed matrix membranes · Composite membranes

S. S. Karim · O. Mehmood · S. Farrukh (✉)
School of Chemical and Materials Engineering (SCME), National University of Sciences and Technology (NUST), Sector H-12, Islamabad, Pakistan
e-mail: sarah.farrukh@scme.nust.edu.pk

S. S. Karim
e-mail: sshujaat.che6scme@student.nust.edu.pk

M. Ayoub
Department of Chemical Engineering, Universiti Teknologi Petronas (UTP), Perak, Malaysia
e-mail: muhammad.ayoub@utp.edu.my

S. Farrukh et al. (eds.), *Facilitated Transport Membranes (FTMs) for CO_2 Capture: Overview and Future Trends*, Green Energy and Technology,
https://doi.org/10.1007/978-3-031-21444-8_5

Nomenclature

1-D	One dimensional
3-D	Three dimensional
6FDA	4, 4´-(Hexafluoroisopropylidene) diphthalic anhydride
α	Selectivity
Ac	Acetate
B(POSS)	$B(OH)^{-}{}_4$ Polyhedral Oligomeric Silsequioxane
Be-NADES	Betaine-based natural deep eutectic solvent
BF_4	Boron fluoride
CH_4	Methane
Cl^-	Chlorine ion
CNC	Crystalline nano-cellulose
cNFC	Carboxymethylated-nanofibrillated cellulose
CO	Carbon monoxide
CO_2	Carbon dioxide
Cr^{+3}	Chromium-III metal ion
D4	Octamethylcyclotetrasiloxane
[Emim]-[OAc]	1-Ethyl-3-methylimidazolium acetate
FFV	Fractional free volume
FSC	Fixed-site carrier
FTM	Facilitated transport membrane
GHG	Greenhouse gas
GO	Graphene oxide
H_2	Hydrogen
H_2O	Water (moisture)
H_2S	Hydrogen sulfide
HMMP-1	High-valence metal-induced microporous polymer
[Hmim]-[NTf_2]	1-Hexyl-3-methylimidazolium bis(trifluoromethylsulfonyl)imide
LDHN	Layer double hydroxide nano-cage
M.C	Mobile carrier
Ma-NADES	DL-malic acid-based natural deep eutectic solvent
MDA	4, 4´-Diaminodiphenylmethane
MMMs	Mixed matrix membranes
mPSF	Modified polysulfone
N_2	Nitrogen
NC	Nano-composite
N-PSiNTs	Amino-modified novel porous organosilicon nanotubes
O_2	Oxygen
P	Permeance (GPU) or Permeability (Barrer)
PCNF	Phosphorylated nano-cellulose fibrils
PDA	Polydopamine
PDCS	Modified polymer-directed chemical synthesis
Pebax	Poly ether block amide

PEI	Polyethyleneimine
PEM	Polymer electrolyte membranes
PGO	Porous graphene oxide
PI	Polyimide
PIM	Polymer of intrinsic micro-porosity
PIP-Py	Pyridine-based porous ionic polymer
PPO	Polyphenylene oxide
ProK	Potassium L-prolinate
PSA	Pressure swing adsorption
PSF	Polysulfone
PSiNTs	Novel porous organosilicon nanotubes
PVA	Polyvinyl Alcohol
PVAm	Polyvinyl amine
PVDF	Poly vinylidene fluoride
QAPSF	Quaternary ammonium polysulfone
RH	Relative Humidity
S.L	Support layer
SHPAA	Poly-N-isobutyl allyl amine
SiNTs	Non-porous organosilicon nanotubes
SLM	Support liquid membrane
sβ-CD	Sulfobutylether-beta-cyclodextrin
Tart	L-(+)-tartaric acid
TiO_2	Titanium dioxide
v	Volume
VOC	Volatile organic compounds
Zn^{2+}	Zinc ion

5.1 Introduction

In this current era, because of the increasing population at an exponential rate, the demand for industrial and agricultural products, along with the consumption of energy, has been escalating every day. As of today, fossil fuels are the world's gigantic source of energy, which fulfills 84% of world's energy requirements. Nevertheless, fossil fuels are simultaneously a huge source of the greenhouse gases (GHG), causing global warming by emitting tonnes of carbon dioxide (CO_2) into the atmosphere. This compels environmentalists and scientists towards the use of renewable energy sources. Therefore, efforts are being made constantly to explore alternative or renewable sources of energy to fulfill the energy needs and to prevail over the exhausting resources of fossil fuels and the growing problem of global warming [1, 2].

Biogas is a renewable source of energy produced from the anaerobic digestion of organic feedstocks involving municipal waste, farm waste, organic food waste, and energy crops as illustrated in Fig. 5.1. The residual digestate, on the other hand, can be utilized for a various purposes, including bio-fertilizer, livestock bedding, and soil amendments [3, 4]. Biogas is predominantly composed of methane CH_4 (50–80%), carbon dioxide CO_2 (20–50%), and a smaller fraction of hydrogen (H_2) up to 5% as represented in Table 5.1. Furthermore, traces of hydrogen sulfide (H_2S), ammonia (NH_3), oxygen (O_2), nitrogen (N_2), carbon monoxide (CO) and siloxane are also present in raw biogas depending upon its source [5, 6]. The presence of a higher concentration of CO_2 reduces the calorific value of biogas. In addition, the burning of CH_4 fuel emits CO_2 per unit of energy. However, the use of biogas as a source of fuel for automobiles can reduce the release of volatile organic compounds (VOC) by 50% and nitrogenous compounds by 25%. Therefore, use of biogas as a substitute fuel is supposed to be carbon negative and promote the reduction of GHG emissions [7].

There are certain applications of biogas where high purity and energy content are required prior to its usage, and it can be achieved by separating the impurities from biogas. Mostly, separation of CO_2 from biogas dramatically increases its calorific value by up to many folds, and upgraded biogas normally has a CH_4 content greater than 95%. For optimizing the CH_4 concentration in biogas fuel, the removal of CO_2 molecules by utilizing less energy intensive and more efficient technology is essential. Numerous conventional technologies are employed for biogas upgradation in-order to utilize it for commercial purposes (e.g. transportation, power station and heating applications). Initially, the removal of water content (H_2O) from the feed is processed by cryogenic condensation. Moreover, adsorption of H_2S on iron

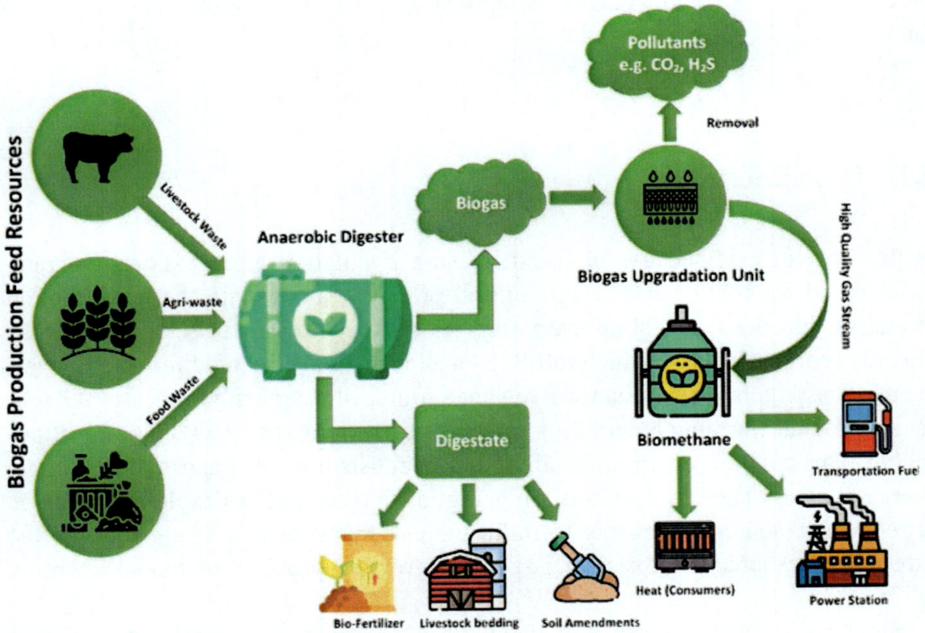

Fig. 5.1 Pictorial representation of biogas production, upgradation and consumption system [3, 4]

Table 5.1 Typical composition data of biogas and the physical properties of its components

Feed gas components	Agricultural waste (%)	Landfills gas (LFG) (%)	Industrial waste (%)	Kinetic diameter (Å)	Molecular weight (g/mol)	Ref.
Methane (CH_4)	50.00–80.00	50.00–80.00	50.00–70.00	3.80	16.00	[5, 6, 10]
Carbon-dioxide (CO_2)	30.00–50.00	20.00–50.00	30.00–50.00	3.30	44.00	[5, 6, 10]
Moisture (H_2O)	Saturation	Saturation	Saturation	2.65	18.00	[5, 6, 11]
Hydrogen (H_2)	0.00–2.00	0.00–5.00	0.00–2.00	2.89	2.00	[5, 6, 11]
Hydrogen sulfide (H_2S)	0.70	0.10	0.80	3.60	34.00	[5, 6, 12]
Ammonia (NH_3)	Traces	Traces	Traces	2.60	17.00	[5, 6, 13]
Carbon monoxide (CO)	0.00–1.00	0.00–1.00	0.00–1.00	3.76	28.00	[5, 6, 12]
Nitrogen (N_2)	0.00–1.00	0.00–3.00	0.00–1.00	3.64	28.00	[5, 6, 10]
Oxygen (O_2)	0.00–1.00	0.00–1.00	0.00–1.00	3.46	32.00	[5, 6, 10]
Siloxane (D4)	Traces	Traces	Traces	~10.00	296.61	[6, 14, 15]

oxide is a well-recognized process for eliminating the trace levels of H_2S from biogas. Whereas, separation of CO_2 from biogas is often performed by physical and chemical absorption, pressure swing adsorption (PSA), cryogenic method, and membrane-based separation processes [8, 9].

Among different separation processes, membrane technology offers numerous benefits, including low-cost, ease of processing, and exceptional reliability. This technology also has been recognized as a green technology because it does not utilize hazardous chemicals and requires no complex process units. However, there is a trade-off between permeability and selectivity when it comes to gas separation using thick polymeric membranes, though attempts are underway to develop new types of membranes to improve gas separation performance [16, 17]. In which facilitated transport membranes (FTMs) outperformed all other membranes in terms of gas separation, due to two key effects: i.e., pure physical diffusion and reversible chemical reaction between the CO_2 molecules and reactive carriers present in membrane matrix. By utilizing the combination of these two mechanisms FTMs has achieved superior CO_2 permeability for biogas purification while maintaining high CO_2/CH_4 selectivity [18].

Figure 5.2 depicts the general process of FTM-based biogas purification. Typically, these membrane materials have a high affinity for CO_2 molecules, enabling CO_2 molecules to diffuse at a faster rate than other component of the raw biogas mixture at elevated feed pressure. Following that, the carriers present in the membrane matrix form a reversible bond with the CO_2 molecules, resulting in complexes that diffuse rapidly across the membrane due to the concentration gradient via fixed-site carriers

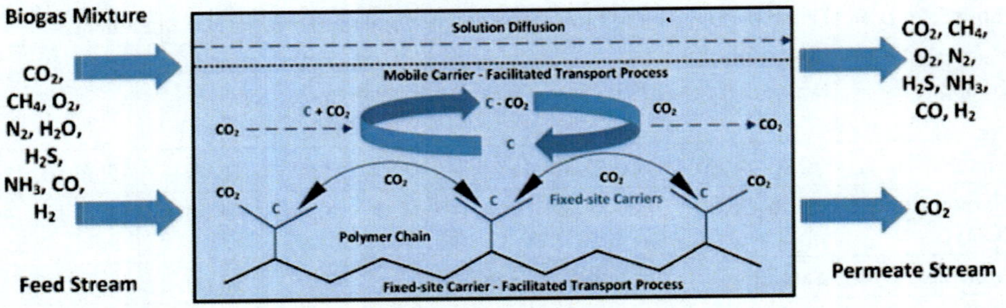

Fig. 5.2 Schematic diagram of CO_2 separation mechanism from biogas mixture stream using FTM [19]

(FSC) through a hopping mechanism or mobile carriers (M.C) using a vehicle mechanism. Whereas this complex bond breaks down on the permeate side at low pressure, releasing the CO_2 molecule while renewing the carrier, which subsequently reacts with the CO_2 molecule on the feed side. While the gas separation through a solution diffusion mechanism that works in combination with the facilitated transport process depends on the difference in diffusivity and solubility of the permeating biogas molecules across membrane materials [18–20].

Furthermore, the raw biogas stream usually has a maximum saturation of moisture content, which can be used to boost the facilitated transport process. Like most of the FSC-based FTMs, per-conditioning of feed is necessary by introducing moisture to keep the CO_2 permeability and CO_2/CH_4 selectivity of the membrane within an acceptable range. Water has the capability of increasing the mobility of both mobile and fixed-site carriers as well as increasing the reaction rate between the carriers and CO_2 molecules, resulting in improved CO_2/CH_4 separation performance of the FTMs in the biogas purification process [21].

5.2 Recent Development of FTMs for Biogas Purification

Over the last decades, many researchers have developed different types of FTMs for efficient upgradation of biogas fuel. A wide range of carriers, nano-fillers, and hybrid fillers that are incorporated into different pristine polymers and their blends based-FTM membranes matrixes have been recently studied for biogas purification are mentioned in Table 5.2. In 2018, Jahan et al. [22], fabricated polyvinyl alcohol (PVA)/crystalline nano-cellulose (CNC) nano-composite (NC) membranes for CO_2/CH_4 separation. Polyvinyl alcohol (PVA) is a water-soluble synthetic polymer that demonstrates high tensile, non-toxic, and bio-compatible features. Mixed matrix membranes (MMMs) made of PVA have been shown to be promising candidates for the removal of CO_2 from biogas mixtures (CO_2/CH_4). CNCs was incorporated into 2 weight percentages (wt%) of PVA in a concentration ranging from 0.5 to 4 wt%. A novel methodology was developed for making equal viscosity solutions with varied CNC concentration using water dilution for viscosity control to achieve equal thickness of NC layer. The homogenous and defect-free PVA/CNC layers solution at a pH value of 9.00 were cast on polysulfone (PSF) support membrane separately using the dip-coating method, followed by heat treatment forming PVA/CNC based FSC-FTM composites, which were then employed for mixed gas permeation testing systems ($CH_4/CO_2 = 60/40$ mol%) at a pressure range of 5–10 bar and temperature of 25 °C under a humidified condition. According to this study, an increase in the CO_2 permeability and selectivity of CO_2/CH_4 trend was seen when the filler loading content of CNC was increased. The CNC-PVA/PSF NC membrane having skin layer thickness of 0.713 μm at a filler loading of 1.5 wt% showed the optimum CO_2/CH_4 separation results at a pressure of 5 bar and at a relative humidity (R.H) range of 30–100%. The PVA/CNCs NC membranes follow the FSC-based FTM mechanism as a main transport mechanism for the separation of CO_2, which showed the CO_2/CH_4 separation factor of 5.96–38.86 and the CO_2 permeability is in the range of 10.62–71.06 Barrer. The permeation results revealed that by increasing the degree of water swelling in the membrane material, CO_2 may be preferentially carried as a bicarbonate anion across the swelled FSC-based FTM matrix. In addition, it was also observed that the FTMs permeability normally drops with an increase in feed pressure compressed the water swollen membrane matrix, which subsequently caused loss of water content. Whereas another reason of this decrement trend is the due to the effect of carrier saturation at high pressure.

Similarly, in another research work of Jahan et al. [23], investigated the CO_2 separation performance as a function of the CNC loading percentage in PVA membrane and the pH of the casting solution. PVA is a hydrophilic synthetic polymer with superior mechanical strength and high moisture absorption properties. Therefore, they developed a novel FSC-based NC FTMs comprised of PVA as a polymer and CNCs as fillers, which were incorporated in the different ranges of filler loading from 0.5 to 2 wt% using solution mixing method at pH value of 6–10. The PVA/CNC were cast to produce a thin, dense selective layer at an average thickness of 0.797–1.450 μm over the PSF flat-sheet support membrane by dip-coating method and then subjected to post heat treatment. The fabricated FTMs were then tested in a mixed gas permeation system ($CH_4/CO_2 = 60/40$ mol%) at a temperature of 25 °C and a pressure range of 5–10 bar, under humidified conditions at an R.H range of 30–100%. The permeation results revealed an increase in CO_2 permeability with an increase in concentration of CNC in PVA from 0.05 to 2.00 wt%. Whereas by increasing the pH value from 6 to 10 has a positive effect on both swelling behaviour and CO_2/CH_4 separation performance of the FTM. These NC membranes demonstrated the coupled mechanism of both carrier mediation and solution diffusion process for the transport of CO_2 molecules. Moreover, by introducing the moisture content in feed the FTM process for the CO_2/CH_4 separation can be further boosted due to the increase in hydroxyl (OH^-) ion resulting more dissolving of CO_2 molecules in the membrane matrix. At 1 wt% CNC filler loading of PVA/PSF NC FSC-based FTM having skin layer thickness of 0.905 μm, which was formed at a higher pH value of 10, exhibited superior CO_2/CH_4 selectivity of 43.00. Furthermore, the corresponding CO_2 permeability of

Table 5.2 Recent advancement data of FTMs for bio-gas purification (CO_2/CH_4)

Materials	Carriers/fillers	Fabrication methods	FTMs types	Module/thickness	Operating conditions	$P(CO_2)$	$P(CH_4)$	$\alpha \frac{(CO_2)}{(CH_4)}$	Ref.
(2 wt%) PVA/PSF	CNC (1.5 wt%)	Solution mixing, dip-coating method and heating treatment [b]	FSC (composites)	Flat-sheet (skin layer: 0.713 μm)	T = 25 °C P = 5 bar pH = 9.00 Humidified RH = 30–100% Mixed gas CH_4/CO_2 (60/40 mol%)	10.62–71.06 (Barrer)	1.78–1.82 (Barrer)	5.96–38.86	[22]
(2 wt%) PVA/PSF	CNC (1 wt%)	Solution mixing, dip-coating method and heat treatment [b]	FSC (composites)	Flat-sheet (Skin layer: 0.905 μm)	T = 25 °C P = 5–10 bar pH = 10.00 Humidified RH = 93% Mixed gas CH_4/CO_2 (60/40 mol%)	97.32–33.58 (Barrer)	2.26–1.01 (Barrer)	43.00–32.97	[23]
Pebax (5 wt%)	SiNTs (0.5 wt%)	Hydrolysis and co-condensation reaction [a] solution mixing and casting method [b]	FSC (MMMs)	Flat-sheet (60–80 μm)	T = 30 °C P = 2 bar Mixed gas CH_4/CO_2 (70/30 mol%)	131.04 (Barrer)	6.28 (Barrer)	20.84	[24]

(continued)

Table 5.2 (continued)

Materials	Carriers/fillers	Fabrication methods	FTMs types	Module/thickness	Operating conditions	$P(CO_2)$	$P(CH_4)$	$\alpha\frac{(CO_2)}{(CH_4)}$	Ref.
Pebax (5 wt%)	PSiNTs (0.5 wt%)	Hydrolysis and co-condensation reaction [a] solution mixing and casting method [b]	FSC (MMMs)	Flat-sheet (60–80 μm)	T = 30 °C P = 2 bar Mixed gas CH_4/CO_2 (70/30 mol%)	152.31 (Barrer)	6.58 (Barrer)	23.11	[24]
Pebax (5 wt%)	N-PSiNTs (1 wt%)	Hydrolysis and co-condensation reaction [a] solution mixing and casting method [b]	FSC (MMMs)	Flat-sheet (60–80 μm)	T = 30 °C P = 2 bar Mixed gas CH_4/CO_2 (70/30 mol%)	164.08 (Barrer)	5.44 (Barrer)	30.15	[24]
Pebax (5 wt%)	SiNTs (0.5 wt%)	Hydrolysis and co-condensation reaction [a] solution mixing and casting method [b]	FSC (MMMs)	Flat-sheet (60–80 μm)	T = 30 °C P = 2 bar Humidified mixed gas CH_4/CO_2 (70/30 mol%)	754.04 (Barrer)	24.14 (Barrer)	31.23	[24]
Pebax (5 wt%)	PSiNTs (0.5 wt%)	Hydrolysis and co-condensation reaction [a] solution mixing and casting method [b]	FSC (MMMs)	Flat-sheet (60–80 μm)	T = 30 °C P = 2 bar Humidified Mixed gas CH_4/CO_2 (70/30 mol%)	912.00 (Barrer)	36.27 (Barrer)	25.14	[24]

(continued)

Table 5.2 (continued)

Materials	Carriers/fillers	Fabrication methods	FTMs types	Module/thickness	Operating conditions	$P(CO_2)$	$P(CH_4)$	$\alpha \frac{(CO_2)}{(CH_4)}$	Ref.
Pebax (5 wt%)	N-PSiNTs (1 wt%)	Hydrolysis and co-condensation reaction [a] solution mixing and casting method [b]	FSC (MMMs)	Flat-sheet (60–80 μm)	T = 30 °C P = 2 bar Humidified Mixed gas CH_4/CO_2 (70/30 mol%)	971.82 (Barrer)	28.15 (Barrer)	34.51	[24]
PIM-1 (19/1 wt%) (solvent/polymer)	PIP-Py-Cl (15 wt%)	Polymerization, chloromethylation, ionization and ion exchange method [a] solution mixing and casting method [b]	MC (MMMs)	Flat-sheet (30 μm)	T = 30 °C P = 2 bar Pure gas	4959.80 (Barrer)	122.60 (Barrer)	40.45	[25]
PIM-1 (19/1 wt%) (solvent/polymer)	PIP-Py-Ac (15 wt%)	Polymerization, chloromethylation, ionization and ion exchange method [a] solution mixing and casting method [b]	MC (MMMs)	Flat-sheet (30 μm)	T = 30 °C P = 2 bar Pure gas	6204.80 (Barrer)	110.50 (Barrer)	56.15	[25]
PIM-1 (19/1 wt%) (solvent/polymer)	PIP-Py-BF_4 (15 wt%)	Polymerization, chloromethylation, ionization and ion exchange method [a] solution mixing and casting method [b]	MC (MMMs)	Flat-sheet (50 μm)	T = 30 °C P = 2 bar Pure gas	5584.30 (Barrer)	140.10 (Barrer)	39.85	[25]

(continued)

Table 5.2 (continued)

Materials	Carriers/fillers	Fabrication methods	FTMs types	Module/thickness	Operating conditions	$P(CO_2)$	$P(CH_4)$	$\alpha_{\frac{(CO_2)}{(CH_4)}}$	Ref.
PIM-1 (19/1 wt%) (solvent/polymer)	PIP-Py-Ac (15 wt%)	Polymerization, chloromethylation, ionization and ion exchange method [a] solution mixing and casting method [b]	MC (MMMs)	Flat-sheet (30 μm)	T = 30 °C P = 2 bar Mixed gas CH_4/CO_2 (1:1 v:v)	5300.20 (Barrer)	143.24 (Barrer)	37.00	[25]
PES (S.L)	GO-POSS-NH_2 (5 wt%)	Modified Hummer's method and crosslinked reaction [a] Vacuum-assisted method [b]	FSC (composite)	Flat-sheet (skin Layer: 0.048–0.050 μm)	T = 30 °C P = 2 bar Humidified RH = 85% Mixed gas CH_4/CO_2 (70/30 vol%)	13.66 (GPU)	0.1833 (GPU)	74.50	[26]
(6 wt%) SHPAA-(4 wt%) PVA/PPO	(0.2 wt%) pG06 + 10 wt% [Emim]-[OAc]	PGO: Sonication: 6 h., exfoliation and Hydrothermally treatment [a] solution mixing and casting method [b]	Hybrid MC (composite)	Hollow fiber (skin layer: 0.20 μm)	T = 35 °C P = 2 bar Humidified Mixed gas CH_4/CO_2 (60/40 vol%)	779.37 (GPU)	38.58 (GPU)	20.20	[27]
(6 wt%) SHPAA-(4 wt%) PVA/PPO	(0.2 wt%) pG06 + 20 wt% ProK	PGO: Sonication: 6 h., exfoliation and Hydrothermally treatment [a] solution mixing and casting method [b]	Hybrid MC (composite)	Hollow Fiber (Skin layer: 0.20 μm)	T = 35 °C P = 2 bar Humidified Mixed gas CH_4/CO_2 (60/40 vol%)	825.40 (GPU)	41.27 (GPU)	20.00	[27]

(continued)

Table 5.2 (continued)

Materials	Carriers/fillers	Fabrication methods	FTMs types	Module/thickness	Operating conditions	$P(CO_2)$	$P(CH_4)$	$\alpha\,\frac{(CO_2)}{(CH_4)}$	Ref.
(2 wt%) PVA/PSF	PCNF (1 wt%)	Solution mixing and dip-coating method [b]	FSC (composites)	Flat-sheet (skin layer: 6 μm)	T = 25 °C P = 5 bar pH = 12.00 Humidified RH = 93% Mixed gas CH_4/CO_2 (60/40 mol%)	266.64 (Barrer)	5.89 (Barrer)	45.20	[28]
Pebax (4 wt%)	[Hmim]-[NTf_2]@LDHN (6 wt%)	Etching and Soaking [a] Solution mixing and casting method [b]	MC (MMMs)	Flat-sheet	T = 25 °C P = 2 bar Humidified Mixed gas CH_4/CO_2 (90/10 vol%)	644.00 (Barrer)	18.94 (Barrer)	34.00	[29]
QAPSF (4 wt%)	s*β*-CD (0.5 wt%)	Slow addition, heating and vacuum drying [a] Solution mixing and casting method [b]	FSC (PEM-MMMs)	Flat-sheet (120–140 μm)	T = 30 °C P = 2 bar Humidified Mixed gas CH_4/CO_2 (70/30 vol%)	1303.00 (Barrer)	33.41 (Barrer)	39.00	[30]
QAPSF (5 wt%)	B(POSS) (2 wt%)	Template etching, ion-exchange and soaking method [a, b]	FSC (PEM-MMMs)	Flat-sheet	T = 25 °C P = 1 bar Humidified Mixed gas CH_4/CO_2 (70/30 vol%)	1113.00 (Barrer)	24.73 (Barrer)	45.00	[31]

(continued)

Table 5.2 (continued)

Materials	Carriers/fillers	Fabrication methods	FTMs types	Module/thickness	Operating conditions	$P(CO_2)$	$P(CH_4)$	$\alpha_{\frac{(CO_2)}{(CH_4)}}$	Ref.
(3 wt%) PVAm/PSF	CNC (1 wt%)	Solution mixing and dip-coating method [b]	FSC (composite)	Flat-sheet (1.5 μm)	T = 25 °C P = 5 bar Humidified RH = 87% Pure gas	7.00 (Barrer)	0.17 (Barrer)	41.00	[32]
PVDF (15 wt%)	Be-Tart-NADESs (equimolar)	Heating and stirring [a] Phase inversion and impregnated method [b]	MC (SILM)	Flat-sheet (250 μm)	T = 25 °C P = 14 bar Pure gas	25.55 (Barrer)	0.50 (Barrer)	51.10	[33]
PVDF (15 wt%)	Be-Ma-NADESs (equimolar)	Heating and stirring [a] Phase inversion and impregnated method [b]	MC (SILM)	Flat-sheet (250 μm)	T = 25 °C P = 14 bar Pure gas	29.33 (Barrer)	0.52 (Barrer)	56.40	[33]
PVDF (15 wt%)	Be-Tart-NADESs (equimolar)	Heating and stirring [a] Phase inversion and impregnated method [b]	MC (SLM)	Flat-sheet (250 μm)	T = 25 °C P = 14 bar Mixed gas CH_4/CO_2 (50/50%)	24.44 (Barrer)	0.48 (Barrer)	50.92	[33]
PVDF (15 wt%)	Be-Ma-NADESs (Equimolar)	Heating and stirring [a] Phase inversion and impregnated method [b]	MC (SLM)	Flat-sheet (250 μm)	T = 25 °C P = 14 bar Mixed gas CH_4/CO_2 (50/50%)	26.43 (Barrer)	0.49 (Barrer)	53.94	[33]

(continued)

Table 5.2 (continued)

Materials	Carriers/fillers	Fabrication methods	FTMs types	Module/thickness	Operating conditions	$P(CO_2)$	$P(CH_4)$	$\alpha \frac{(CO_2)}{(CH_4)}$	Ref.
(0.1 wt%) PVAm/ mPSF	HMMP-1 (44.4 wt%)	Modified PDCS method [a] Radical polymerization, solution mixing and casting method [b]	FSC (composite)	Flat-sheet (0.18 μm)	T = 25 °C P = 1.5 bar Humidified Mixed gas CH_4/CO_2 (90/10 vol%)	337.00 (Barrer)	3.24 (Barrer)	104.00	[34]
(27.5 wt%) PVAm/(27.5 wt%) cNFC	L-arginine (45 wt%)	Solution mixing and casting method [b]	Hybrid MC (MMMs)	Flat-sheet (50–60 μm)	T = 35 °C Humidified RH = 100% Pure gas	340.00 (Barrer)	17.00 (Barrer)	20.00	[35]
PI 6FDA-MDA (12.5 wt%)	TiO_2-PDA-Zn^{2+} (7.5 wt%)	Hydrolysis, coating, and ion loading [a] Polymerization, solution mixing and casting method [b]	FSC (MMMs)	Flat-sheet (190 μm)	T = 36 °C P = 6 bar Mixed gas CH_4/CO_2 (90/10 vol%)	108.36 (Barrer)	2.19 (Barrer)	49.48	[36]

[a] Fabrication method for carrier
[b] Fabrication method for FTMs

97.32 Barrer was obtained at a lower pressure of 5 bar under a humidified condition at a R.H value of 93%.

In 2019, Yang et al. [24], investigated the CO_2/CH_4 separation performance of Polyether Block Amide (Pebax) based FSC-based FTM (MMMs) by embedding three different organosilicon nanotubes in the 5 wt% Pebax membrane matrix, which included non-porous, novel porous, and amino-modified novel porous organosilicon nanotubes—SiNTs, PSiNTs, and N-PSiNTs, respectively, as filler at a variable loading concentration of 0.2–2 wt% for the application of biogas purification. The fillers were synthesized by the hydrolysis and co-condensation methods, and then MMMs were fabricated using the solution mixing and casting method at a thickness range of 60–80 μm represented as Pebax-N-PSiNTs, Pebax-PSiNTs, and Pebax-SiNTs. These MMMs were then employed for the mixed gas permeation system (CH_4/CO_2 = 70/30 mol%) at a pressure of 5 bar and a temperature of 30 °C under dry and humidified conditions, separately. According to the gas permeation results, the Pebax-based MMMs demonstrated outstanding performance in terms of CO_2/CH_4 permeability and selectivity. The main transport of CO_2 molecules in the case of Pebax-based SiNTs occurred using three paths: (1) free volume present in the Pebax membrane matrix, (2) the one-dimensional (1-D) channels provided by SiNTs, and (3) the interface between polymer and filler. The Pebax-PSiNTs and Pebax-N-PSiNTs showed similar behaviour in terms of their permeability and selectivity, which initially demonstrated an increment trend and then a decrement in performance when filler loading concentration was increased from 0.5 to 2 wt%. Furthermore, both these Pebax-based MMMs provide an additional path for permeation of CO_2 molecules through the micro-pores of the PSiNTs wall, which significantly improved their CO_2/CH_4 separation performance as compared to the Pebax-SiNTs based MMM. Under dry conditions, at filler loadings of 0.5 and 1.0 wt%, the Pebax-PSiNTs and Pebax-N-PSiNTs showed the most optimum CO_2/CH_4 separation performance of 23.11 and 30.15, with CO_2 permeability of 152.31 Barrer and 164.08 Barrer, respectively. Whereas, in the case of humidified conditions, the intensification of the facilitated transport process occurred, causing these membranes to show remarkable enhancement in CO_2 permeability of up to 912 and 971.82 Barrer with corresponding CO_2/CH_4 selectivity of 25.14 and 34.51, respectively. Among the other two fillers, Pebax-N-PSiNTs MMMs achieved the superior CO_2/CH_4 separation performance because of the presence of an amino group in the N-PSiNTs filler, which facilitates the rapid transport of CO_2 across the membrane and demonstrated CO_2/CH_4 separation performance that even surpassed the revised Robeson upper bound limit (2008).

Wang et al. [25], studied the novel method of CO_2 separation from biogas using innovative bionic channels for the rapid facilitated transport of CO_2 molecule across the membrane. For this purpose, filtering porous ionic polymer (PIP) based fillers were initially fabricated by a three step synthesis process including polymerization of triptycene and formaldehyde dimethyl acetate, followed by chloromethylation, and pyridine ionization, and finally anion exchange, in which pyridine-based PIPs with varied anions of PIP-Py-Cl, PIP-Py-Ac, and PIP-Py-BF_4 were produced. Afterward, a series of PIM-1 based M.C-FTMs (MMMs) were fabricated by embedding these fillers into the PIM-1 matrix using a solution mixing and casting method at a thickness of 30, 50 and 30 μm, respectively. Then a permeation study was conducted using a pure and mixed gas system ($CH_4/CO_2 = 1:1$ v:v) at a pressure of 2 bar and a temperature of 25 °C to investigate the potential of these membranes for CO_2 separation efficiency from biogas mixture. For pure gas testing system, at filler loading of 15 wt%, PIM-PIP-Py-Cl, PIM-PIP-Py-Ac, and PIM-PIP-Py-BF_4 exhibited the maximum CO_2/CH_4 selectivity of 40.45, 56.15, and 39.85 with the corresponding CO_2 permeability of 4959.80, 6204.80, and 5584.30 Barrer, respectively. However, in the case of a mixed gas permeation system, the above-mentioned membranes showed a slight decrement in the CO_2/CH_4 separation performance because of the competitive permeation process. Whereas, according to the results, all the MMMs demonstrated higher perm-selectivity than pristine PIM-1 membranes far surpassing the Robeson upper bound line (2008). Therefore, pyridine-based PIP fillers can be used as potential filler candidate for the fabrication of high separation performance MMMs for the biogas upgradation application.

Yang et al. [26], investigated the modification of graphene oxide (GO) membranes containing two-dimensional interlaminar channels, demonstrating unique aspects in gas separation where regulating the size and interlayer interface of 2D channels can enhance the gas separation performance. Therefore, for its development, GO was initially prepared by the modified hummer method, followed by integrating a distinct type of nano-cage intercalator, amino-functionalized POSS-NH_2 into the GO nano-sheet using pre-crosslinked reaction, forming a GO-POSS-NH_2 hybrid filler. The POSS-NH_2 has a rigid structure with eight amino groups on its side chain, regulating the channel structure of the GO interlayer by reducing the channel size between 9.1 and 4.7 Å, thus forming a promising environment for the efficient separation of CO_2 from biogas mixture (CO_2/CH_4). These GO-POSS-NH_2 hybrid fillers at a concentration of 1–10 wt% suspension were then prepared separately and deposited in the polyether sulfone (PES) support layer using a vacuum-assisted method, forming different FSC-FTM based composite membranes having a skin layer thickness of 0.048–0.050 μm. The gas separation performance of composite membranes was evaluated by a mixed gas composition of $CH_4/CO_2 = 70/30$ vol% at a pressure of 2 bar and a temperature of 30 °C in a humidified condition of R.H = 85%. With the concentration of POSS-NH_2 increasing from 1 to 10 wt%, the membranes showed a first increasing and then decreasing trend of CO_2/CH_4 selectivity. At POSS-NH_2-GO filler loading of 5 wt%, the highest CO_2/CH_4 selectivity of 74.50 with a CO_2 permeance of 13.66 were obtained, which is 1.4 times greater than the pristine GO membrane because of the smaller channel size of GO-POSS-NH_2. The CO_2 transport in these composite membranes mainly occurred due to the facilitated transport process in combination with the solution diffusion mechanism. The crosslinking between GO and POSS-NH_2 caused decreased in channel size. However further increment in POSS-NH_2-GO filler concentration obstructed the gas molecular transport. Whereas, in humidified conditions, the water molecule adsorption in filler channels further enhances the facilitated transport of CO_2 molecules, which resulted in the improvement in CO_2/CH_4 separation performance.

In 2020, Saravanan et al. [27], developed a thin film composite hollow fiber membrane with a defect-free, ultrathin selective layer at a thickness of 0.20 μm. The hybrid facilitated transport membranes (HFTMs) comprises of three main phases for CO_2 separation, which include a host polymeric blend membrane matrix made up of 6 wt% Poly-N-isobutyl allyl amine (SHPAA) and 4 wt% Polyvinyl alcohol (PVA) with amine-based FSC, a modified porous graphene oxide (pGO)-based 2D nano-filler, and a CO_2-philic M.C e.g. 1-Ethyl-3-methylimidazolium acetate [Emim]-[OAc] and Potassium L-prolinate (ProK). The GO 2D nano fillers were prepared by the simultaneous exfoliation method to produced a controlled size of GO, followed by sonication and hydrothermal treatment processes to introduce non-selective pores into GO, resulting in pGO-based 2D nano-fillers. Whereas the HFTMs were fabricated using the solution mixing and casting method, in which polyphenylene oxide (PPO) was used as a porous support membrane. The fabricated HFTMs were then investigated for gas permeation performance by utilizing humidified mixed gas permeation setup, where the composition of the feed was 40/60 vol% of CO_2/CH_4 mixture at a pressure range of 2–20 bar and a temperature of 35 °C. Which revealed that HFTM with 0.2 wt% pGO-06 prepared at a 6 h sonication and 10 wt% [Emim]-[OAc] as a filler had the highest CO_2/CH_4 separation factor of 20.20 and a corresponding CO_2 permeance of 779.82 GPU at a pressure of 2 bar. Under similar conditions, the HFTM exhibited a maximum CO_2 permeance of 825.40 GPU with a CO_2/CH_4 separation of 20.00 at a filler loading of 0.2 wt% pGO-06 and 10 wt% [Emim]-[OAc]. In these optimized HFTMs, the superior CO_2/CH_4 separation performance is due to the combination of three main factors: size-modified pGO, amine group as FSC, along with high M.C mobility, which formed a reversible bond with CO_2 molecules to form carbonate/carbene-CO_2 for improved performance of CO_2 permeation across the membrane matrix. Furthermore, the addition of pGO nano-filler and the relative increase in CO_2-philic species resulted in an increase in carrier saturation resistance, which provide stability of CO_2/CH_4 separation performance even at an elevated pressures up to 20 bar.

Jahan et al. [28], investigated a novel phosphorylated nano-cellulose fibril (PCNF)/polyvinyl alcohol (PVA) NC-based FSC FTM for biogas purification. The PCNF is a cellulose nanofiber with a phosphoryl group attached to the surface. Furthermore, because it is hydrophilic in nature and possesses significant mechanical strength, PCNF is considered a viable choice as an additive in PVA polymer for the preparation of composite membranes. For this purpose, different thin dense layers comprised of 2 wt% of PVA with varying concentrations of PCNF ranging from 1 to 4 wt% were cast on a PSF-based flat-sheet support membranes at a thickness of 0.38 to 6 μm by dip coating technique followed by heat treatment to prepare different PCNF/PVA/PSF FTMs. These membranes were then tested in a mixed gas testing setup with a feed mixture of $CH_4/CO_2 = 60/40$ mol% at a pressure of 5 bar and a temperature of 25 °C, in a humidified condition with an R.H value of 93%. The result showed that the addition of filler content in PCNF/PVA membranes, enhanced both CO_2 permeability and CO_2/CH_4 selectivity as compared to pristine PVA membranes. The critical value of PCNF at 1 wt% prepared at a casting solution value of 12 pH exhibited the most optimum CO_2/CH_4 separation value of 45.20 and a CO_2 permeability of 266.64 Barrer. However, a further increase in filler loading resulted in a decrement in CO_2/CH_4 separation performance and an increment in the crystallinity of the membrane. In these membranes the facilitated transport mechanism is more prominent for the transport of CO_2 molecules as compared to the solution diffusion mechanism, which is further boosted at higher humidity conditions.

Ding et al. [29], developed high-performance FTM-based MMMs for efficient CO_2 gas removal from biogas mixtures by incorporating ionic decorated layered double hydroxide nano cages ([Hmim]-[NTf_2]@LDHN) at filler loading range from 4 to 10 wt% into a 4 wt% Pebax-based matrix using solution mixing and casting method. The ionic liquid-based heterogeneous filler was synthesized by two step method: (1) etching ZIF-67 to prepare LDHN nano cage, (2) followed by soaking it in [Hmim]-[NTf_2] ionic liquid. The CO_2/CH_4 separation performance experiment was carried out with feed pressure at 2–8 bar and an operating temperature of 25–65 °C. In order to study the separation performance of MMMs in real conditions, the mixed-gas experiments ($CH_4/CO_2 = 90/10$ vol%) were carried out under humidified conditions. In this study, it was observed that the MMMs demonstrated superior CO_2 separation performance when the [Hmim]-[NTf_2]@LDHN content was less than 6 wt%, and the CO_2/CH_4 separation performance dropped as the [Hmim][NTf_2]@LDHN concentration was raised to 10 wt%. Whereas, at a filler loading of 6 wt%, the Pebax/[Hmim]-[NTf_2]@LDHN-6 MMM exhibited the maximum CO_2 permeability of 644 ± 25 Barrer and CO_2/CH_4 selectivity of 34 ± 0.53. In these MMMs, the synergistic effect of [Hmim]-[NTf_2] and LDHN was the key factor for the membrane's outstanding CO_2 separation performance. First, the [Hmim]-[NTf_2] on the LDHN's surface was responsible for attracting CO_2 molecules while increasing interfacial compatibility. Secondly, the anions in the LDHN inter-lamellar region (CO_3^{2-} and [NTf_2]) demonstrated affinity for CO_2, caused boosting in CO_2 transport. Finally, the [Hmim]-[NTf_2] enclosed in the interior hollow core of LDHN has an ultrahigh CO_2 capacity, resulting improved CO_2/CH_4 separation performance across these MMMs.

Guo et al. [30], proposed the novel idea of developing a hybrid cluster in the shape of a polymer electrolyte membrane (PEM) for CO_2 separation from biogas mixture. Nano-sized sulfo-butylether-beta-cyclodextrin (sβ-CD) with sulfonic acid groups on the exterior surface and an internal hollow cavity was employed as the cluster core that effectively induces the quaternary ammonium groups on the polysulfone to form hybrid clusters via electrostatic interaction. Under humidified conditions, the gas separation performance of the membrane was evaluated using a mixed gas ($CH_4/CO_2 = 70/30$ vol%) testing setup at a pressure of 2 bar and a temperature of 25 °C. Which revealed that the CO_2 permeability across the QAPSf/sβ-CD based PEM membrane increased at first and then decreased with the increment of sβ-CD concentration. At an optimum filler loading of 0.5 wt% the MMM attained the maximum CO_2 permeability of 1303 Barrer with a CO_2/CH_4 selectivity of 39. The reason for this superior CO_2/CH_4 separation performance is due to three main factors: First, the hybrid clusters, which are made up of QAPSf chains and sβ-CD, are important in the CO_2/CH_4 separation performance because they use a combination of multiple gas transport mechanisms. In the QAPSf, hydrated chlorine ion—(Cl^-) works as a CO_2-facilitated transport carrier by undergoing a reversible chemical reaction with CO_2 in a humidified condition. Meanwhile, the CO_2-philic cationic quaternary ammonium groups serve as continuous electrostatic interaction sites, allowing HCO_3^- to move freely among hybrid clusters. Both solution diffusion and facilitated transport mechanisms in the membranes are improved because of the cluster structure induced by sβ-CD. Secondly, the internal cavity of the sβ-CD provides increased fractional free volume (FFV) to the hybrid cluster, which facilitates CO_2 diffusion. Furthermore, the presence of humid environment caused swelling of the hybrid clusters and salting out effects, which provide an interconnected pathway for rapid transport of CO_2 molecules during the CO_2/CH_4 separation process.

Qu et al. [31], proposed an innovative approach to develop a tunable hollow CO_2-philic cluster through a sacrificial template method in quaternized polysulfone (QAPSf) membrane for attaining compulsory micro-phase separation in order to achieve rapid transport of CO_2 molecules for biogas purification applications. The octa (tetramethylammonium) POSS (octaTMA-POSS), having a molecular-level size of 1–3 nm, has been selected as a template in this research work. The dimension of template-induced CO_2-philic clusters can be controlled by modifying the template content. Then the templates are etched out to produce hollow cavities after the CO_2-philic clusters have developed on the surface of the templates, which efficiently enhances the free volume in membranes, hence generates pathways for rapid diffusion of CO_2 molecules. Furthermore, the counter ions are replaced by $B(OH)^-{}_4$, which increases the CO_2-facilitating capability of the hollow clusters. Due to these factors the fabricated QAPSf-B(POSS) based PEMs demonstrate superior CO_2/CH_4 performance and long-term running stability for biogas upgradation. The separation performance of polyelectrolyte membranes was evaluated in the case of mixed gas ($CH_4/CO_2 = 70/30\%$) at a low feed pressure of 1 bar and a temperature of 25 °C by custom-built testing apparatus using a constant pressure/volume technique under humidified conditions. The results showed that with the increase of filler concentration, the CO_2 permeability and CO_2/CH_4 selectivity of QAPSf/POSS membranes first increased and then decreased gradually. The QAPSf/POSS based PEM at a filler loading of 2 wt% exhibited a maximum CO_2 permeability of 1113 Barrer and an optimum CO_2/CH_4 selectivity of 45, which is 334% and 73% higher than the pure QAPSf based PEM, respectively. The main reason of this increment in perm-selectivity is because of the CO_2-philic clusters present in these MMM. The development of CO_2-philic clusters reduce the distance between nearby clusters, thus enabling the development of interconnected transport pathways in humidified conditions. This interlinked hydrated CO_2-philic clusters, network provides greater transport sites for HCO_3^-, which mainly decreases the tortuosity of CO_2 transport channels, causing an enhancement of CO_2 permeability. The greater degree of aggregation of hydrated $-N^+ (CH_2CH_3)_3$ with strong CO_2 sorption enthalpy can successfully enhance CO_2 sorption in membranes, thereby enhancing the CO_2/CH_4 selectivity.

Saeed et al. [32], performed research to enhance the mechanical properties and water affinity of polyvinyl amine (PVAm) membranes in order to improve

CO_2 separation at relatively high pressures for biogas purification applications. The CNC composite-based FSC FTMs have been integrated into a 3 wt% PVAm membrane matrix using solution mixing and then cast on a PSF flat-sheet support membrane using the dip-coating method at a thickness of 1.12–2.16 μm in order to prepare different CNC-PVAm/PSF composite-based FSC FTMs. These fabricated membranes were examined for CO_2/CH_4 separation performance using a pure gas testing system under humidified conditions at a high pressure range of 5–15 bars. It was observed that the addition of CNC filler loading in the PVAm matrix up to 1 wt%, enhanced the CO_2/CH_4 perm-selectivity of the membrane. However, a further increment of filler content showed a decrement in gas separation performance. Similarly, the moisture uptake percentage also increased up to the addition of 1 wt% CNC and started decreasing beyond this concentration. Where, CNC(1)-PVAm/PSF composite FTM showed a maximum swelling degree of 75.88 ± 1.1%. Furthermore, it is highly recommended to work at maximum swollen conditions to get the maximum benefit from FSC-based facilitated transport of CO_2. It was seen that CNC(1)-PVAm/PSF composite FTM showed the highest CO_2 permeance and selectivity of 0.0216 m^3 (STP)/m^2 bar h and 41, respectively, at a lower pressure of 5 bar.

In 2021, Saeed et al. [33], studied the role of Betaine-based natural deep eutectic solvents (NADESs), a novel class of green solvents immobilized into a porous polyvinylidene fluoride (PVDF) membrane matrix for the separation of CO_2 gas from biogas mixture. Be-based NADES was preferred because of its availability in natural, low-cost, nonhazardous, and renewable source. Two forms of Be-Ma-NADES and Be-Tart-NADES were produced by mixing an equimolar amount of betaine with malic acid and tartaric acid at 80 °C for 2 h, respectively, which were then impregnated into a 15 wt% PVDF porous membrane matrix (250 μm) to form two different support liquid membranes (SLMs): Be-Ma-NADES SLM and Be-Tart-NADES SLM. The performance of NADESs SLMs was examined by both single permeation of pure CO_2 and CH_4 gas and mixed gas (CO_2/CH_4 = 50/50%) testing system at a pressure of 14 bar and a temperature of 25 °C. In case of the pure gas system, the CO_2 permeability and CO_2/CH_4 selectivity of Be-Ma-NADES SLM were found to be 29.33 Barrer and 56.40 respectively. While Be-Tart-NADES SLM showed relatively lower CO_2/CH_4 separation performance of 51.10 and CO_2 permeability of 25.55 Barrer. However, in the mixed gas permeation system, a slight decrement in gas separation performance is due to the high saturation of CO_2 molecules in NADES SILMs. In these SLMs, higher CO_2 permeability was caused by strong affinity for the carboxylate group (-COO) of Be towards CO_2 molecules, which resulted in high CO_2 solubility in the NADES. Whereas, the lower viscosity of the Be-Ma NADES relative to the Be-Tart-NADES may explain the superior CO_2/CH_4 separation performance of the Be-Ma NADES SLM over the Be-Tart-NADES SLM.

Yaun et al. [34], reported a novel porous polymer high-valence metal-induced microporous polymer (HMMP-1) produced by applying a modified polymer-directed chemical synthesis (PDCS) method. Unlike metal-induced microporous polymers (MMP), which have a 1:1 ratio of free amines to chlorine atoms in small linkers, HMMP-1 was synthesized using a high-valence chromium (Cr^{+3}) metal ion rather than a divalent ion. These high-valence metal ion groups considerably enhance the alkaline stability of HMMP-1 nanoparticles, allowing them to interact with more active free amine groups and to maintain their stability even in the presence of a strong alkaline environment. Furthermore, the polyethylene imine (PEI) polymer that has been embedded into the HMMP-1 framework exhibits improved interfacial compatibility between the polymer matrix and filler content, assuring the stability of the amine-rich nano-channels within the membrane matrix, which provides better adsorption and monomolecular surface diffusion of CO_2 molecules. The optimum HMMP-1 nanoparticles filler loading of 44.4 wt% was incorporated into 0.1 wt% of PVAm and then coated on the modified PSF support layer at a thickness of 0.18 μm using a solution casting method. The CO_2/CH_4 separation of the HMMP-1-PVAm/PSF composite FSC-based FTM was tested using a mixed gas testing setup (CH_4/CO_2 = 90/10 vol%) at a pressure range of 1.1 to 5 bar and a temperature of 25 °C under humidified conditions At a pressure of 1.5 bar, the HMMP-1-PVAm/PSF based FTM achieved the highest CO_2 permeability of 337 Barrer and CO_2/CH_4 selectivity of 104. The CO_2 transport in the MMM exhibited the usual attributes of facilitated transport. This is because amine functional groups are present in the PVAm matrix and HMMP-1 nano channels, which act as CO_2 fixed carriers under humidified conditions, and they tend to get saturated when feed gas pressure rises.

Casadei et al. [35], fabricated different FTMs using a blend of PVAm, L-arginine, and carboxymethylated nanocellulose fibers (cNFC) and studied them for water sorption and gas separation performance. These FTMs have a thickness of 50–60 μm and are composed of a PVAm/cNFC blend of 50/50 wt% as a polymeric matrix, in which different amounts of L-arginine as M.C filler were incorporated into a range of 0–45 wt% by using the solution mixing and casting method. In this study, sorption tests were carried out at a temperature of 35 °C and an R.H value of up to 90%, while gas permeation was evaluated at an R.H close to 100% using a pure gas testing system to investigate the CO_2/CH_4 separation performance of these MMMs at the same temperature. The sorption data revealed that PVAm, being the most hydrophilic of the membrane's components, governs water absorption; hence, solubility follows the proportion of this polymer in the blend. Whereas, the permeation results revealed that incorporation of MC was responsible for the increase in CO_2/CH_4 separation performance of the PVAm/cNFC based FTMs. The maximum CO_2/CH_4 separation value of 20 and CO_2 permeability of 340 Barrer was obtained when the MC loading of L-arginine was increased up to 45 wt% in the PVAm/cNFC membrane. Furthermore, PVAm also act as an FSC in this FTM process, which is one of the key factors in the increased gas separation performance, while addition of cNFC improved the membrane matrix's mechanical properties and stability under humidified conditions.

Zhang et al. [36], investigated MMMs with zinc (Zn^{2+}) ion functionalized spherical titanium oxide (TiO_2) nano-fillers for efficient facilitated transport of CO_2 molecules. By loading Zn^{+2} ions at around 3.43% on a Polydopamine (PDA) pre-coated TiO_2 nano-sphere, a three-dimensional (3D) zinc ion functionalized TiO_2 spherical nano-filler (TiO_2-PDA-Zn^{2+}) was synthesized in this research work. Different flat-sheet MMMs were fabricated at a thickness of 190 ± 5 μm by incorporating this TiO_2-PDA-Zn^{2+} nano-filler content at a range of 0–10 wt% into a 4, 4'-(Hexafluoroisopropylidene) diphthalic anhydride (6FDA)-based polyimide (PI) matrix using a solution mixing and casting method. These FTM (MMMs) were then tested using a mixed gas testing system (CO_2/CH_4 = 10/90 vol%) in humidified conditions at a pressure of 6 bar and a temperature of 36 °C. According to the gas permeation results, CO_2 permeability improved with increasing TiO_2-PDA-Zn^{2+} concentration, reaching a maximum value of 145.37 Barrer with a corresponding CO_2/CH_4 selectivity of 49.98 at 7.5 wt% filler loading, which is close to the revised upper bound line (2008). However, after exceeding this threshold concentration up to 10 wt%, a decrement in CO_2 separation performance was observed. The integration of TiO_2-PDA-Zn^{2+} into the polymer matrix inhibits the packing of polymer chains, resulting in a decreased crystallinity of MMMs. Whereas the Zn^{2+} ion provided a fixed site carrier which facilitates the rapid transport of CO_2 molecules due to a reversible π-interaction between Zn^{2+} and CO_2, which led to the superior CO_2/CH_4 separation performance.

5.3 Trade-Off Robeson Plot of FTMs for Biogas Purification—CO_2/CH_4 Separation

The empirical bound relationship between the CO_2 permeability and CO_2/CH_4 separation for polymeric membranes was first published in 1991 and updated in 2008 given the large volume of new data reported [37]. The polymeric-based FTMs Robeson upper bound plot is depicted in Fig. 5.3, which summarizes the recently reported literature of different FTMs for biogas purification (CO_2/CH_4) separation, based on the data in Table 5.2. According to the data obtained from recent research studies, it can be clearly seen in Fig. 5.3 that the majority of polymeric based FTMs exceeded the upper bound limit, while some of the membranes were in close proximity to the upper bound line, indicating that FTMs outperformed pristine dense polymeric membranes in terms of biogas purification (CO_2/CH_4) separating performance.

The PVAm/mPSF/HMMP-1 composite-based FTMs for biogas purification stand out as the most optimized membranes as compared to different FTMs, surpassing the 2008 upper bound limit in terms of CO_2/CH_4 selectivity, having a value of 104 with a CO_2 permeability of 337 Barrer. Furthermore, in terms of CO_2 permeability, the PIM-1/PIP-Py-Ac-15 MMM-based MC FTMs with an optimized CO_2/CH_4 selectivity of 37 exhibited the highest CO_2 permeability of 5300 Barrer among other FTMs. To sum up, the Robeson plot clearly demonstrates and highlights that the latest FTM-based technologies, including composites, MMMs, and hybrid-based FTMs, are one of the most promising candidates for biogas purification applications and which can be further optimize in terms of CO_2 permeability and CO_2/CH_4 selectivity.

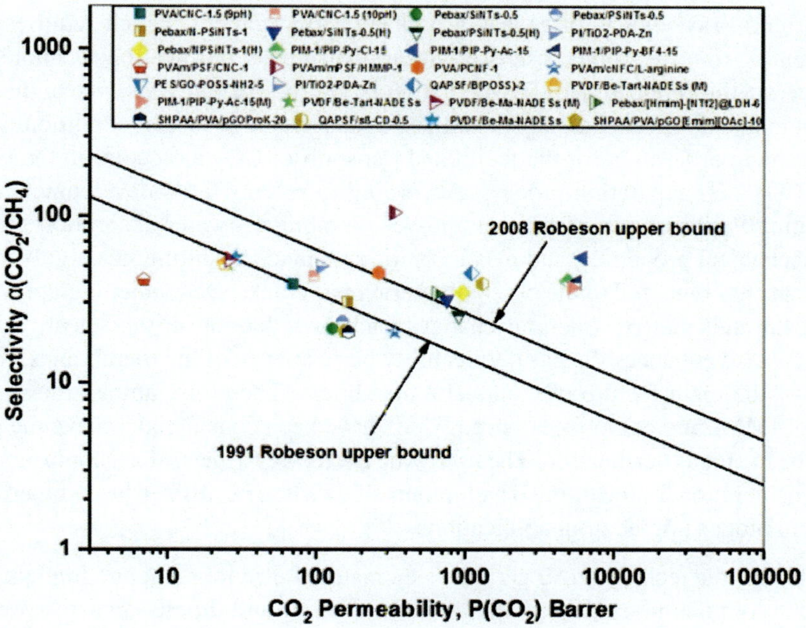

Fig. 5.3 Robeson's upper bound graph illustrating the Biogas Purification (CO_2/CH_4) selectivity for different FTMs versus CO_2 permeability from Table 5.2 data

5.4 Conclusion and Future Perspectives

This chapter elaborated the application of polymeric-based FTMs for biogas purification, which has been proven to be a remarkable and efficient technology for removing pollutants, especially CO_2 gas, from raw biogas at low partial pressures, hence enhancing the calorific value of the bio-fuel for commercial applications. A wide range of composites, MMMs, and hybrid-based FTMs that can separate CO_2 from biogas using various fillers and carriers, such as M.C, FSC, and its hybrid, have recently been reported are thoroughly reviewed in terms of their gas transport mechanism, membrane and carriers fabrication methods, membrane modules and CO_2/CH_4 separation performance in different operating conditions. A Robeson's graph has also been plotted, which depicts the CO_2 permeability and CO_2/CH_4 selectivity values of various FTMs that have recently been reported, which indicates that majority of FTMs showed outstanding gas separation performance that exceeded the revised trade-off limitation of polymeric membrane for CO_2/CH_4 separation performance, while some of them coincided close to the upper bound limit. The main reason of this excellent gas separation performance of FTMs is mainly due to the coupled mechanism of carrier mediation process and solution diffusion mechanism. Whereas, in some cases (e.g. hybrid FTMs) multiple transport mechanism such surface diffusion and molecular sieve effects are also observed along with the coupled mechanism of FTMs, demonstrating rapid CO_2 transportation in CO_2/CH_4 separation process.

Furthermore, raw biogas often has a high moisture content, which requires pre-treatment to remove in the case of conventional biogas purification technologies. However, in the application of FTMs (particularly FSC-based FTMs), the high water content in the biogas stream may be utilized as an advantage for pre-conditioning the membrane in order to boost the facilitated transport of CO_2 molecules in the FTM-based CO_2/CH_4 separation process. According to recent FTMs development data, integrating various types of fillers improves membrane's water absorption properties, mechanical properties, and plasticization resistance, resulting in a highly stable membrane as compared to the pure polymeric membranes even under high pressure. While, through matrix structural changes such as reduction in the selective layer thickness also enhances the gas permeability performance of the membranes, which is one of the main feature of composite membrane. Therefore, among the various FTMs, MMMs and composites-based FTMs have been considered to have the great potential in biogas purification. The following are the key aspects that should be taken into consideration for the future development of FTMs to be utilized be commercially utilize in biogas upgradation applications.

- To bridge the technological gap between material development and implementation in commercial-scale gas separations, modelling and simulation with extensive field testing is require.
- For commercial up scaling, new modules such as high-performance hollow fiber-based FTMs for biogas purification process must be developed, along with the various challenges related to module designing, coating, and sealing must also be overcome.
- The membrane's endurance and carrier stability when exposed to raw biogas, which usually contains other contaminants such as H_2S gas should be investigated and documented for further improvement of materials.

References

1. Weiland P (2010) Biogas production: current state and perspectives. Appl Microbiol Biotechnol 85(4):849–860
2. Favre E, Bounaceur R, Roizard D (2009) Biogas, membranes and carbon dioxide capture. J Membr Sci 328(1–2):11–14
3. Iskov H, Kvist T, Bruun J (2019) Biogas, Biomethane and electro-methane cost comparison. University of Gävle Gävle, Sweden
4. Khalil M, Berawi MA, Heryanto R, Rizalie A (2019) Waste to energy technology: the potential of sustainable biogas production from animal waste in Indonesia. Renew Sustain Energy Rev 105:323–331
5. Huertas J, Giraldo N, Izquierdo S (2011) Removal of H_2S and CO_2 from biogas by amine absorption. Mass Trans Chem Eng Processes 307
6. Chen XY, Vinh-Thang H, Ramirez AA, Rodrigue D, Kaliaguine S (2015) Membrane gas separation technologies for biogas upgrading. RSC Adv 5(31):24399–24448
7. Abbasi T, Tauseef S, Abbasi S (2012) Biogas and biogas energy: an introduction. In: Biogas energy. Springer, Berlin, pp 1–10

8. Bauer F, Persson T, Hulteberg C, Tamm D (2013) Biogas upgrading–technology overview, comparison and perspectives for the future. Biofuels Bioprod Biorefin 7(5):499–511
9. Adnan AI, Ong MY, Nomanbhay S, Chew KW, Show PL (2019) Technologies for biogas upgrading to biomethane: a review. Bioengineering 6(4):92
10. Karim SS, Farrukh S, Hussain A, Noor T, Younas M (2021) A comprehensive overview of dual-layer composite membrane for air (O_2/N_2) separation. Polym Polym Compos 29(9_suppl):S1630–S1640
11. Ismail AF, Khulbe KC, Matsuura T (2015) Gas separation membranes. Switz Springer 10:978–973
12. Matteucci S, Yampolskii Y, Freeman BD, Pinnau I (2006) Transport of gases and vapors in glassy and rubbery polymers. Mater Sci Membr Gas Vap Sep 1:1–2
13. Breck DW, Breck DW (1973) Zeolite molecular sieves: structure, chemistry, and use. Wiley
14. Workplace Environmental Exposure Level: Octamethylcyclotetrasiloxane (D4) (2017). Toxicol Ind Health (SAGE Journals) 33(1):2–15. https://doi.org/10.1177/0748233716670061
15. Yu M, Gong H, Chen Z, Zhang M (2013) Adsorption characteristics of activated carbon for siloxanes. J Environ Chem Eng 1(4):1182–1187
16. Vrbová V, Ciahotný K (2017) Upgrading biogas to biomethane using membrane separation. Energy Fuels 31(9):9393–9401
17. Baena-Moreno FM, le Saché E, Pastor-Perez L, Reina T (2020) Membrane-based technologies for biogas upgrading: a review. Environ Chem Lett 18(5):1649–1658
18. Tong Z, Ho WW (2017) Facilitated transport membranes for CO_2 separation and capture. Sep Sci Technol 52(2):156–167
19. He X (2018) A review of material development in the field of carbon capture and the application of membrane-based processes in power plants and energy-intensive industries. Energy Sustain Soc 8(1):1–14
20. Deng L, Hägg M-B (2010) Techno-economic evaluation of biogas upgrading process using CO_2 facilitated transport membrane. Int J Greenhouse Gas Control 4(4):638–646
21. Rafiq S, Deng L, Hägg MB (2016) Role of facilitated transport membranes and composite membranes for efficient CO_2 capture–a review. ChemBioEng Rev 3(2):68–85
22. Jahan Z, Niazi MBK, Hägg M-B, Gregersen ØW (2018) Decoupling the effect of membrane thickness and CNC concentration in PVA based nanocomposite membranes for CO_2/CH_4 separation. Sep Purif Technol 204:220–225
23. Jahan Z, Niazi MBK, Hägg M-B, Gregersen ØW (2018) Cellulose nanocrystal/PVA nanocomposite membranes for CO_2/CH_4 separation at high pressure. J Membr Sci 554:275–281
24. Yang L, Zhang S, Wu H, Ye C, Liang X, Wang S, Wu X, Wu Y, Ren Y, Liu Y (2019) Porous organosilicon nanotubes in pebax-based mixed-matrix membranes for biogas purification. J Membr Sci 573:301–308
25. Wang C, Guo F, Li H, Xu J, Hu J, Liu H, Wang M (2020) A porous ionic polymer bionic carrier in a mixed matrix membrane for facilitating selective CO_2 permeability. J Membr Sci 598:117677
26. Yang L, Liu X, Wu H, Wang S, Liang X, Ma L, Ren Y, Wu Y, Liu Y, Sun M (2020) Amino-functionalized POSS nanocage intercalated graphene oxide membranes for efficient biogas upgrading. J Membr Sci 596:117733
27. Janakiram S, Espejo JLM, Høisæter KK, Lindbråthen A, Ansaloni L, Deng L (2020) Three-phase hybrid facilitated transport hollow fiber membranes for enhanced CO_2 separation. Appl Mater Today 21:100801
28. Jahan Z, Niazi MBK, Hagg M-B, Gregersen ØW, Hussain A (2020) Phosphorylated nanocellulose fibrils/PVA nanocomposite membranes for biogas upgrading at higher pressure. Sep Sci Technol 55(8):1524–1534
29. Ding S, Li X, Ding S, Zhang W, Guo R, Zhang J (2020) Ionic liquid-decorated nanocages for cooperative CO_2 transport in mixed matrix membranes. Sep Purif Technol 239:116539
30. Guo Z, Qu Z, Wu H, Zhao R, Wu Y, Liu Y, Yang L, Ren Y, Ye C, Jiang Z (2020) Polymer electrolyte membranes with hybrid cluster network for efficient CO_2/CH_4 separation. ACS Sustain Chem Eng 8(17):6815–6825

31. Qu Z, Zhao R, Wu H, Ren Y, Liu Y, Guo Z, Wu Y, Yang L, Liang X, Jiang Z (2020) Polyelectrolyte membranes with tunable hollow CO_2-philic clusters via sacrificial template for biogas upgrading. J Membr Sci 612:118445
32. Saeed U, Jahan Z, Niazi MBK, Pervaiz E, Sher F (2020) Biogas upgrading with novel cellulose nano-crystals and polyvinyl amine nanocomposite membranes. Polym Test 91:106867
33. Saeed U, Khan AL, Gilani MA, Aslam M, Khan AU (2021) CO_2 separation by supported liquid membranes synthesized with natural deep eutectic solvents. Environ Sci Pollut Res 28(26):33994–34008
34. Yuan Y, Qiao Z, Xu J, Wang J, Zhao S, Cao X, Wang Z, Guiver MD (2021) Mixed matrix membranes for CO_2 separations by incorporating microporous polymer framework fillers with amine-rich nanochannels. J Membr Sci 620:118923
35. Casadei R, Firouznia E, Baschetti MG (2021) Effect of mobile carrier on the performance of PVAm–nanocellulose facilitated transport membranes for CO_2 capture. Membranes 11(6):442
36. Zhang B, Yan Q, Chen G, Yi C, Qi S, Yang B (2021) Fabrication of mixed matrix membranes with zinc ion loaded titanium dioxide for improved CO_2 separation. Sep Purif Technol 254:117472
37. Robeson LM (2008) The upper bound revisited. J Membr Sci 320(1–2):390–400

Chapter 6
Facilitated Transport Membranes (FTMs) for Syngas Purification (CO_2/H_2)

Syed Shujaat Karim, Ovaid Mehmood, Sarah Farrukh, and Muhammad Ayoub

Abstract The most widely used method for hydrogen (H_2) production are gasification and reforming methods, which in combination with the water–gas shift reaction produces a gaseous mixture, primarily composed of hydrogen (H_2) and carbon dioxide (CO_2). For the separation of CO_2 from this product stream, membrane technology is considered to be one of the most efficient process as compared to conventional technologies due to its high energy efficiency, compact module, and ease of operation. Among different types of membranes, facilitated transport membranes (FTMs) have emerged as the most promising technology for CO_2 capturing processes because of their remarkable gas separation selectivity and high permeability. In this chapter, different syngas production methods combined with water–gas shift reaction and their product stream purification through FTMs were concisely described. A wide range of recently reported FTMs for the application of CO_2/H_2 separation in syngas purification have been critically reviewed, which include mobile carrier (MC) FTMs, fixed-site carrier (FSC) FTMs, ionic liquid (IL) membranes, poly (ionic-liquid) (PIL) IL membranes, hybrid filler based MMMs and their composite based FTMs. Moreover, a Robeson's permeability-selectivity trade-off graph has been discussed in this literature for comparing the CO_2/H_2 separation performances of various FTMs. Lastly, future research directions have also been proposed for further development of FTMs in order to utilize this technology for CO_2 capture in industrial scale H_2 production processes.

S. S. Karim · O. Mehmood · S. Farrukh (✉)
School of Chemical and Materials Engineering (SCME), National University of Sciences and Technology (NUST), Sector H-12, Islamabad, Pakistan
e-mail: sarah.farrukh@scme.nust.edu.pk

S. S. Karim
e-mail: sshujaat.che6scme@student.nust.edu.pk

M. Ayoub
Department of Chemical Engineering, Universiti Teknologi Petronas (UTP), Perak, Malaysia
e-mail: muhammad.ayoub@utp.edu.my

S. Farrukh et al. (eds.), *Facilitated Transport Membranes (FTMs) for CO_2 Capture: Overview and Future Trends*, Green Energy and Technology,
https://doi.org/10.1007/978-3-031-21444-8_6

Keywords CO_2-selective facilitated transport membranes · Syngas purification · Mixed matrix membranes (MMMs) · Composite membranes · Hydrogen (H_2) purification · CO_2/H_2 separation

Nomenclature

$\alpha_{i/j}$	Selectivity (ith/jth) components
AAO	Anodic alumina oxide
AF-MWCNTs	Amino-functionalized multi-walled carbon nanotubes
AIBA	2-Aminoisobutyric acid
APTES	3-Aminopropyltriethoxysilane
Ar	Argon
[BMIM][BF_4]	1-Butyl-3-methylimidazolium tetrafluoroborate
[C_2mim][C(CN)$_3$]	1-Ethyl-3-methylimidazolium tricyanomethanide
[C_2mim]-[FSI]	1-Ethyl-3-methylimidazolium bis(fluorosulfonyl)imide
[C_2mim][NTf_2]	1-Ethyl-3-methylimidazolium bis(trifluoromethylsulfonyl)imide
[C_4mim]-[Tf_2N]	3-Butyl-1-methyl-1H-imidazol-3-iumbis((trifluoromethyl) sulfonyl)amide
[C_4mpyr][NTf_2]	1-Butyl-1-methylpyrrolidinium bis(trifluoromethylsulfonyl)imide
[EMIM]-[DCA]	1-Ethyl-3-methylimidazolium dicyanamide
C_2H_4	Ethylene
CH_4	Methane
CO	Carbon monoxide
CO_2	Carbon dioxide
COF	Covalent organic framework
COF_L	COF-LZU1 modified by 1,3,5-triformylbenzene
COF_M	COF-LZU1 modified by 1,4-diaminobenzene
COF_P	COF-LZU1 modified by glacial acetic acid
$-COO^-$	Carboxylate group
Cu^{2+}	Copper ion
$Cu_3(BTC)_2$	Copper Benzene-1,3,5-tricarboxylate
FSC	Fixed-site carrier
FTM	Facilitated transport membrane
GO	Graphene oxide
GO-IL-NH_2	1-(3-Aminopropyl)-3-methylimidazolium bromide ionic liquid covalent functionalized Graphene oxide
H_2	Hydrogen
H_2O	Water (Moisture or steam)
H_2S	Hydrogen Sulfide
ILs	Ionic liquids
Im-TB (*o*)-PA	Ortho Imidazolium-mediated TB containing polyamide ionene polymers

Im-TB (*p*)-PA	Para Imidazolium-mediated TB containing polyamide ionene polymers
K.D	Kinetic diameter
KOH	Potassium hydroxide
MC	Mobile carrier
MIL-53	Materials Institute of Lavoisier-53
MOF	Metal organic framework
mPSF	Modified polysulfone (coated with crosslinked PDMS)
N_2	Nitrogen
NBR	Nitrile-butadiene rubber
$-NH_2$	Amine group
NH_3	Ammonia
NMR	Nuclear magnetic resonance
PDADMQ-F	Poly(diallyldimethylammonium fluoride)
PDAMA-F	Poly(diallyldimethylammonium fluoride)
PDMS	Polydimethylsiloxane
Pebax-1657	Polyether block amide Pebax® MH 1657
PEO	Polyethylene Oxides
P_i	Permeability in Barrer of ith component
PIL	Poly (ionic) liquid
Poly ([Pyr_{11}])	Poly-Pyrrolidinium
Poly [Pyr_{11}][Tf_2N]	Poly(diallylmethylammonium) bis(trifluoromethylsulfonyl)
POM	Partial oxidation of methane
POSS	Polyhedral Oligomeric Silsesquioxane
PSF	Poly sulfone
PTMSP	Poly(1-trimethylsilyl-1-propyne)
PVA	Polyvinyl alcohol
PVAm	Poly-(vinyl amine)
PVDF	Polyvinylidene fluoride
QA	Quaternary ammonium
SILM	Support ionic liquid membrane
SILP	Supported ionic liquid phase
sIPN	Ionic liquid-based semi-interpenetrating polymer network
SMR	Steam methane reforming
SOFC	Solid oxide fuel cell
STP	Standard Temperature and Pressure
TB	Tröger's base
TFBA	Tetrafluoroboric acid
[Tf_2N]	Bis(trifluoromethylsulfonyl) imide
TGA	Thermogravimetric analysis
TMAF	Tetramethylammonium fluoride tetrahydrate
TMAOH	Tetra-methyl-ammonium hydroxide
TMQOH	Tetramethylquaternaryammonium hydroxide
WSGR	Water gas shift reaction
XL-PVA	Crosslinked poly (vinyl) alcohol

XLPVA-POS	Crosslinked poly vinyl alcohol—polysiloxane
ZIF-8	Zeolitic imidazolate frameworks-8
Zn^{2+}	Zinc ion

6.1 Introduction

Synthesis gas, which is also known as syngas, is a flammable fuel gas mainly made up of a mixture of carbon monoxide and hydrogen ($CO + H_2$). Syngas can be produced from the conversion of a variety of carbonaceous fuel sources, including natural gas, coal, petroleum coke, heavy oil, bitumen, and biomass, mainly by using gasification process and reforming methods, which are then primarily used to produce H_2 gas and a variety of chemicals for commercial purposes, such as ammonia, methanol, and synthetic fuels, as depicted in Fig. 6.1 [1, 2]. The gasification process is typically used to produce syngas from solid feedstocks like coal, pet coke, and biomass in the presence of air, steam (H_2O), oxygen (O_2), carbon dioxide (CO_2), or a mixture of gasifying agents at a high temperature range of 600–1000 °C and moderate pressure of up to 70 barg [3].

The steam methane reforming (SMR) process is the endothermic process, in which natural gas is reacted with high-temperature steam (700–1000 °C) at a pressure of 3 to 25 bar and in the presence of a catalyst to produce syngas mixture [4]. Whereas, partial oxidation is a considerably faster process because it is an exothermic process in which natural gas and other light hydrocarbons fuel are oxidized at a temperature range of 700–900 °C in the presence of catalyst with a limited amount oxygen (O_2) from air to produced syngas mixture and heat. However, as compared to the SMR approach, this method produces less amount of H_2 gas [5]. The following equations are the total reactions that summarized the gasification process and reforming methods for the syngas production process.

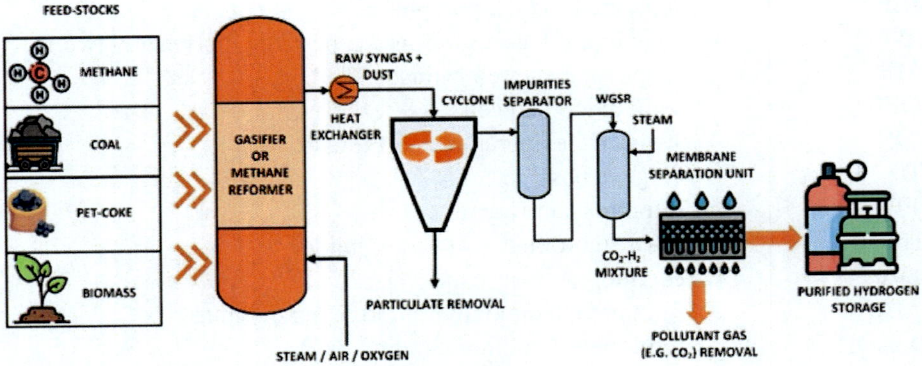

Fig. 6.1 Pictorial representation of syngas production, purification and utilization system [6, 7]

(a) **Gasification process (partially oxidizing) reaction** [2]:

$$[2CH]_{feedstock} + O_2 \rightarrow 2CO + H_2 \tag{6.1}$$

(b) **Steam methane reforming (SMR) reaction** [4]:

$$CH_4 + H_2O\,(\text{Steam}) \rightarrow CO + 3H_2 \tag{6.2}$$

(c) **Partial oxidation of methane (POM) reaction** [5]:

$$CH_4 + \frac{1}{2}\,O_2 \rightarrow CO + 2H_2 \tag{6.3}$$

The chemical composition of raw syngas comprises of 20–40% hydrogen (H_2), 35–40% of carbon monoxide (CO), 25–35% carbon dioxide (CO_2), 0–15% methane (CH_4), with a lower proportion of other gases [8]. However, the composition of syngas can varies based on the types of raw materials (feedstocks), operating conditions, and the technology employed to produce it [9]. The different syngas composition with its physical properties produced by converting some of different raw materials using gasification and SMR processes are represented in Table 6.1.

After the production of syngas, the hot reformed gaseous mixture from the gasifier or methane reformer is then allowed to cool down through heat exchangers. Later on, through a cyclone, the separation of solid impurities i.e. particulates would be carried out from the effluent stream and scrubbed with water to eliminate constant impurities, i.e. NH_3, halides and tar. Due to availability of wastewater treatment facility at site the scrubber water is purged and treated constantly. Then CO is allowed to react with steam at a temperature range of 150–300 °C in the presence of a catalyst for its conversion into CO_2 gas and more production H_2 gas through water–gas shift reaction (WGSR), as shown in the following equation [20].

$$CO + H_2O\,(\text{Steam}) \rightarrow CO_2 + H_2 \tag{6.4}$$

This WGS reaction is a crucial step in the production of H_2 gas in different industries, such as in hydrogen, ammonia, and methanol plants that employ natural gas or coal as a feed-stock. However, in the product stream, H_2 gas is mixed with CO_2, which must be separated before being used commercially. At this stage, pressure swing adsorption (PSA) and amine absorption processes are two common methods utilized for H_2 gas purification. Aside from these conventional technologies, different types of membranes are also being developed that would enable more efficient gas separation performance [9, 21].

Table 6.1 Typical composition data of raw syngas obtained from different production methods and the physical properties of its components

Feed gas components	Coal [a] (vol%)	Pet coke[a] (mol%)	Biomass[a] (vol%)	Natural gas[b] (vol%)	Kinetic diameter (Å)	Molecular weight (g/mol)	Ref.
Methane (CH_4)	0.00–5.00	0.00–1.00	1.00–6.20	0.00–9.00	3.80	16.00	[10–16]
Carbon dioxide (CO_2)	5.00–16.00	8.00–20.00	8.00–20.00	5.00–10.00	3.30	44.00	[10–16]
Moisture (H_2O)	2.00–62.00	1.00–2.00	5.00–35.00	30.00–40.00	2.65	18.00	[10–17]
Hydrogen (H_2)	13.00–34.00	22.00–30.00	5.00–22.00	30.00–40.00	2.89	2.00	[10–17]
Hydrogen Sulfide (H_2S)	0.00–1.50	1.00–2.00	N.S	N.S	3.60	34.00	[10–18]
Ammonia (NH_3)	0.00–0.30	N.S	N.S	N.S	2.60	17.00	[10–19]
Carbon monoxide (CO)	5.00–64.00	39.00–48.00	9.00–22.00	5.00–10.00	3.76	28.00	[10–19]
Nitrogen (N_2)	0.00–4.50	1.00–2.00	0.00–62.00	0.00–16.00	3.64	28.00	[10–16]
Argon (Ar)	0.00–1.10	0.00–1.00	N.S	0.00–0.20	3.40	40.00	[10–16]
Ethylene (C_2H_4)	0.00–0.20	N.S	0.00–2.00	N.S	3.90	28.00	[10–18]

[a] Feed-stock for the gasification process
[b] Feed-stock for the reforming process
N.S Not specified

6.2 Membrane Technology for Hydrogen (H_2) Purification

Membrane technology has a number of advantages over traditional technologies, including lower equipment and energy costs, as well as a lightweight, compact module that requires less installation area. Additionally, membrane technology can be used to purify H_2 gas at both high and low pressures processes. There are two main types of H_2 purification membranes: H_2-selective membranes and CO_2-selective membranes. In case of H_2-selective membranes, H_2 gas permeates across the membrane, while other gas molecules are retained, resulting in a low-pressure H_2 stream and a high-pressure CO_2 stream. Whereas, CO_2-selective membranes, allows CO_2 molecules to pass through it while retaining other gas molecules, resulting in a high-pressure H_2 stream and a low-pressure CO_2 stream. The CO_2-selective membranes are the most suitable technology for H_2 purification procedures when high pressure H_2 streams are needed on a commercial scale [22].

6.2.1 Facilitated Transport Membranes (FTMs) for CO_2/H_2 *Separation*

CO_2-selective FTMs has recently gained interest in both academia and commercial sectors for H_2 purification due to their advantages of higher separation selectivity and productivity. They have also offers superior durability and mechanical strength at the moderate and higher operating conditions. Furthermore, higher CO_2/H_2 selectivity of FTMs makes the purification process efficient, requiring less driving force to achieve separation efficiency, which results in lowering the cost of separation system. Similarly, because of the higher CO_2 flux, the process requires less membrane area, reducing the total capital cost of the membrane based separation system [21–23].

As shown in Fig. 6.2, in these type of FTMs, the reactive carriers are generally small molecules that acts as mobile carriers (MCs) or functional groups attached to the polymeric chain as fixed site carriers (FSCs), reacts reversibly with the CO_2 molecule to form a CO_2-carrier complex as a product. Some of the common functional group and ions that can be used as carriers for the transport of CO_2 molecules are amines groups ($-NH_2$), carboxylate group ($-COO^-$), and metallic ions (Cu^{2+}, Zn^{2+}) etc. In MC-FTMs, the CO_2 molecules are freely transported across the membrane in the form of a CO_2-carrier complex (product) at a faster rate due to the concentration gradient. While, in the case of FSC-FTMs, the CO_2 molecules diffuse across the membrane by a hopping mechanism. Following that, the CO_2-carrier complex bond breaks down, releasing the CO_2 molecule on the permeate side at a lower partial pressure than on the feed side [21–23].

Moreover, inert gas molecules (such as H_2, N_2, and CH_4) cannot react with reactive carriers and can only pass through the membrane via the solution diffusion mechanism, in which the permeability of the gas molecules depends on their solubility and

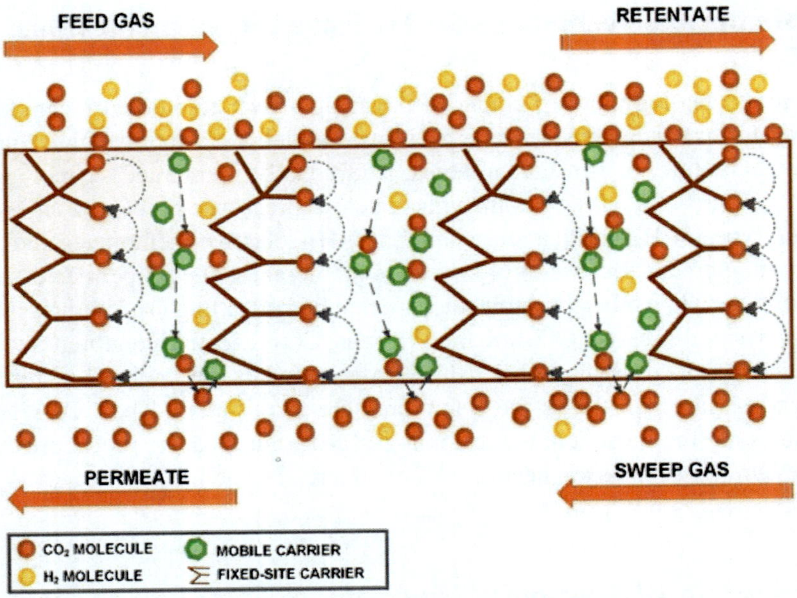

Fig. 6.2 Illustration of CO_2-selective FTM mechanism for hydrogen purification [22]

diffusivity within the polymeric membrane [24]. Whereas, in the case of CO_2 transport, the FTMs utilize the coupled mechanisms, which comprise a reversible reaction–diffusion process as well as a solution diffusion mechanism, resulting in superior CO_2/H_2 separation performance as compared to the dense polymeric membranes [21–24].

6.2.2 *Recent Advancement of FTMs for Syngas Purification (CO_2/H_2)*

Recently, CO_2-selective FTMs have gained a lot of interest in the field of H_2 purification because of their outstanding CO_2/H_2 separation performance. Throughout the years, many groups of researchers have fabricated and explored a wide range of FTMs having different carriers, nanofillers, and hybrid nanofillers to achieve high CO_2/H_2 separation performance. The recent progress details of these FTMs, as represented in Table 6.2, are discussed in this section.

In 2018, Ying et al. [25], studied the ionic liquids for gas separation applications because of their low volatilities, outstanding chemical and thermal stability, and high viscosities. Generally, ionic liquids (ILs) have more affinity towards CO_2 gas molecules than H_2, CH_4, and N_2. However, membranes made from direct ionic liquids (ILs) are difficult to fabricate because of their liquid state as well as high material and production costs. Therefore, it must be supported by a porous, solid membrane matrix with high mechanical strength. There are two main types of immobilized ionic

Table 6.2 Recent advancement data of FTMs for syngas purification (CO_2/H_2)

Materials	Carriers/fillers	Fabrication method	FTMs types	Module/thickness	Operating conditions	$P(CO_2)$	$P(H_2)$	$\alpha_{\frac{(CO_2)}{(H_2)}}$	Ref.
AAO (what man—Pore size: 0.2 μm)	GO/[BMIM][BF_4] (0.05 mg/mL)	Hummers' method, mixing and sonication [a] Solution mixing and vacuum filtration [b]	Hybrid-MC (SILM)	Flat-sheet (1.05 μm)	T = 25–50 °C P = 1 bar Pure gas	71.93 (Barrer)	3.03 (Barrer)	23.70	[25]
Poly ([Pyr_{11}] [$C(CN)_3$]) (6 wt%)	[C_2mim]-[$C(CN)_3$] (60 wt%)	Solvent casting method [b]	Hybrid-MC (PIL-IL) (MMM)	Flat-sheet (70–210 μm)	T = 20–35 °C P = 1 bar Pure gas	438.00–505.00 (Barrer)	29.10–40.30 (Barrer)	15.10–12.50	[26]
Poly ([Pyr_{11}] [NTf_2]) (12 wt%)	[C_4mpyr]-[NTf_2] (60 wt%)	Solvent casting method [b]	Hybrid-MC (PIL-IL) (MMM)	Flat-sheet (70–210 μm)	T = 20–35 °C P = 1 bar Pure gas	232.00–288.00 (Barrer)	29.80–46.00 (Barrer)	7.80–6.30	[26]
Poly ([Pyr_{11}] [NTf_2]) (12 wt%)	[C_2mim]-[NTf_2] (40 wt%)	Solvent casting method [b]	Hybrid-MC (PIL-IL) (MMM)	Flat-sheet (70–210 μm)	T = 20–35 °C P = 1 bar Pure gas	214.00–287.00 (Barrer)	26.20–43.80 (Barrer)	8.20–6.50	[26]
(2 wt%)—Pebax-1657/(2 wt%)—PTMSP/PVDF	GO-IL-NH_2 (0.2 wt%)	Improved Hummers', mixing, and epoxide ring-opening reaction [a] Solution mixing and casting [b]	Hybrid-FSC \(composite)	Flat-sheet (skin layer: 0.1 μm)	T = 25 °C P = 4 bar Pure gas	143.00 (Barrer)	10.40 (Barrer)	13.80	[27]
(2 wt%)—Pebax-1657/(2 wt%)—PTMSP/PVDF	GO-IL-NH_2 (0.2 wt%)	Improved Hummers', mixing, and epoxide ring-opening reaction [a] Solution mixing and casting [b]	Hybrid-FSC (composite)	Flat-sheet (Skin layer: 0.1 μm)	T = 25 °C P = 4 bar Mixed gas (CO_2/H_2) (30/70 vol%)	118.60 (Barrer)	12.40 (Barrer)	9.50	[27]

(continued)

Table 6.2 (continued)

Materials	Carriers/fillers	Fabrication method	FTMs types	Module/thickness	Operating conditions	$P(CO_2)$	$P(H_2)$	$\alpha_{\frac{(CO_2)}{(H_2)}}$	Ref.
Poly [Pyr_{11}] [Tf_2N] (0.6 g)	[C_4mpyr] [Tf_2N] (0.4 g) / MIL-53 (30 wt%)	Solvent evaporation method [b]	Hybrid-MC (PIL-IL) (MMMs)	Flat-sheet	T = 30 °C P = 0.7 bar Pure gas	89.01 (Barrer)	6.70 (Barrer)	13.28	[28]
Poly [Pyr_{11}] [Tf_2N] (0.6 g)	[C_4mpyr] [Tf_2N] (0.4 g) / $Cu_3(BTC)_2$ (30 wt%)	Solvent evaporation method [b]	Hybrid-MC (PIL-IL) (MMM)	Flat-sheet	T = 30 °C P = 0.7 bar Pure gas	77.15 (Barrer)	12.30 (Barrer)	6.27	[28]
Poly [Pyr_{11}] [Tf_2N] (0.6 g)	[C_4mpyr] [Tf_2N] (0.4 g)/ZIF-8 (30 wt%)	Solvent evaporation method [b]	Hybrid-MC (PIL-IL) (MMM)	Flat-sheet	T = 30 °C P = 0.7 bar Pure gas	97.15 (Barrer)	21.35 (Barrer)	4.55	[28]
PVA (0.5 g)	[EMIM]-[DCA] (53 wt%)	Solution mixing and spin-coating method [b]	MC (MMM-IL)	Flat-sheet (104.40 μm)	T = 25 °C P = 2 bar Pure gas	66.90 (Barrer)	8.50 (Barrer)	7.73	[29]
(1.0 wt%) PVAm/mPSF	COF_L (10 wt%)	COF-LZU1 synthesis and modification [a] Solution mixing and casting [b]	FSC (Composite)	Flat-sheet (Skin layer: 0.152 μm)	T = 25 °C P = 1.5 bar Humidified Mixed gas (CO_2/H_2) (40/60 vol%)	221.10 (Barrer)	13.73 (Barrer)	16.10	[30]
(1.0 wt%) PVAm/mPSF	COF_M (10 wt%)	COF-LZU1 synthesis and modification [a] Solution mixing and casting [b]	FSC (Composite)	Flat-sheet (0.165 μm)	T = 25 °C P = 1.5 bar Humidified Mixed gas (CO_2/H_2) (40/60 vol%)	234.60 (Barrer)	13.64 (Barrer)	17.20	[30]

(continued)

Table 6.2 (continued)

Materials	Carriers/fillers	Fabrication method	FTMs types	Module/thickness	Operating conditions	$P(CO_2)$	$P(H_2)$	$\alpha_{\frac{(CO_2)}{(H_2)}}$	Ref.
(1.0 wt%) PVAm/mPSF	COF_P (10 wt%)	COF-LZU1 synthesis and modification [a] Solution mixing and casting [b]	FSC (Composite)	Flat-sheet (0.154 μm)	T = 25 °C P = 1.5 bar Humidified Mixed gas (CO_2/H_2) (40/60 vol%)	201.30 (Barrer)	9.06 (Barrer)	22.20	[30]
Im-TB (*o*)-PA (0.795 mmol–1g)	$[C_4mim]$-$[Tf_2N]$ (1.59 mmol)	Menshutkin reaction and Solution mixing and casting [a, b]	MC (IL-MMM)	Flat-sheet (100–130 μm)	T = 20 °C P = 3.0 bar Pure gas	47.19 (Barrer)	10.78 (Barrer)	4.38	[31]
Im-TB (*p*)-PA (0.795 mmol–1g)	$[C_4mim]$-$[Tf_2N]$ (1.59 mmol)	Menshutkin reaction and Solution mixing and casting [a, b]	MC (IL-MMM)	Flat-sheet (100–130 μm)	T = 20 °C P = 3.0 bar Pure gas	21.13 (Barrer)	5.43 (Barrer)	3.89	[31]
Crosslinked (20.6 wt%) PVA-POS/PSF	APTES-(5.5 wt%)/KOH-(19.4 wt%)/AIBA-K—(31.7 wt%)/PVAm-(22.8 wt%)/AF-MWCNTs-(10 wt%)	AF-MWCNTs: Oxidation and reaction with APTES [a] Sol–gel reaction, solution mixing, coating method and crosslink reaction [b]	MC-FSC hybrid (composite)	Flat-sheet (Skin layer: 17 μm)	T = 107 °C P = 2 bar R.H = 59.20% Humidified Pure gas	3196.00 (Barrer)	15.59 (Barrer)	205.00	[32]
(52.6 wt%)—PDAMA-F/(25.3 wt%)—XL-PVA/PSF	TMAF (22.1 wt%)	Solution mixing and coating method [b]	MC (composite)	Flat-sheet (skin layer: 15 μm)	T = 120 °C P = 0.10 bar R.H = 54% Humidified Mixed gas ($CO_2/H_2/CO$) (59.5/26.5/14%)	108.00 (GPU)	1.01 (GPU)	106.00	[33]

(continued)

Table 6.2 (continued)

Materials	Carriers/fillers	Fabrication method	FTMs types	Module/thickness	Operating conditions	$P(CO_2)$	$P(H_2)$	$\alpha_{\frac{(CO_2)}{(H_2)}}$	Ref.
Poly([Pyr_{11}]-[$C(CN)_3$]	[C_2mim]-[$C(CN)_3$] (60 wt%)	Solvent casting method [b]	Hybrid-MC (PIL-IL) (MMM)	Flat-sheet (86–122 μm)	T = 35 °C P = 1 bar Mixed gas ($CO_2/H_2/N_2$) (40/57.1/2.9 vol%)	324.70 (Barrer)	28.30 (Barrer)	11.40	[34]
Poly([Pyr_{11}]-[NTf_2]	[C_4mpyr]-[NTf_2] (60 wt%)	Solvent casting method [b]	Hybrid-MC (PIL-IL) (MMM)	Flat-sheet (86–122 μm)	T = 35 °C P = 1 bar Mixed gas ($CO_2/H_2/N_2$) (40/57.1/2.9 vol%)	254.20 (Barrer)	38.30 (Barrer)	6.60	[34]
Poly([Pyr_{11}]-[NTf_2]	[C_2mim]-[NTf_2] (40 wt%)	Solvent casting method [b]	Hybrid-MC (PIL-IL) (MMM)	Flat-sheet (86–122 μm)	T = 35 °C P = 1 bar Mixed gas ($CO_2/H_2/N_2$) (40/57.1/2.9 vol%)	201.60 (Barrer)	29.00 (Barrer)	6.90	[34]
PEO/NBR (60/40 w/w) (sIPN)	[C_2mim]-[NTf_2] (66 wt%)	In-situ polymerization and post-swelling method [b]	Hybrid-MC (sIPN) (MMM)	Flat-sheet (238–387 μm)	T = 35 °C P = 1 bar Pure gas	609.00 (Barrer)	74.80 (Barrer)	8.10	[35]
PEO/NBR (60/40 w/w) (sIPN)	[C_2mim]-[$C(CN)_3$] (66 wt%)	In-situ polymerization and post-swelling method [b]	Hybrid-MC (sIPN) (MMM)	Flat-sheet (238–387 μm)	T = 35 °C P = 1 bar Pure gas	633.00 (Barrer)	58.10 (Barrer)	10.90	[35]

(continued)

Table 6.2 (continued)

Materials	Carriers/fillers	Fabrication method	FTMs types	Module/thickness	Operating conditions	$P(CO_2)$	$P(H_2)$	$\alpha_{\frac{(CO_2)}{(H_2)}}$	Ref.
PEO/NBR (60/40 w/w) (sIPN)	$[C_2mim]$-[FSI] (66 wt%)	In-situ polymerization and post-swelling method [b]	Hybrid-MC (sIPN) (MMM)	Flat-sheet (238–387 μm)	T = 35 °C P = 1 bar Pure gas	727.00 (Barrer)	66.80 (Barrer)	10.90	[35]
Pebax-1657 (6.0 wt%)	Core shell ZIF-8@ZIF-67-templated GO (5.0 wt%)	GO: modified Hummer's, GO/ZIF-8@ZIF-67: ZIF-8 seed-mediated deposition [a] Solution mixing and casting method [b]	FSC (MMM)	Flat-sheet (100 μm)	T = 35 °C P = 4 bar Pure gas	173.20 (Barrer)	14.90 (Barrer)	11.60	[36]

[a] Fabrication method for carrier
[b] Fabrication method for FTMs

liquid membranes (ILMs) such as supported ionic liquid phase (SILP) and supported ionic liquid membrane (SILM) are widely used for CO_2 separation applications. In the case of SILP, the porous surface of the substrate is coated with a thin layer of IL, while in the preparation of SILM, IL is confined into the pores of the support substrate.

In this study, the researchers synthesized a two-dimensional (2D) nano channel of laminated graphene oxide (GO) by the modified Hummer's method. Following that, 0.05 mg/mL of immobilized 1-butyl-3-methlyimidazolium tetrafluoroborate [BMIM]-[BF_4] IL was incorporated into the 2D nano channel of laminated GO membranes to fabricate GO-[BMIM][BF_4] based SILMs for CO_2/H_2 separation using solution mixing and vacuum filtration method. The anodic alumina oxide (AAO) membrane having a pore size of 0.2 μm was used as the support layer in this research work. The CO_2/H_2 separation performance of optimized GO-[BMIM][BF_4]-SILM was then evaluated using a pure gas permeation testing system at a pressure of 1 bar and temperature range of 25–50 °C. According to the results, the fabricated membrane demonstrated a rapid CO_2 permeance and high CO_2/H_2 selectivity, which is almost seven times higher than the pristine GO membranes. In the case of pristine GO membranes, the selectivity of CO_2/H_2 is only 4.00, which is too close to Knudsen selectivity owing to the large, wrinkled nano-channels of the GO 2D nano-sheets. GO-[BMIM][BF_4]-SILM having a thickness of 1.05 μm, on the other hand, achieved a higher CO_2/H_2 selectivity of 23.70 with a CO_2 permeability of 71.93 Barrer (CO_2 permeance = 68.5 GPU) [25].

The reason for this superior gas separation performance as compared to conventional SILM is the nano confinement of IL in the 2D channels of GO membranes, which have a low thickness of just 1.05 μm. Furthermore, the molecular dynamics (MD) simulations also revealed that the comparison between the binding energies of the following gas species, i.e., CO_2, and H_2 and the confined [BMIM][BF_4] is marginally higher than that of the bulk [BMIM][BF_4], due to which dissolution of gas molecules in the nano-confined [BMIM][BF_4] IL into the GO channel is dynamically more preferable as compared to the bulk form. This also resulted in the enhancement of CO_2 solubility and selectivity of CO_2/H_2 separation. The effect of GO-SILM membrane thickness on separation performance was also investigated, which indicated a decreasing trend of gas permeance as membrane thickness increases, while CO_2/H_2 selectivity improves. Whereas, the effect of temperature on CO_2 permeance was also explored, which revealed that CO_2 permeance increased with a rise in temperature in the case of [BMIM][BF_4]-GO-SILM [25].

Gouveia et al. [26], fabricated the high-performance polymer ionic liquids (PIL)-ionic liquid (IL) based composite membranes for the purification of H_2 gas for use in fuel cell applications. In this research work, solvent casting method was used to fabricate composite membranes at thickness of 70–210 μm composed of poly pyrrolidinium-based PILs—Poly ([Pyr_{11}]) with 6 wt% of tricyanomethanide [$C(CN)_3$] and 12 wt% of Bis(tri-fluoromethylsulfonyl)imide [NTf_2] anions, respectively, in which different concentration of free ILs, including

1-ethyl-3-methylimidazolium tricyanomethanide [C_2mim][C(CN)$_3$], 1-butyl-1-methylpyrrolidinium bis(tri-fluoro-methyl-sulfonyl) imide [C_4mpyr][NTf_2], and 1-ethyl-3-methylimidazolium bis(tri-fluoro-methyl-sulfonyl) imide [C_2mim][NTf_2] are incorporated, separately. The permeability, solubility, and diffusivity of pure CO_2 and H_2 gases were examined at two different temperatures of 20 and 30 °C under the influence of a transmembrane pressure differential of 1 bar at the feed side and an initial vacuum pressure of 1 bar at the downstream side.

The gas permeation experiment results revealed that the CO_2 gas has a much higher permeability than H_2 gas, with a tendency to increase in permeability with a rise in temperature. Furthermore, the solubilities of both CO_2 and H_2 were enhanced with the incorporation of a higher concentration of IL in the PIL-based composite membranes. At a temperature of 20 °C, the solubilities of both CO_2 and H_2 gas increased significantly when the concentration of [C_2mim][C(CN)$_3$] was increased from 40 to 60%. Moreover, by increasing the temperature up to 35 °C, the increment in CO_2 and H_2 gases solubilities was found to be 39% and 5%, respectively. Whereas, the gas diffusivities follows the order of $CO_2 < H_2$ based on their kinetic diameter (K.D). When the temperature was increased from 20 to 35 °C, the [C_2min]-[C(CN$_3$)]–PIL-C(CN)$_3$ composite membrane showed an increase in CO_2 permeability from 438 to 505 Barrer with a decrease in selectivity of 15.10–12.50 at an optimum concentration of 60 wt%. Similarly, at 60 wt% of [C_4mpyr]-[NTf_2] and 40 wt% [C_2mim]-[NTf_2] the PIL-IL composite membranes also demonstrated a decrement in CO_2/H_2 separation performances of 7.80–6.30 and 8.20–6.50, respectively, with the increment in corresponding CO_2 permeabilities of 232–288 Barrer and 214–287 Barrer, accordingly. Moreover, these results also clearly show that PIL-IL composite membranes containing $[C(CN)_3]^{-1}$ anion are a better candidate for CO_2/H_2 separation applications than membranes containing $[NTf_2]^{-1}$ anion [26].

Huang et al. [27], fabricated a hybrid IL-filler based polyether block amide (Pebax)—1657 composite membrane for the application of CO_2/H_2 separation. For the preparation of these composite membranes, GO was initially synthesized using the improved Hummer's method, Then 1-(3-aminopropyl)-3-methylimidazolium bromide ionic liquid (IL-NH_2) was covalently functionalized with GO through an epoxy ring-opening reaction, which formed a GO-IL-NH_2 hybrid filler. Afterward, different concentration range from 0.01 to 0.5 wt% of GO-IL-NH_2 were mixed in 2 wt% Pebax—1657 solution, followed by coating on the Poly(1-trimethylsilyl-1-propyne) (PTMSP)/Polyvinylidene fluoride (PVDF) support layer at a thickness of 0.1 μm using solution casting method.

The mixed gas (CO_2/H_2 = 30/70 vol%) and pure gas permeation experiments were conducted at a pressure of 4 bar and a temperature of 25 °C to evaluate the CO_2/H_2 separation performance of the fabricated hybrid (GO-IL-NH_2) composite membranes. According to the results, Pure Pebax showed a high CO_2 permeability but low selectivity for CO_2/H_2 separation process. The use of GO fillers into Pebax membranes improves gas selectivity without substantially effecting the CO_2 permeability. Even at very moderate filler loading of GO-IL-NH_2 increased the permeability and selectivity of the membrane relative to pure Pebax and GO-based Pebax membranes, exceeding the 2008 Robeson's upper bound line. At an optimized filler loading of 0.2% GO-IL-NH_2, composite membrane exhibited the CO_2/H_2 selectivity of 13.80 and CO_2 permeability of 143 Barrer in pure gas testing system, which is much higher than that of a GO based Pebax membrane with the same filler content having CO_2/H_2 selectivity of 8.0 with a corresponding CO_2 permeability of 88 Barrer. While in the case of mixed gas testing system, the composite membranes showed a slightly lower CO_2/H_2 selectivity of 9.50 with CO_2 permeability of 118.60 Barrer due to the competitive interaction of both gases and non-ideal effects [27].

The three main reasons for this superior gas separation performance of hybrid composite membranes are: First, the addition of IL-NH_2 enhanced the interfacial interaction by strengthening hydrogen bonding. Secondly, it also improved the interface compatibility between GO nanosheets and the Pebax-1657 polymeric matrix. Lastly, the presence of IL-NH_2 on the interspaces of GO nanosheets facilitated the rapid transport of CO_2 via a reversible reaction. These findings also suggest that the GO-IL-NH_2 nanosheet has the potential to be used as a nanofiller in the development of highly selective thin-film composite membranes for industrial scale H_2 gas purification processes [27].

In 2019, Nabais et al. [28], developed three different hybrid PIL-IL/Inorganic filler based mixed matrix membranes (MMMs) using the solvent evaporation method for CO_2 separation from syngas stream. These membranes were composed of 0.6 g of pyrrolidinium-based PIL, poly(diallyl-methyl-ammonium) bis(tri-fluro-methyl-sulfonyl) imide (poly[Pyr_{11}][Tf_2N]), 0.4 g of an IL having same structural anion 1-butyl-3-methlypyrrolidinium bis(tri-fluro-methyl-sulfonyl) imide [C_4mpyr][Tf_2N], and a varied concentration of 10–30 wt% of metal organic framework (MOFs), which include MIL-53(AI), $Cu_3(BTC)_2$ and ZIF-8. The CO_2/H_2 separation performance of the fabricated membranes was then examined by employing pure gas testing equipment at a transmembrane pressure of 0.7 bar and a temperature of 30 °C.

According to the gas permeation analysis results, hybrid MMMs demonstrated an improvement in CO_2/H_2 separation performance despite the fact that H_2 molecules have smaller K.D of 2.8 Å than the CO_2 having the K.D of 3.3 Å. At a low filler content of 10 wt%, the MMMs with $Cu_3(BTC)_2$ and ZIF-8 showed the lower CO_2 permeabilities than the pure PIL/IL based membrane. This is likely due to an increase in tortuosity caused by the low concentration of MOFs particles, which may inhibit gas diffusion through the membrane. However, with a continuous increase in MOF loading up to 30 wt%, this obstacle was overcome, resulting in increased CO_2 permeabilities, which surpassed the value obtained for the pure PIL/IL based membrane [28].

At an optimized filler loading of 30 wt%, MIL-53(Al)-based PIL-IL MMMs exhibited the highest increase in ideal gas selectivity of 13.28 with the CO_2 permeability of 89.01 Barrer. Whereas, at same filler content, ZIF-8-based hybrid MMMs showed the maximum CO_2 permeability of 97.15 Barrer with a low ideal CO_2/H_2 selectivity of 4.55, while $Cu_3(BTC)_2$-based hybrid MMM containing 30 wt% of filler loading, demonstrated an intermediate results of ideal CO_2/H_2 selectivity of 6.27 and CO_2 permeability of 77.15 Barrer. The main reasons for this outstanding results are due to the combination of high ionic content of PIL-IL, which has an affinity for CO_2 molecules and high adsorption characteristics of the MOFs in these membranes, which overcomes the revised Robeson's upper bound limit in term of CO_2/H_2 separation performance. In addition, the incorporation of high content of MOFs also boosted the mechanical strength and thermal stability of the membranes, making them a viable option for CO_2/H_2 separation applications within the temperature range of the syngas product stream [28].

In 2020, Klepic et al. [29], investigated the effect of 1-ethyl-3-methylimidazolium dicyanamide [EMIM][DCA] as water-soluble IL in polyvinyl alcohol (PVA) membranes for efficient H_2 recovery from syngas product stream (CO_2/H_2). For these membrane preparation, a variable concentration range of 9–53 wt% [EMIM][DCA] was blended with the PVA polymer and then spin-coated on the stainless plate to produce PVA-[EMIM][DCA] MMMs with thicknesses ranging from 75.30 to 104.40 μm. Following that, the MMMs were tested using a pure gas permeation testing system at a pressure of 2 bar and a temperature of 25 °C, which revealed that MMMs with lower IL concentrations ranging from 9 to 14 wt% had significantly higher H_2 permeability than CO_2 gas and exhibited H_2/CO_2 selectivities ranging from 3.00 to 3.45. This indicates that the gas transport was regulated by a solution diffusion mechanism, which is a typical property of the dense polymeric membrane.

However, when the IL content was increased above 20 wt%, the PVA-[EMIM][DCA]-based MMMs remarkably demonstrated a "reverse selectivity" of CO_2/H_2 separation, which indicated that the CO_2 was permeated much faster than the H_2 gas. The IL loading was optimized at 53 wt%, in which the membrane showed the highest CO_2/H_2 selectivity of 7.73 with a maximum CO_2 permeability of 66.90 Barrer. The higher CO_2-philicity of the IL was the main reason for this improvement in gas separation performance, which caused a complete shift in membrane characteristics from a diffusion-controlled process to a solubility-controlled process, resulting in superior CO_2/H_2 perm-selectivity of PVA-[EMIM][DCA]-based MMMs with a high IL concentration that exceeded the revised Robeson's upper bound limitation of CO_2/H_2 separation. This finding also opens up a new way of employing IL to change the gas separation capacities of membrane materials [29].

Cao et al. [30], prepared three different high-performance covalent organic frameworks (COFs) based composite fixed-site carrier (FSC) FTMs for CO_2/H_2 separation applications. In this study, polyvinyl amine (PVAm) was used as the polymer matrix because of its amino group rich structure, which acts as fixed-site carrier (FSC) in the facilitated transport process. Whereas, COF-LZU1 was selected as the nano-filler because of its water stability, excellent compatibility with PVAm polymer, and large pore structure. For the membrane preparation, initially the COF-LZU1 was modified by glacial acetic acid, 1,3,5-triformylbenzene, and 1,4-diaminobenzene to produced COF_P, COF_L, and COF_M, respectively, which further enhanced its compatibility with PVAm polymeric matrix due to the presence of PVAm segment or groups on the surface of modified COFs. Afterwards, 10 wt% of each modified COFs were mixed with the 1.0 wt% of PVAm and 0.1 wt% of PVA to produce the COFs/PVAm MMMs coating solution. Meanwhile, the PSf substrates were coated with crosslinked polydimethylsiloxane (PDMS) to prepare a modified polysulfone (mPSf) substrates, which acts as a gutter layer in-order to prevent the coating solutions from penetrating the pores. Each COFs/PVAm MMM solution was then coated on these mPSf layers to fabricate COF_P-PVAm/mPSf, COF_L-PVAm/mPSf, and COF_M-PVAm/mPSf composite FSC FTMs with skin layer thicknesses of 0.152, 0.165, and 0.154 μm, respectively.

These membrane were then tested in a mixed gas permeation testing setup with a feed mixture of $CO_2/H_2 = 40/60$ vol% at pressure of 1.5 bar and a temperature of 25 °C under humidified condition. The result showed that the COF_P-PVAm/mPSf composite membrane exhibited the most optimum CO_2/H_2 selectivity value of 22.20 with a CO_2 permeability of 201.30 Barrer. While the highest CO_2 permeability was achieved by the COF_M-PVAm/mPSf composite membrane having a value of 234.60 Barrer with the CO_2/H_2 selectivity of 17.20. However, the COF_L-PVAm/mPSf composite membrane showed the lowest gas separation performance as compared to the other modified COFs-based composite membranes, having a CO_2/H_2 selectivity of 16.10 and a CO_2 permeability of 221.10 Barrer [30].

In these composite membranes, the introduction of modified COFs was primarily responsible for this superior gas separation performance. For starters, the porous nature of COFs might act as gas transport routes while also increasing the free volume of the PVAm membrane matrix. The gas transport resistance in the nano channels was lower than that in the polymer matrix because the pore diameters of the nano channels were greater than the gaps present between the packed polymeric chains of the membrane matrix, allowing more CO_2 molecules to pass through the nano channels of COFs, resulting in a substantial boost in CO_2 permeability. Meanwhile, the integration of COFs as a filler in the membrane matrix could promote the polymer coating solution to penetrate into the nano pores of the COFs filler, which reduced the diameter of COFs pores to an optimal size as well as the abundant amino group due to the presence of PVAm molecules suitable for high absorption of CO_2 molecules while preventing the H_2 molecules from passing through them. As a result, a significant improvement in CO_2/H_2 separation performance was observed. Lastly, the strong intermolecular interaction between modified COFs surfaces and the PVAm membrane matrix also enhanced the composite membranes' interfacial compatibility and lowered the polymeric chain's mobility under elevated pressure. As a result, these composite membranes showed high plasticization resistance, which allowed them to perform well even under high pressure conditions [30].

Kammakakam et al. [31], investigated the synthetic route for the preparation of imidazolium-Tröger's base containing polyamide (Im-TB-PA) ionene polymeric membranes and its IL-based MMMs for the facilitated transport of CO_2 for the purification of H_2 from syngas streams. In this study, two novel Im-TB-ortho (*o*)-PA and Im-TB-para (*p*)-PA ionenes polymers were optimized using a Menshutkin reaction and then two equivalent amounts (0.667 g) of free IL-[C_4mim]-[Tf_2N] were mixed into each 1 g of Im-TB-PA ionene polymeric solution to prepare MMMs solutions. Afterwards, these solutions were casted on a Teflon plates to prepared Im-TB-ortho (*o*)-PA ionene-[C_4mim]-[Tf_2N] and Im-TB-para (*p*)-PA ionene-[C_4mim]-[Tf_2N] MMMs using solution casting method at a controlled thickness of 100–130 μm.

The fabricated MMMs were then tested for the CO_2/H_2 separation performance using a pure gas permeation testing system at a feed pressure of 3 bar against the downstream vacuum pressure of less than 6.89×10^{-4} bar and a temperature of 20 °C. The result revealed that both pristine and their free IL-based Im-TB-PA ionenes MMMs demonstrated a much higher CO_2 permeability than the smaller H_2 molecule, which indicated the CO_2-philic behaviour of these membranes. The pristine membranes exhibited similar gas separation performance regardless of their ortho or para connectivity, having the CO_2/H_2 selectivities of 2.65 and 2.69 with the CO_2 permeabilities of 5.27 and 4.87, respectively. While, their free IL MMMs demonstrated superior CO_2/H_2 selectivities of 4.38 and 3.89 with the CO_2 permeabilities of 47.19 and 21.13 Barrer, respectively, which reached the reverse Robeson's upper limit of CO_2/H_2 separation [31].

The incorporation of free IL caused irregularity in the inter-chain spacing that affects the flexibility of the polymeric chain in these MMMs, which resulted in a considerable increase in diffusivity. Whereas, due to a stronger affinity of IL groups towards CO_2 molecules, these membranes had much higher solubility for CO_2 than H_2 molecules, resulting in a solubility controlled mechanism, which led to a higher reverse CO_2/H_2 perm-selectivity. Furthermore, these membranes also showed excellent physical characteristics such as enhanced mechanical strength, thermal stability, and shape memory properties at an elevated temperature of 60 °C [31].

Yang et al. [32], developed amine-based hybrid composite (MC-FSC) FTMs in which amino-functionalized multi-walled carbon nanotubes (AF-MWCNTs) were embedded in the polymeric membrane matrix for high performance CO_2/H_2 separation applications. For the synthesis of AF-MWCNTs, the pure multiwall carbon nanotubes (MWCNTs) were initially aggressively oxidized in an acidic environment. Then they reacted with the 3-Aminopropyltriethoxysilane (APTES) to form AF-MWCNTs. Afterwards, these nano-fillers at a concentration range of 0–10 wt% were mixed with crosslinked polyvinyl alcohol-polysiloxane (PVA-POS) solution, which was prepared using the sol–gel reaction method. Then a mixture of potassium hydroxide (KOH) and AIBA-K (deprotonated) and Lupamin® 9095 was added to this MMM coating solution to prevent the gelation of the solution. After that, the final solution is coated on a nano porous PSF support substrate to fabricate each hybrid (MC-FSC) composite FTM with a 17 μm skin layer thickness. In these membranes, KOH and AIBA-K act as MCs, while PVAm, present in the Lupamin® 9095 mixture, serves as a FSC in the rapid facilitated transport of CO_2 molecules.

The fabricated membranes were then tested for CO_2/H_2 separation performance using a pure gas testing system at a feed gas pressure range of 2–15 bars and a temperature of 107 °C under humidified conditions having a relative humidity (R.H) value of 59.20–100.0%. In addition, the sweep gas was also used in the gas permeation system with an R.H of 47.40–72.50% at a constant pressure of 1 bar. The results revealed that the CO_2/H_2 selectivity and CO_2 permeability decreased steadily as the nano-filler concentration increased from 0 to 6 wt%. At 6 wt% of AF-MWCNTs, the FTM exhibited a lower CO_2/H_2 selectivity of 193 and CO_2 permeability of 2907 Barrer at a pressure of 2 bar due to the concentration dilution of active carriers available for the facilitated transport of CO_2 molecules when the filler loading increased in the membrane matrix. However, as the concentration of AF-MWCNTs increased beyond 6 to 10 wt%, the CO_2 permeability began to increase and eventually reached up to 3196 Barrer with a high CO_2/H_2 selectivity value of 205, which surpassed the revised Robeson's upper bound limit of CO_2/H_2 separation, making them a suitable candidate for industrial syngas purification processes [32].

This remarkable enhancement in gas separation performance can be explained by two main factors. First, AF-MWCNTs disturbed the polymer-chain packing by embedding between the spaces present in the polymer chains, which increased the free volume of the polymeric membrane and resulted in an improvement in gas diffusivity. Secondly, CO_2 diffusivity within AF-MWCNTs is exceptionally high, indicating that their tubular structure provides tunnels for the diffusion of gas molecules. In addition, these nano-fillers also enhanced the compaction resistance of the membrane. However, above 10 wt% nano-filler loading, the membrane showed a sudden drop in CO_2/H_2 separation performance due to the nano-filler agglomeration caused inhomogeneous structure and rigidification of the polymeric chain of the membrane. Furthermore, at a high feed pressure of 15 bar, the optimized 10 wt% nano-filler loading of FTM also showed a decrement in gas separation performance, having the CO_2/H_2 selectivity value of 31 with a CO_2 permeability of 776 Barrer due to the condition of carrier saturation that occurred at an elevated pressure [32].

Chen et al. [33], fabricated a quaternary ammonium (QA)—tetramethylammonium fluoride tetrahydrate (TMAF)-based composite FTM for the separation of CO_2 from the anode exhaust stream of the solid oxide fuel cell (SOFC) to recycle the residual H_2 and CO mixture back into the fuel cell for the production of more energy. In this investigation, the TMAF was chosen as an alternative to the tetramethyl-ammonium hydroxide (TMAOH) in order to enhance the CO_2/H_2 separation performance and to improve the thermal stability of the membrane. For the preparation of this composite membrane, an optimized MMM solution was obtained by blending 52.60 wt% Polydiallyl-dimethyl-ammonium fluoride (PDMA-F) with 25.3 wt% of crosslinked polyvinyl alcohol (XL-PVA) and 22.1 wt% of TMAF using a solution mixing method and then coated on a PSF support layer to produce a selective layer thickness of 17 μm. After that, the composite membrane was allowed to complete the crosslinking reaction for 6 h in the oven at a temperature of 120 °C. The fabricated composite membrane was then examined for the CO_2/H_2 separation performance using a mixed gas permeation testing system having the feed gas composition of $CO_2/H_2/CO = 59.50/26.50/14.00\%$ at a feed pressure of 0.10 bar with a permeate pressure of 0.068 bar and a temperature of 120 °C under humidified conditions with an R.H value of 54%. The sweep gas (air) was used on the permeate side with an R.H value of 30%. Whereas, the thermal stability of the membrane was evaluated using a thermogravimetric analysis (TGA) and nuclear magnetic resonance (NMR) characterization method.

According to the gas permeation result, the TMAF-based composite FTM showed excellent gas separation performance, having a CO_2/H_2 selectivity of 106 with a CO_2 permeance of 108 GPU (CO_2 permeability = 1620 Barrer) due to the presence of TMAF, which served as M.C in the facilitated transport process of CO_2 molecules. Whereas, other gas molecules, e.g. H_2, which have no affinity with the carrier species, can only permeate through the solution diffusion mechanism, resulting in a lower gas permeation rate as compared to the CO_2 molecules. Furthermore, the fabricated membrane also exhibited a superior thermal stability of approximately four times as compared to the TMAOH-based membrane at a temperature of 130 °C, making the TMAF-based composite membrane a better option for capturing CO_2 in the SOFC process [33].

In 2021, Gouveia et al. [34], investigated the PIL-IL-based FTMs for CO_2/H_2 separation performance using a mixed gas testing system. These membranes were prepared using solvent casting method at a thickness of 86–122 μm and were composed of pyrrolidinium-based PILs carrying $[NTF_2]^-$ and $[C(CN)_3]^-$ anions and different concentration range (40–60 wt%) of corresponding ILs, such as $[C_2mim][C(CN)_3]$, $[C_4mpyr][NTf_2]$, and $[C_2mim][NTf_2]$. The gas permeation test was conducted using a ternary gas mixture having the composition of ($CO_2/H_2/N_2$ = 40.00/57.10/2.90) at temperatures ranging from 20 to 80 °C and a different feed pressure of 1–4 bar with a constant sweep gas (Helium) flowrate of 20 cm^3 (STP). The gas permeation result revealed that, when the temperature was raised from 20 to 80 °C, the fabricated PIL-IL-based MMMs showed a considerable increase in gas permeability with a decrease in CO_2/H_2 selectivity. However, at a T = 35 °C, an optimum balance between the gas permeability and selectivity was achieved in all cases. Whereas, with the increased of feed pressure, the gas permeabilities and CO_2/H_2 selectivities of the PIL-IL based MMMs containing up to 40 wt% IL remained constant, while in the case of higher IL concentration of 60 wt%, the membrane showed slight decrease in gas separation performance.

At an optimized concentration of 60 wt% the PIL $[C(CN)_3]$-60$[C_2mim][C(CN)_3]$ MMM demonstrated the highest CO_2/H_2 selectivity of 11.40 with an CO_2 permeability of 324.70 Barrer, as compared to the PIL $[NTf_2]$-60$[C_4myr][NTf_2]$ and PIL $[C_2mim]$-40$[NTf_2]$ MMMs, which showed an lower gas separation performance having the CO_2/H_2 selectivities value of 6.60 and 6.90 with an CO_2 permeabilities of 254.20 Barrer and 201.60 Barrer, respectively at a constant feed pressure of 1 bar and an optimum temperature condition of 35 °C. Furthermore, minor variations were observed between the mixed and pure gas CO_2/H_2 separation performances, which showed that these membranes can maintain their gas separation efficiency despite the competitive effect that occurred during the mixed gas experimentations and also surpassed the revised Robeson's upper bound of CO_2/H_2 separation [34].

In another research work, Gouveia et al. [35], explored the IL-based semi-interpenetrating polymer network (sIPN) FTMs for the application of CO_2/H_2 separation. In this study, sIPNs were prepared by the in-situ polymerization method between the polyethylene oxide (PEO) networks and linear nitrile butadiene rubber (NBR) at a 60/40 weight ratio. Meanwhile, depending on the ILs' nature and their concentration, the ILs were integrated at a concentration range of 50–66 wt% using two different methods. In the case of 1-ethyl-3-methylimidazolium bis(tri-fluoro-methyl-sulfonyl) imide $[C_2mim][NTf_2]$, the IL was directly mixed as a co-solvent in the preparation of PEO/NBR based sIPN. While other ILs such as, 1-ethyl-3-methylimidazolium tricyanomethanide $[C_2mim][C(CN_3)]$ and 1-ethyl-3-methylimidazolium bis (fluoro sulfonyl) imide $[C_2mim][FSI]$, were incorporated into sIPN via a post-swelling method due to the solubility constraints between the IL and the solvent mixture. In this method, the membrane matrix was fully immersed in each IL until the sIPN achieved the desired content of IL in the membrane matrix. The fabricated PEO/NBR-based sIPN/IL FTMs having a thickness in the range of 238–387 μm were then evaluated for CO_2/H_2 separation performance using a pure gas testing setup at a feed pressure of 1 bar with a downstream vacuum pressure of less than 0.001 bar and a temperature range of 20–35 °C.

The gas permeation results revealed that all the fabricated FTMs showed a substantial increment in both CO_2 and H_2 permeabilities with the increase in IL content in the membrane matrix. Similarly, the rise in operating temperature from 20 to 35 °C also promoted a significant improvement in CO_2 permeability as compared to the pure sIPN membrane. In addition, the fabricated PEO/NBR-based sIPN membranes with high content IL showed remarkable CO_2/H_2 separation performance that surpassed the revised Robeson's upper bound limitation. At a maximum temperature of 35 °C and a higher IL content of 66 wt% [C_2mim][FSI] in PEO/NBR sIPN matrix, the FTM demonstrated the highest CO_2 permeability of 725 Barrer with optimum CO_2/H_2 selectivity of 10.90. Similarly, at the same conditions and IL content, the PEO/NBR sIPN—[C_2mim][C(CN)$_3$] exhibited an identical CO_2/H_2 separation with lower CO_2 permeability of 633 Barrer. While PEO/NBR sIPN—[C_2mim][NTf$_2$] at IL content of 66 wt% showed the lowest CO_2/H_2 selectivity of 8.1 with an CO_2 permeability of 609 Barrer [35].

Liu et al. [36], synthesized the Pebax MMMs based FTMs incorporated with graphene oxide (GO)/core shell ZIF-8@ZIF-67 hybrid nano-fillers to achieve high CO_2/H_2 separation performance. In this study, they used a ZIF-8 seed-mediated deposition method to prepare the core shell ZIF-8@ZIF-67-templated GO hybrid nano-fillers at an optimized GO concentration of 0.5 wt%, which has a sandwich like structure. After that, at a concentration range of 2.5–10 wt%, these nano-fillers were embedded into a 6 wt% Pebax membrane matrix using a solution mixing and casting method to fabricate different GO/core shell ZIF-Pebax MMMs based FTMs at a thickness of 100 μm. The CO_2/H_2 separation performance of these FTMs was then evaluated using a pure gas permeation testing system at a pressure of 4 bar and a temperature of 35 °C.

The gas permeation analysis revealed that, the MMMs exhibited a higher CO_2 permeability than the H_2, despite the fact that the H_2 molecule is a smaller molecule with a K.D of 2.8 Å as compared to the CO_2 molecule, which has a bigger K.D of 3.3 Å. This attributed that these MMMs have strong affinity towards CO_2 gas, and the gas transport is mainly dependent on the solubility of the gas rather than the diffusivity process. The incorporation of GO/core shell ZIF-8@ZIF-67 from 2.5 to 5.0 wt% in the Pebax matrix showed an increasing trend of CO_2/H_2 separation performance. At a filler loading of 5.0 wt%, GO/core shell ZIF-Pebax MMM demonstrated the highest CO_2/H_2 separation of 11.60 and optimum CO_2 permeability of 173.20 Barrer, which were 20.80% and 97.70%, respectively, higher than the pristine Pebax membrane, exceeding the revised Robeson's (2008) upper bound limitation [36].

This remarkable improvement in gas separation performance was explained by enhanced diffusivity and solubility of CO_2 molecules due to the synergistic effects of two main factors. Firstly, the high CO_2 adsorption capacity of optimized GO/core shell ZIF-8@ZIF-67 hybrid nanofiller is because of the strong electrostatic interaction and pi (π)—complexation between the unsaturated metallic sites of the nanofiller such as Zn–O or Co–O, nitrogen-containing sites, polar oxygen carrying functional groups, and CO_2 molecules. Secondly, the high porosity and flexibility of the embedded hybrid nano-filler provide micro-voids at the polymer-nanofiller interface and additional free volume, which facilitated the diffusion process of CO_2 molecules. However, further addition of nano-fillers up-to 10 wt% caused aggregation and void formation in the Pebax membrane matrix, which resulted in a decrement of CO_2/H_2 perm-selectivity [36].

6.3 Trade-Off Robeson Plot of FTMs for Syngas Purification—CO_2/H_2 Separation

The Robeson upper bound (trade-off graph), illustrate the relationship between membrane gas permeability (P_i) in Barrer and selectivity ($\alpha_{i/j}$) was initially published in 1991 and updated in 2008 given the large volume of newly reported research data of the modified gas separation membranes. The two empirical line in the graph represent the gas separation performance trade-off limit of the polymeric membranes, which served as benchmark for the membrane selection. Different upper bound correlations for various CO_2 gas mixture separation such as CO_2/CH_4, CO_2/N_2, H_2/CO_2, and He/CO_2 were determined by L. M. Robeson's in their literatures [37, 38]. However, a Robeson upper bound graph for plotting CO_2/H_2 selectivity against CO_2 permeability does not exist. Therefore, in this case, the Robeson's H_2/CO_2 upper bounds [37, 38], were inverted to represent the data of CO_2/H_2 selectivities of different membranes against their CO_2 permeabilities. Based on the recently reported data in Table 6.2, a Robeson plot has been developed as illustrated in Fig. 6.3, which showed the comparison between the CO_2/H_2 separation performances of various FTMs. According to this Figure, all the modified polymeric-based CO_2-selective FTMs have exceeded the revised (2008) Robeson's upper bound limit of CO_2/CH_4 separation, which showed that the FTMs outperformed the pristine dense polymeric membranes in syngas purification processes.

The crosslinked PVA-POS/PSF hybrid MC-FSC based FTMs containing AF-MWCNTs at a 10 wt% filler loading stood out as the best CO_2-selective membrane in these FTMs, exhibiting the highest gas separation performance in terms of both CO_2/H_2 selectivity and CO_2 permeability having the values of 205 and 3196 Barrer, respectively. While TMAF-PDMA-F/XL-PVA/PSF composite FTM at an MC concentration of 22.10 wt%, which came in second, also displayed an exceptional CO_2/H_2 perm-selectivity had a separation value of 106 and a CO_2 permeability of 1620 Barrer (CO_2 Permeance = 108 GPU). Whereas, other FTMs based on MMMs demonstrated a mid-range CO_2/H_2 perm-selectivities as compared to the hybrid MC-FSC composite FTM. To summarize it, the Robeson graph clearly reveals that the most recent FTM-based technologies, in which hybrid carrier based

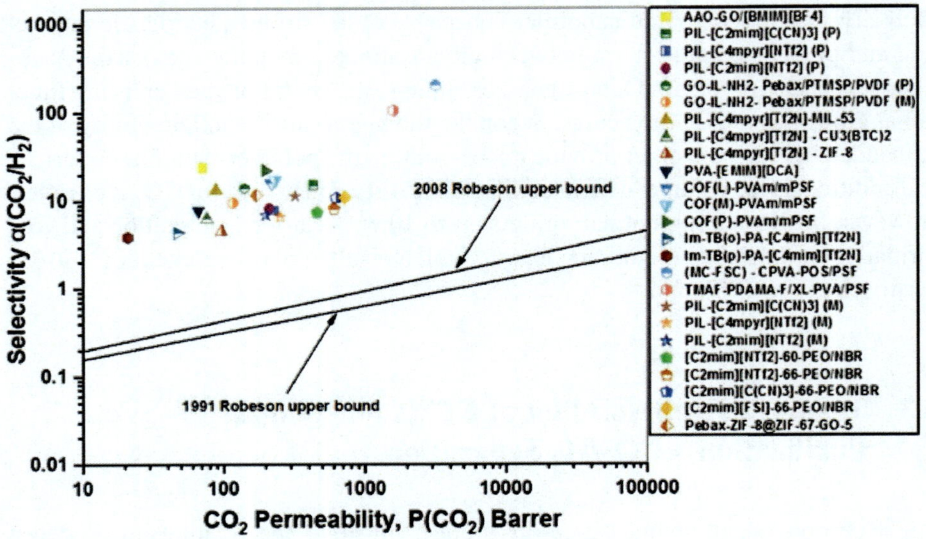

Fig. 6.3 Robeson's upper bound graph illustrating the syngas purification (CO_2/H_2) selectivity for different FTMs versus CO_2 permeability from Table 6.2 data

composite FTMs, are one of the most viable membranes for CO_2/H_2 separation processes as compared to the other type of membrane technology and have the potential to be utilized in commercial scale syngas purification applications, which can replace the current conventional technologies and also formed a hybrid system for efficient gas separation processes.

6.4 Conclusion and Future Recommendations

This chapter has critically reviewed recent advancements in CO_2-selective FTMs for the application of syngas (CO_2/H_2) purification. A concise introduction to syngas has been summarized, including syngas production through different methods, raw syngas composition details from various sources, water gas shift reaction for the conversion of syngas into CO_2 and H_2 gaseous mixture, have been explained. The comparison between the conventional methods and membrane technology for the purification of capturing CO_2 gas from the syngas mixtures has also been included in this literature. In the second phase of the chapter, an overview of the past years' developmental progress of different FTMs, which include pristine, MMMs, composite, and hybrid-based FTMs for the syngas purification (CO_2/H_2) separation application, was thoroughly summarized in detail. The CO_2/H_2 separation performance of different types of FTMs has been discussed with emphasis on the different types of membrane material, the role of nano-fillers, carriers (e.g. MC and FSC), CO_2-selective ILs, hybrid fillers and their optimization for synergistic effects, fabrication method of both membrane and nano-fillers or carriers, CO_2/H_2 selectivity and permeability at operating conditions.

Furthermore, a Robeson upper bound graph has also been plotted to illustrate the comparison between the CO_2 permeability and CO_2/H_2 selectivity values of different recently reported FTMs, which clearly showed that the CO_2-selective FTMs demonstrated a superior gas separation performance that has surpassed the current trade-off limitation of polymeric membranes that only depend on the solution diffusion mechanism for CO_2/H_2 separation applications. In these recently reported FTMs, the combination of CO_2-selective ILs with a PIL membrane showed a remarkable gas separation performances and enhanced membrane stabilities. Whereas, in comparison to these FTMs, the incorporation of hybrid/mixed carriers in polymeric membrane showed a superior CO_2/H_2 separation performance, which opened up a new research area for the FTMs, that needs to further explored and optimized for real operating conditions to evaluate its CO_2 separation performance and mechanical stability in commercial scale. In addition, the adhesion between the composite layers and functionalizing polymeric membrane materials for improved membrane stability, perm-selectivity, and mechanical strength needs to be considered for effective performance of FTMs for CO_2 separation applications.

References

1. El-Nagar RA, Ghanem AA (2019) Syngas production, properties, and its importance. Sustain Altern Syngas Fuel 1–408
2. Speight J (2019) Unconventional gas. Natural gas. In: Speight JG (ed) Gulf Professional Publishing, Boston, pp 59–98
3. Torres MT, Kameid NR (2019) Syngas fuel production from carbonaceous feedstocks using hybrid porous media. In: Sustainable alternative syngas Fuel. IntechOpen
4. Das HS, Chowdhury MFF, Li S, Tan CW (2021) Fuel cell and hydrogen power plants. In: Hybrid renewable energy systems and microgrids. Elsevier, pp 313–349
5. Tanios C, Labaki M (2022) Catalytic reforming: a sustainable technology for hydrogen production. In: Recent advances in renewable energy technologies. Elsevier, pp 199–247
6. Ismail AF, Khulbe KC, Matsuura T (2015) Application of gas separation membranes. In: Gas separation membranes. Springer, Berlin, pp 241–287
7. Kurkela E, Kurkela M, Hiltunen I (2021) Pilot-scale development of pressurized fixed-bed gasification for synthesis gas production from biomass residues. Biomass Convers Biorefinery 1–22
8. Energy C (2016) Syngas cogeneration/combined heat & power. Clarke Energy Website
9. Liu K, Song C, Subramani V (2010) Hydrogen and syngas production and purification technologies. Wiley
10. Cayan FN, Zhi M, Pakalapati SR, Celik I, Wu N, Gemmen R (2008) Effects of coal syngas impurities on anodes of solid oxide fuel cells. J Power Sources 185(2):595–602
11. EG, Services GT, Inc ETS, Energy USDo (2016) Fuel cell handbook, 7th edn. Lulu.com
12. Murthy BN, Sawarkar AN, Deshmukh NA, Mathew T, Joshi JB (2014) Petroleum coke gasification: a review. Can J Chem Eng 92(3):441–468
13. Heaven DL (1996) Gasification converts a variety of problem feedstocks and wastes. Oil and Gas J 94 (22)
14. Couto N, Rouboa A, Silva V, Monteiro E, Bouziane K (2013) Influence of the biomass gasification processes on the final composition of syngas. Energy Procedia 36:596–606

15. Karim SS, Farrukh S, Hussain A, Noor T, Younas M (2021) A comprehensive overview of dual-layer composite membrane for air (O_2/N_2) separation. Polym Polym Compos 29(9_suppl):S1630–S1640
16. Jess A, Wasserscheid P (2020) Chemical technology: from principles to products. Wiley
17. Ismail AF, Khulbe KC, Matsuura T (2015) Gas separation membranes. Switz Springer 10:973–978
18. Matteucci S, Yampolskii Y, Freeman BD, Pinnau I (2006) Transport of gases and vapors in glassy and rubbery polymers. Mater Sci Membr Gas Vapor Sep 1:1–2
19. Breck DW, Breck DW (1973) Zeolite molecular sieves: structure, chemistry, and use. Wiley
20. Houston R, Labbé N, Hayes D, Daw CS, Abdoulmoumine N (2019) Intermediate temperature water–gas shift kinetics for hydrogen production. React Chem Eng 4(10):1814–1822
21. Huang J, Zou J, Ho WW (2008) Carbon dioxide capture using a CO_2-selective facilitated transport membrane. Ind Eng Chem Res 47(4):1261–1267
22. Salim W, Ho WW (2018) Hydrogen purification with CO_2-selective facilitated transport membranes. Curr Opin Chem Eng 21:96–102
23. Bai H, Ho WW (2011) Carbon dioxide-selective facilitated transport membranes for hydrogen purification. Prod Purif Ultraclean Transp Fuels 115–141
24. Baker RW (2012) Membrane technology and applications. Wiley
25. Ying W, Cai J, Zhou K, Chen D, Ying Y, Guo Y, Kong X, Xu Z, Peng X (2018) Ionic liquid selectively facilitates CO_2 transport through graphene oxide membrane. ACS Nano 12(6):5385–5393
26. Gouveia AS, Ventaja L, Tomé LC, Marrucho IM (2018) Towards biohydrogen separation using poly (ionic liquid)/ionic liquid composite membranes. Membranes 8(4):124
27. Huang G, Isfahani AP, Muchtar A, Sakurai K, Shrestha BB, Qin D, Yamaguchi D, Sivaniah E, Ghalei B (2018) Pebax/ionic liquid modified graphene oxide mixed matrix membranes for enhanced CO_2 capture. J Membr Sci 565:370–379
28. Nabais AR, Martins AP, Alves VD, Crespo JG, Marrucho IM, Tomé LC, Neves LA (2019) Poly (ionic liquid)-based engineered mixed matrix membranes for CO_2/H_2 separation. Sep Purif Technol 222:168–176
29. Klepić M, Setničková K, Lanč M, Žák M, Izák P, Dendisová M, Fuoco A, Jansen JC, Friess K (2020) Permeation and sorption properties of CO_2-selective blend membranes based on polyvinyl alcohol (PVA) and 1-ethyl-3-methylimidazolium dicyanamide ([EMIM][DCA]) ionic liquid for effective CO_2/H_2 separation. J Membr Sci 597:117623
30. Cao X, Xu H, Dong S, Xu J, Qiao Z, Zhao S, Wang J, Wang Z (2020) Preparation of high-performance and pressure-resistant mixed matrix membranes for CO_2/H_2 separation by modifying COF surfaces with the groups or segments of the polymer matrix. J Membr Sci 601:117882
31. Kammakakam I, Bara JE, Jackson EM (2020) Synthesis and characterization of imidazolium-mediated Tröger's base containing poly (amide)-ionenes and composites with ionic liquids for CO_2 separation membranes. Polym Chem 11(46):7370–7381
32. Yang Y, Han Y, Pang R, Ho W (2020) Amine-containing membranes with functionalized multi-walled carbon nanotubes for CO_2/H_2 separation. Membranes 10(11):333
33. Chen KK, Salim W, Han Y, Gasda M, Ho WW (2020) Fluoride-and hydroxide-containing CO_2-selective membranes for improving H_2 utilization of solid oxide fuel cells. J Membr Sci 612:118484
34. Gouveia AS, Yanez M, Alves VD, Palomar J, Moya C, Gorri D, Tomé LC, Marrucho IM (2021) CO_2/H_2 separation through poly (ionic liquid)–ionic liquid membranes: The effect of multicomponent gas mixtures, temperature and gas feed pressure. Sep Purif Technol 259:118113
35. Gouveia AS, Bumenn E, Rohtlaid K, Michaud A, Vieira TM, Alves VD, Tomé LC, Plesse C, Marrucho IM (2021) Ionic liquid-based semi-interpenetrating polymer network (sIPN) membranes for CO_2 separation. Sep Purif Technol 274:118437
36. Liu N, Cheng J, Hou W, Yang X, Zhou J (2021) Pebax-based mixed matrix membranes loaded with graphene oxide/core shell ZIF-8@ ZIF-67 nanocomposites improved CO_2 permeability and selectivity. J Appl Polym Sci 138(23):50553

37. Robeson LM (1991) Correlation of separation factor versus permeability for polymeric membranes. J Membr Sci 62(2):165–185
38. Robeson LM (2008) The upper bound revisited. J Membr Sci 320(1–2):390–400

Chapter 7
Facilitated Transport Membranes (FTMs) for CO_2 Separation from Flue Gas (CO_2/N_2)

Syed Shujaat Karim, Huma Shaikh, Sarah Farrukh, Saeed Ahmed Memon, Tahira Qureshi, and Shahabuddin Memon

Abstract Membrane technology is broadly considered as one of the best options for mitigating CO_2 emissions from the combustion or utilization of fuels that are major sources of flue gas. Various membranes with different compositions and functionalities have already been reported that enable extremely selective CO_2 transport by a facilitated transport mechanism. These membranes are called Facilitated Transport Membranes (FTMs). FTMs have been broadly used in a variety of industrial fields due to their extraordinary abilities for separation of targeted species. In spite of extensive research on FTMs, the trade-off relationship between their stability and separation efficiency has always been a challenge. However, it is worth mentioning that the trade-off between membrane permeability and selectivity for CO_2 separation from flue gas has surpassed the upper bound limit given in the Robeson graph. This progress has been achieved due to the fabrication of functionalized FTMs based on a variety of polymers, copolymers, nanofillers, mobile and fixed-site carriers, etc. Membrane technologists are also striving towards more robust FTMs for the separation of CO_2 so that they can be utilized commercially. In this chapter, a wide variety of FTMs are discussed based on their composition, functional properties, stability, operating conditions, and separation efficiency for CO_2 separation from the flue gas streams.

S. S. Karim · S. Farrukh (✉)
School of Chemical and Materials Engineering (SCME), National University of Sciences and Technology (NUST), Sector H-12, Islamabad, Pakistan
e-mail: sarah.farrukh@scme.nust.edu.pk

S. S. Karim
e-mail: sshujaat.che6scme@student.nust.edu.pk

H. Shaikh · S. A. Memon · S. Memon
National Center of Excellence in Analytical Chemistry (NCEAC), University of Sindh, Jamshoro, Sindh, Pakistan
e-mail: huma.shaikh@usindh.edu.pk

S. Memon
e-mail: shahabuddin@usindh.edu.pk

T. Qureshi
Government College University, Hyderabad, Sindh, Pakistan

S. Farrukh et al. (eds.), *Facilitated Transport Membranes (FTMs) for CO_2 Capture: Overview and Future Trends*, Green Energy and Technology,
https://doi.org/10.1007/978-3-031-21444-8_7

Keywords Facilitated transport membranes (FTMs) · Flue gas separation · CO_2/N_2 separation CO_2 capture · Mixed matrix membranes (MMMs) · Composite membranes

Nomenclature

AAO	Anodic alumina oxide
AFSNP	Amine surface-functionalized silica nanoparticles
[BMIM][BF_4]	1-Butyl-3-methylimidazolium tetrafluoroborate
Ca	Calcium
CAn	Carbonic anhydrase
CCS	Carbon capture storage
CMC	Carboxymethyl chitosan
CNT	Carbon nanotube
CO	Carbon monoxide
CO_2	Carbon dioxide
CS	Chitosan
DA	Dopamine
DAmBS	Sodium 3,5-diaminobenzoate
DETA	Diethylenetriamine
DGBAmE	Diethylene glycol bis(3-aminopropyl) ether
EDA	Ethylene diamine
EDC	1-Ethyl-3-(3-dimethylaminopropyl) carbodiimide
f-GO	Aminosilane functionalized graphene oxide
FSC-FTM	Fixed-site carrier facilitated transport membrane
FTM	Facilitated transport membrane
GA	Glutaraldehyde (crosslinking agent)
GO	Graphene oxide
GO-PA-Mg^{2+}	Graphene oxide-Phosphoric acid-Magnesium (metallic) ion
GPU	Gas permeance unit
H_2O	Water (Moisture content)
HCl	Hydrochloric acid
HDC	Hexamethylenetetramine
HT	Hydrotalcite
IL	Ionic liquid
IP	Interfacial polymerization
KOH	Potassium hydroxide
MC-FTM	Mobile carrier facilitated transport membrane
MEDA	N-methyldiethanolamine
MFC	Micro-fibrillated cellulose
mGO	Modified graphene oxide
MK	Amorphous layered metakaolin
MMM	Mixed matrix membrane

MMM	Mixed matrix membrane
MWCNTs	Multi-wall carbon nanotubes
MWCNTs-NH_2	Amine-modified multi walled carbon nanotubes
[m-2HEA](Pr)	N-methyl-2-hydroxyethylammonium propionate
N_2	Nitrogen
NaOH	Sodium hydroxide
NCs	Nano-capsules
NHS	N-hydroxysulfosuccinimide
NIPS	Non-solvent induced phase separation
NO_x	Nitrogen oxides
O_2	Oxygen
OH	Hydroxyl Group
P	Feed pressure
P(AAS-co-AAm)	Poly (sodium acrylate-co-acrylamide)
P(CO_2)	CO_2 Permeability (Barrer)/Permeance (GPU)
P(N_2)	N_2 Permeability (Barrer)/Permeance (GPU)
PAA	Poly(allylamine)
PAMAM dendrimer	Poly(amidoamine) dendrimer
PAN	Polyacrylonitrile
PCC	Post-combustion capture
PDA	Polydopamine
P_{DS}	Downstream pressure
Pebax	Polyether block amide
PEG	Poly ethylene glycol
PEG4	4Arm-poly(ethylene glycol)-NH_2
PEI	Poly(ethylene imine)
PEO	Polyethylene oxide
PES	Polyethersulfone
PM	Particulate matter
PNVF-*co*-VAm	Poly(Nvinylformamide-co-vinylamine)
Poly (VA-*co*-VAm)	Poly(vinylalcohol-*co*-vinylamine)
P_p	Permeate pressure
ppm	Parts per million
ppmw	Parts per million by weight
PSF	Polysulfone
PSS	Polystyrene sulfonate
PTFE	Polytetrafluoroethylene
PTFE	Polytetrafluoroethylene
PTMSP	Poly(1-trimethylsilyl-1-propyne)
P_{UPS}	Upstream stream pressure
PUU	Polyurethane urea (elastomer)
P_v	Vacuum pressure
PVA	Poly (vinyl) alcohol
PVA-*g*-POEM	Poly(vinyl alcohol)-g-poly(oxyethylene methacrylate)
PVDF	Polyvinylidene Fluoride

PVI	Poly (N-vinylimidazole)
PVP7-*co*-VAc3	Poly(1-vinylpyrrolidone-co-vinyl acetate) (7:3 mol ratio VP:VAc)
PZEA-Ala	2-(1-Piperazinyl)ethylamine-Alanine
PZEA-Lys	2-(1-Piperazinyl)ethylamine-Lysine
PZEA-Pro	2-(1-Piperazinyl)ethylamine-Proline
PZEA-Sar	2-(1-Piperazinyl)ethylamine sarcosinate
PZEA-Sar	2-(1-Piperazinyl)ethylamine-Sarcosine
R.H	Relative humidity
RNH_2	Primary amine group
SDS	Sodium dodecyl sulfate
SHPAA	Sterically hindered polyallylamine
SILM	Supported ionic liquid membrane
SL	Selective layer
SO_2	Sulfur dioxide
SO_x	Sulphur oxides
T	Temperature
TEPA	Tetraethylenepentamine
TETA	Triethylenetetramine
TiO_2	Titanium dioxide
TMC	Trimesoyl chloride
VAm-VOH	Vinyl amine-vinyl alcohol copolymer
wt%	Weight percent
ZIF	Zeolitic imidazolate framework
ZIF-8-d-MK	ZIF-8-decorated metakaolin hybrid
Zn	Zinc
α	Gas separation selectivity
α-C	α-Cellulose (0.05 mol of anhydrous glucose unit)

7.1 Introduction

Flue gas is a mixture of air pollutants that is released when fossil fuels such as oil, natural gas, coal, or wood are burned for heat and power generation. The average furnace output temperature of flue streams is often around 1200 °C, which steadily drops along the heat transfer pathway. The temperature of the flue gas streams flowing into the stack is generally up to 150 °C. It mainly contains carbon dioxide (CO_2), nitrogen (N_2), oxygen (O_2), moisture content (H_2O), and small amount of nitrogen oxides (NO_x), sulphur oxides (SO_x), carbon monoxide (CO), mercury content and particulate matter (PM). The major sources of flue gas are power plants and industrial facilities such as large refineries, natural gas plants, heavy oil plants, coal-fired plants, chemical plants, petrochemicals, and waste incineration plants, etc. Typically, the mercury content of various fuels ranges from 0.01 to 2 μg/g, producing flue gases with mercury concentrations of 1–200 $\mu g/m^3$ [1, 2]. Whereas, the composition of the different components present in the flue gas stream may vary with respect to its production source, as represented in Table 7.1 [2–4].

The flue gas greatly affects the quality of regional air. The treatment of flue gas gathered the attention of environmentalists back in the nineteenth century due to rising concerns about sulfates in the environment. Thus, the twentieth century was marked by the development of sophisticated devices and advanced methods to remove pollutants spread by flue gas. The flue gas pollution control treatment could efficiently remove different contaminate such as, PM, SO_x and NO_x. However, the environmental remediation from mercury content and CO_2 is still in the developing state. Deteriorating climate change is posing a serious threat to the world. Today, we are bound to stop energy production using fossil fuels to save ourselves from extinction. However, this will lead to a major economic and environmental crisis. Prevention of these crises is very difficult because the world's major energy supplies are fulfilled by fossil fuels. CO_2 is an unavoidable product produced from fossil fuels. As long as fossil fuels cannot be replaced, there is an urgent need to stabilize and limit the concentration of CO_2 in the atmosphere via advanced carbon capture and storage (CCS) methodologies. The amount of energy consumed by modern society is very large, which ultimately impacts on the environment and disrupts natural cycles. Therefore, it is important to recognize these impacts while engineering energy infrastructures. The most prominent impacts of energy consumption range from thermal pollution of rivers to acid rain and smog [8–10].

Recently, the release of CO_2 has also been added to the long list of environmental impacts caused by fossil energy resources. The concerns of environmentalists have resulted in the advancement of technology that ultimately led to the reduction in the pollution from fossil fuel-based power generation. However, the promising environmental remediation from large-scale production of CO_2 has not been achieved yet. The pretreatment of flue-gas is required for optimal performance of the post-combustion capture (PCC) of CO_2. However, this step is critical as all the PCC technologies are prone to damaging impacts from untreated flue gas [10]. Amine-based absorption is the most common method used for post-combustion CO_2 capture. Whereas, other conventional PPC technologies are also used for CO_2 separation from flue gas, which include pressure swing adsorption (PSA), temperature swing adsorption (TSA), and chemical looping methods. However, there are a number of drawbacks to these conventional technologies, such as these processes requires high energy, have large footprints and are difficult to scale-up due to their high capital and operational costs. In comparison to conventional PPC technologies for CO_2 separation, using membrane technology have several advantages, for example, it has low energy consumption, low operational cost, and requires less area as compared to the conventional PPC methods, It provide more flexibility in adjusting the process variable and can easily scalable. Furthermore, it can also formed a hybrid system by combining with other separation technologies. It is a clean technology, which does not requires any mass transfer agent for the CO_2 separation processes [11, 12].

In membrane technology, the facilitated transport membranes (FTMs) are gaining attention due to their excellent capability of efficient and selective separation of

Table 7.1 Typical composition data of flue gas streams produced from different sources and the physical properties of its components

Flue gas components	Natural gas plant (vol%)	Coal-fired plant (vol%)	Heavy oil plant (vol%)	Incineration plant (vol%)	Kinetic diameter (Å)	Molecular weight (g/mol)	References
Carbon dioxide (CO_2)	8–10	12–14	9–12	6–12	3.30	44	[2–5]
Moisture (H_2O)	18–20	8–10	8–10	10–18	2.65	18	[2–5]
Oxygen (O_2)	2–3	3–5	4–8	7–14	3.46	32	[2–5]
Nitrogen (N_2)	67–72	72–77	75–76	Balance	3.64	28	[2–5]
Sulfur dioxide (SO_2)	N.S	N.S	150–400 (ppm)	200–1500 (ppmw)	3.60	64	[2–6]
Nitrogen Oxides (NO_x)	N.S	N.S	100–200 (ppm)	200–500 (ppmw)	3.17 [a]	30[a]	[2–7]
Carbon monoxide (CO)	N.S	N.S	400–1000 (ppm)	0.001–0.06	3.76	28	[2–7]

[a] Value of nitrogen oxide (NO)

NS not specified

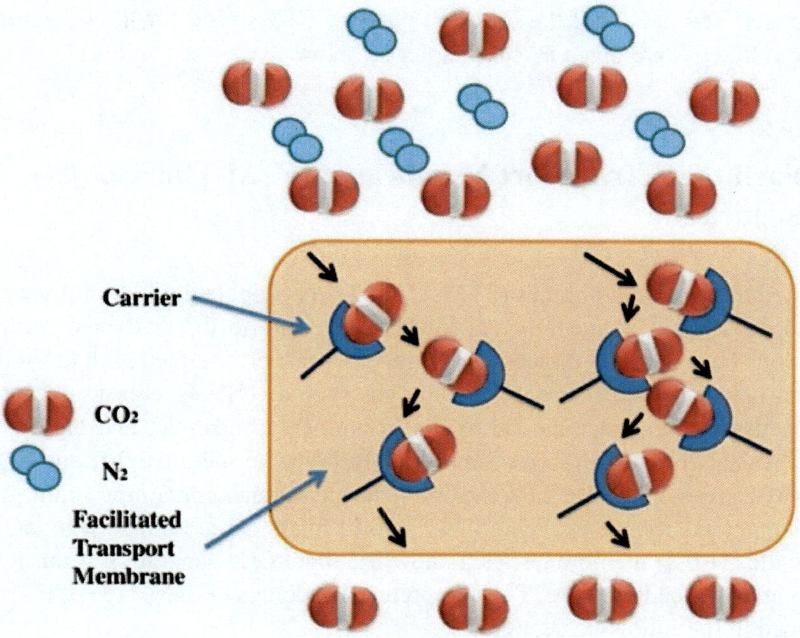

Fig. 7.1 Schematic representation of CO_2 separation from flue gas using FTM

target species from the gaseous mixture. They have been applied successfully in various industrial fields such as wastewater treatment, mining process, gas separation, and trace chemical detection, etc. [13–18]. The principle of FTMs is based on the presence of functional moieties that are also known as carriers. These carriers are able to interact with the target species specifically and reversibly to form transient complexes. The transient complexes are transported to the permeate side of the membrane and the carriers regenerated by releasing the target species via reversible reactions. Figure 7.1 shows the CO_2/N_2 separation process and mechanism of CO_2 permeation through FTM. The reaction kinetics at the membrane interface along with the transport rate of the transient complexes through the membrane mainly governs the CO_2 separation efficiency of FTMs. It is believed that the gas transport rate is much slower than the interfacial reaction rates, suggesting that the transport of the transient complexes is a critical factor in CO_2 separation process [19–22]. The FTMs typically undergo two different transport mechanisms in combination with the solution diffusion mechanism, based on their carrier mobility and physicochemical properties. These are (i) free motion of the complexes by mobile carriers and (ii) hopping of the targeted molecules via fixed carriers. The membranes that follow the former mechanism are supported liquid membrane (SLM) systems, and the membranes that follow the latter mechanism are polymer inclusion membrane (PIM) systems [23, 24]. Due to their outstanding properties, FTMs have been considered a most viable options in membrane systems for the CO_2 separation from flue gas streams. The details of the different types of FTMs used for the separation of CO_2 from flue gas are given in this chapter as below.

7.2 Facilitated Transport Membranes (FTMs) for CO_2/N_2 Separation

Facilitated transport membranes (FTMs) for CO_2/N_2 applications exhibit very high CO_2 selectivity and permeability at low pressure due to the swift and reversible interactions between CO_2 molecules and reactive carriers coupled with the solution diffusion mechanism. Other gases in flue gas such as N_2, O_2, etc. do not interact with reactive carriers such as carboxylic or amines groups, etc. and only depend on the physical diffusion process. Subsequently, high CO_2 permeability and CO_2/N_2 selectivity can be achieved at the same time. The membranes containing amine groups as carriers, metal complexes, nanofillers, hybrid fillers, ionic liquids (ILs) and poly ionic liquids (PILs) membranes, etc., show excellent CO_2 separation from flue gas streams at a low feed pressure. The comprehensive details of these type of FTMs are mentioned in the following sections.

7.2.1 Amine-Based Pure and Blended FTMs for CO_2/N_2 *Separation*

A series of polymeric FTMs have been fabricated by membrane researchers that contain amino groups such as polyethylenimine (PEI), polyallylamine (PAA), Chitosan (CS), and polyvinyl amine (PVAm). Among these, PVAm showed excellent results for the separation of CO_2 under low pressure due to the highest amino group density in its structure [25]. The problems like high carrier saturation and crystallinity restrict PVAm performance in CO_2 separation. By blending of PVAm with other polymers or small molecules results in multi-permselective FTMs with excellent gas separation performance. The multi-permselective membranes simultaneously offer diffusivity-selectivity, solubility-selectivity, and reactivity-selectivity for the separation of CO_2 from flue gas. These membranes show extraordinary selectivity and permeability for the CO_2 molecules. However, the crystallinity of the prepared FTMs can be effectively reduced by physical blending [26, 27].

The pure PVAm membrane at a thickness of 12 μm exhibits an excellent CO_2/N_2 selectivity of 48.50 with the CO_2 permeability of 213.90 Barrer at 0.10 feed pressure and temperature of 57 °C under humidified condition, in which Argon (Ar) was used as a sweep gas [28]. The membrane technology is advancing towards multi-permselective membranes. For instance, Wang et al. fabricated a number of composite membranes by adjusting structural properties using small molecules with copious amino groups that strengthened their multi-permselectivity. These membranes showed high performance for CO_2 separation [29, 30]. In order to overcome the drawback of carrier saturation, membranes with steric hindrance were also fabricated and evaluated [28, 31]. This strategy doubled the loading capacity of CO_2. It is because one CO_2 molecule interacts with one carrier (amine group) in membranes having steric hindered amine groups, while in membranes containing steric unhindered amine groups, one molecule of CO_2 interacts with two carriers and leads to saturation of the membrane [31]. The PVAm based FTMs are broadly divided into two categories on the basis of carrier type, such as: (i). Amine containing fixed site carriers (FSC) FTMs and (ii). Amine containing mobile site carriers (MC) FTMs. Moreover, the most advanced form of membranes is contain both fixed-site and mobile carriers. The CS, PAA, and PEI are typically used as secondary polymers because of their ability to form membranes efficiently.

One of the most dominant secondary polymer is polyvinyl alcohol (PVA) and its crossed-linked polymer, prepared by reacting the PVA with the glutaraldehyde (GA). The PVA contains abundant hydroxyl (OH) groups that allow defect-free formation of thin membranes and improved water uptake that ultimately benefits the amine-CO_2 reaction [31]. Polyethylene glycol (PEG) and its other derivatives were reportedly utilized as block copolymers with polyamines. Liu et al. [32], reported a blend of polyether block amide (Pebax MH-1657) and CS at a mass ratio of 1:1 that resulted in hydrogel membranes at a thickness of 42.40 μm, which showed superior CO_2 permeability of 2884 Barrer and a CO_2/N_2 selectivity of 65.30 at a 2 bar pressure and temperature of 85 °C under humidified condition at a relative humidity (R.H) of greater than 98%. The N_2 gas was used as a sweep gas in this gas permeation test. In another research by Yuan et al. [33] studied the effect of blending ethylene diamine (EDA) into the PVAm polymer and then coated on a porous polysulfone (PSF) support membrane. The fabricated composite membrane at a mass ratio (m_{EDA}/m_{PVAm}) of 3 having a selective layer thickness of 0.15 μm showed remarkable CO_2/N_2 selectivity of 106 with a CO_2 permeance of 607 GPU (CO_2 permeability = 91.05 Barrer) under a humidified mixed gas permeation testing system (CO_2/N_2 = 20/80 by volume) at room temperature and partial pressure of 0.2 bar, in which hydrogen (H_2) gas was used as a sweep gas.

Yuan et al. [34], used interfacial polymerization (IP) method for the preparation of TFC-FTMs, in which N-methyldiethanolamine (MEDA) and TMC were selected as the two monomers and interfacially polymerized them on the modified crosslinked PDMS layer coated on the PSF support membrane. In this research work, they optimized the growth process of film by changing the skin layer formation conditions, which was very important due to its significant role in the membrane performance for CO_2/N_2 separation process. The fabricated membranes were tested using the humidified mixed gas permeation testing system (CO_2/N_2 = 15/85 by volume) at a temperature of 25 °C and a variable feed pressure of 1.1–20 bar, in which humidified H_2 gas was used as a sweep gas. According to the results, the TFC FTM having a thickness of 0.528 μm, which was prepared with the 0.0615 mol L^{-1} MEDA and 0.0250 mol L^{-1} TMC showed the highest CO_2/N_2 selectivity of 87 with a CO_2 permeance of 1035 GPU (CO_2 permeability = 546.48 Barrer) at 0.165 bar CO_2 partial pressure. Whereas, at same operating condition, TFC FTM with a thickness of 0.439 μm synthesized with 0.0615 mol L^{-1} MEDA and 0.0100 mol L^{-1} TMC exhibited the maximum CO_2 permeance of 2905 GPU (CO_2 permeability = 1275.30 Barrer) and CO_2/N_2 selectivity of 64. Furthermore, due to the crosslinking, these membrane also showed an enhancement in plasticization resistance.

7.2.2 Other Types of Carriers-Based FTMs for CO_2/N_2 *Separation*

Although the majority of the research on FTMs for CO_2 separation and capture focused on amine-based carriers. However, the other carrier types such as carboxylate groups for the capture of CO_2 from flue gas streams have also gained a considerable attention. Wang et al. [35] reported a novel composite FTM having carboxylate groups as fixed carriers for CO_2 separation from flue gas streams. This composite membrane was synthesized by casting a 0.5 weight percent (wt%) poly (sodium acrylate-co-acrylamide)—P(AAS-co-AAm) solution of active layer on the PSF support membrane. The fabricated membrane exhibited excellent permselectivity for CO_2/N_2 separation with antioxidative properties, and was appropriate for CO_2 separation from flue gas streams containing oxidant such as O_2 gas. The resultant membrane at a thickness of 0.257 μm, showed a high CO_2/N_2 selectivity of around 74.00 and CO_2 permeability of 85.30 GPU (CO_2 permeability = 21.92 Barrer) at a temperature of 25 °C and pressure of 3 bar under humidified mixed gas permeation testing system (CO_2/N_2 = 15/85 by volume). In this process, water saturated H_2 gas was also used as a sweep gas in-order to prevent the increase in partial pressure of the gas at permeate side and also facilitate in the increase in driving force of CO_2/N_2 separation process.

In another research by Wang et al. [36], they prepared high performance composite FTMs by using the IP method for the application of the CO_2/N_2 separation process, which also have both high antioxidative as well as acid resistant properties. Trimesoyl chloride (TMC) and a mixture of sodium 3,5-diaminobenzoate (DAmBS) and Diethylene glycol bis(3-aminopropyl) ether (DGBAmE) at different concentration were interfacially polymerized on a crosslinked polydimethylsiloxane (PDMS) coated PSF substrate to produce TFC FTMs. The addition of DGBAmE provides ether oxygen group, which is a high affinity CO_2-philic group and can also help in the adjustment of the polymeric chain flexibility of the membrane. The optimized membrane at concentration of 0.004 mol L^{-1} DAmBS, 0.006 mol L^{-1} DGBAmE, and 0.003 mol L^{-1} TMC with the thickness of 0.817 μm exhibited an excellent CO_2 permeance of 5831 GPU (CO_2 permeability = 4763.92 Barrer) and a superior CO_2/N_2 selectivity of 86 at 1.1 bar feed pressure and temperature of 25 °C under a humidified mixed gas permeation testing system (CO_2/N_2 = 15/85 by volume), in which humidified H_2 gas was used as a sweep gas.

Carbonic anhydrase (CAn) an enzyme and its derivatives can also be utilized in FTM as they can catalyze the reversible hydration of CO_2 rapidly due to its active site consists of Zinc (Zn) (II) ions and have been applied for CO_2 separation process. Yao et al. [37], studied the effect of novel poly (N-vinylimidazole) (PVI)—Zn (II) complex blends by variating the PVI/Zn (II) mole ratio from 5 to 20 for the CO_2/N_2 separation application. These FTMs were evaluated using a mixed gas permeation testing system ($CO_2/N_2 = 15/85$ by volume) at a temperature of 25 °C and pressure range of 1–5 bar. H_2 was employed as the sweep gas, saturated with water vapour. The optimized membrane prepared at PVI/Zn (II) mole ratio of 10 with an pH of 5, having a thickness of 0.12 μm showed an high CO_2/N_2 selectivity of 83 with CO_2 permeance of around 1122 GPU (CO_2 permeability = 134.64 Barrer) at a pressure of 1 bar and temperature of 25 °C.

7.2.3 Recent Advancements in FTMs for CO_2/N_2 Separation

Recently, FTMs with high gas selectivity have attracted the interest of the scientific community for the separation of CO_2 from different industrial gas mixture streams. Among them, the most energy-intensive processes is the separation of CO_2 from N_2 from flue gas streams. However, to make these FTMs highly efficient solutions for industrial separation processes, their permeability is also needs to be addressed equally. It is because FTMs with excellent selectivity but with compromised permeability are not cost-effective. Different researches have been ongoing by the membrane scientists to enhance the CO_2 separation performance of FTMs in-order to utilize it for industrial applications. Some of the recently reported FTMs along with their CO_2/N_2 separation performance, material composition, operating conditions, module/configuration types and fabrication methods are represented in Table 7.2. In these membranes, FTMs having amine or amide functionalities have gained significant importance due to their outstanding performance in CO_2/N_2 separations.

The membranes made of less expensive polymers, such as poly (ethylene oxide) (PEO), can also produce good CO_2 separation results. However, their high molecular weight leads to the formation of defected membranes. In 2018, Lee et al. [38] proposed the fabrication of Flat-sheet PEO based mixed matrix FSC FTMs, that were composed of 1,000,000 g mol^{-1} molecular weight of PEO matrix, in which amine-branched poly-(amidoamine) (PAMAM) dendrimers were added as a nanofillers at filler loading range of 2.5–20 wt% using the solution mixing and casting method. The nanofillers were synthesized by using a two-step polymerization process: Michael addition and amidation reaction. The pure gas testing procedure was used to examine the PEO/PAMAM membrane's capabilities to separate CO_2/N_2 at a temperature of 35 °C with 1.01 bar of upstream pressure and 0.002 bar of downstream pressure. The CO_2 permeability and CO_2/N_2 selectivity of the pristine PEO membrane were found to be 34.20 Barrer and 7.5, respectively. With increasing PAMAM concentration up to 10 wt%, the CO_2 permeability decreased slightly while the CO_2/N_2 selectivity increased gradually. However, with a further increment in PAMAM concentration up

Table 7.2 Recent advancement data of FTMs for CO_2 capture from flue gas (CO_2/N_2)

Materials	Carriers/fillers	Fabrication method	FTMs types	Module/thickness	Operating conditions	$P(CO_2)$	$P(N_2)$	$\alpha \frac{(CO_2)}{(N_2)}$	References
PEO	PAMAM dendrimer (RNH_2) (2.5 wt%)	Michael addition polymerization and amidation reaction [a] Solution mixing and casting method [b]	FSC-FTM (MMM)	Flat-sheet	T = 35 °C P_{UPS} = 1.01 bar P_{DS} = 0.002 bar Pure gas	32.30 (Barrer)	0.77 (Barrer)	42.00	[38]
(15 wt%)—PVA-*g*-POEM/(0.5 wt%)—PTMSP/PSF	DETA (10 wt%)	Free radical polymerization, solution mixing and casting method [b]	FSC-FTM (Composite)	Flat-sheet (SL: 0.20–0.35 μm)	T = 25 °C P = 1–5 bar Pure gas	402.50–435.90 (GPU) 110.68–119.87 (Barrer)	7.70–7.80 (GPU) 2.11–2.14 (Barrer)	52.10–55.70	[39]
(10 wt%) Crosslinked PVA/PEG (1: 0.25)/PTFE	25 wt%—PEI/15 wt%—TETA/3.34 wt%—Silica	Silica nanoparticles: In situ sol–gel technique[a] Solution mixing and casting method with crosslinking reaction[b]	FSC-FTM (Composite)	Flat-sheet (27.40 μm)	T = 100 °C P = 2.5 bar Humidified Mixed gas CO_2/N_2 (20/80 vol%)	400.00 (Barrer)	1.90 (Barrer)	210.00	[40]
(70 wt%)—CS/PES	PAA (RNH_2) (30 wt%)	PAA:KOH mixing[a] Solution mixing and casting method[b]	FSC-FTM (Composite)	Flat-sheet (SL: 4.50 μm)	T = 90 °C P = 2 bar Humidified R.H = < 80% Mixed gas CO_2/N_2 (20/80%)	39.00 (GPU) 175.50 (Barrer)	0.15 (GPU) 0.675 (Barrer)	260.00	[41]
AAO (What man–Pore size: 0.2 μm)	GO/[BMIM][BF_4] (0.05 mg/mL)	Hummers' method, mixing and sonication[a] Solution mixing and vacuum filtration[b]	MC-Hybrid (SILM)	Flat-sheet (1.05 μm)	T = 50 °C P = 1 bar Pure gas	71.93 (Barrer)	0.188 (Barrer)	382.00	[42]

(continued)

Table 7.2 (continued)

Materials	Carriers/fillers	Fabrication method	FTMs types	Module/thickness	Operating conditions	$P(CO_2)$	$P(N_2)$	$\alpha_{\frac{(CO_2)}{(N_2)}}$	References
Pebax® MH 1657 (4 wt%)	NCs (20 wt%)	Stöber method[a] Solution mixing and casting method[b]	FSC-FTM (MMM)	Flat-sheet (75–95 μm)	T = 25 °C P = 2 bar Humidified Pure gas	1431.00 (Barrer)	20.74 (Barrer)	69.00	[43]
Pebax® MH 1657 (4 wt%)	NCs (30 wt%)	Stöber method[a] Solution mixing and casting method[b]	FSC-FTM (MMM)	Flat-sheet (75–95 μm)	T = 25 °C P = 2 bar Humidified Pure gas	1993.00 (Barrer)	33.21 (Barrer)	60.00	[43]
(14.4 wt%)—PNVF-*co*-VAm/PES	(3.9 wt%)—PVP7-*co*-VAc3/ (3 wt%)—MWCNTs/(81.7 wt%)—PZEA-Sar	M.C: PZEA reaction with base FSC: Acid treatment, amino-functionalization, polymer wrapping[a] Purification, solution mixing and casting method[b]	FSC-MC-hybrid FTM (Composite)	Flat-sheet (SL: 0.17 μm)	T = 57 °C P = 1–5 bar P_v = 0.20 bar Humidified (Water vapor: 17.2%) Mixed gas CO_2/N_2 (20/80%)	970.00–874.00 (GPU) 164.90–148.58 (Barrer)	5.95–5.98 (GPU) 1.011–1.016 (Barrer)	163.00–146.00	[44]
(14.4 wt%)—PNVF-*co*-VAm/PES	(3.9 wt%)—PVP7-*co*-VAc3/ (3 wt%)—MWCNTs/(81.7 wt%)—PZEA-Sar	M.C: PZEA reaction with base FSC: Acid treatment, amino-functionalization, polymer wrapping[a] Purification, solution mixing and casting method[b]	FSC-MC-Hybrid FTM (Composite)	Flat-sheet (SL: 0.17 μm)	T = 57 °C P = 4.05 bar P_v = 0.20 bar Mixed gas CO_2/N_2 (20/80%)	910.00 (GPU) 154.70 (Barrer)	5.94 (GPU) 1.009 (Barrer)	153.00	[44]

(continued)

Table 7.2 (continued)

Materials	Carriers/fillers	Fabrication method	FTMs types	Module/thickness	Operating conditions	$P(CO_2)$	$P(N_2)$	$\alpha\ \frac{(CO_2)}{(N_2)}$	References
α-C (1.05 g)	Ca–Zn (Ca: 2.38%) (Zn: 22.23%)	Facile and green solvent system, solution mixing and casting method[b]	FSC-FTM (MMM)	Flat-sheet (41.00 μm)	T = 25 °C P = 2–4 bar Pure gas	155.00–166.20 (Barrer)	5.70–5.50 (Barrer)	27.20–30.20	[45]
Pebax® MH 1657 (4.5 wt%)	f-GO (0.7 wt%)	Hummer's method and functionalization[a] Solution mixing and casting method[b]	FSC-FTM (MMM)	Flat-sheet (70–90 μm)	T = 35 °C P = 2 bar Pure gas	172.00 (Barrer)	3.80 (Barrer)	45.20	[46]
Pebax® MH 1657 (4.5 wt%)	f-GO (0.9 wt%)	Hummer's method and functionalization[a] Solution mixing and casting method[b]	FSC-FTM (MMM)	Flat-sheet (70–90 μm)	T = 35 °C P = 2 bar Humidified Mixed gas CO_2/N_2 (20/80 vol%)	934.30 (Barrer)	13.14 (Barrer)	71.10	[46]
(5.0 wt%)—Pebax® MH 1657/PVDF	DA-PEI-TiO_2 (3.0 wt%)	Non-aqueous solvothermal process, facile chelation and Schiff base reaction[a] Solution mixing and casting method[b]	FSC-FTM (Composite)	Flat-sheet	T = 20 °C P = 3 bar Pure gas	67.70 (Barrer)	0.67 (Barrer)	101.00	[47]
Poly(VA-*co*-VAm)/PSf	MWCNTs-NH_2 (2.0 wt%)	Poly(VAm-VOH): Free-radical polymerization, basic and acidic hydrolysis, and ion exchange method Composite: Solution mixing and casting method[b]	FSC-FTM (Composite)	Flat-sheet (SL: 3 μm)	T = 28 °C P = 2 bar Humidified Mixed gas CO_2/N_2 (20/80 vol%)	132.00 (GPU) 396.00 (Barrer)	1.78 (GPU) 5.34 (Barrer)	74.00	[48]

(continued)

Table 7.2 (continued)

Materials	Carriers/fillers	Fabrication method	FTMs types	Module/thickness	Operating conditions	$P(CO_2)$	$P(N_2)$	$\alpha_{\frac{(CO_2)}{(N_2)}}$	References
(90 wt%)—CMC/PES	PAMAM dendrimer (RNH_2) (10 wt%)	Solution mixing and casting method[b]	FSC-FTM (Composite)	Flat-sheet (SL: 2.64 μm)	T = 90 °C P = 2 bar Humidified Mixed gas CO_2/N_2 (20/80 vol%)	98.00 (GPU) 258.72 (Barrer)	0.65 (GPU) 1.716 (Barrer)	149.00	[49]
(2.5 wt%)—(SHPAA/PVA) (9: 1)/PVDF	(30 wt%)—MFC-PEG4/(40 wt%)—PZEA-Sar	Purification, carboxymethylation, EDC/NHS activation reaction[a] Solution mixing and casting method[b]	MC Hybrid-FTM (Composite)	Flat-sheet (SL: 0.30 μm)	T = 35 °C P_{UPS} = 1.7 bar Humidified R.H = 100% Mixed gas CO_2/N_2 (90/10 vol%)	652.00 (GPU) 195.60 (Barrer)	15.78 (GPU) 4.734 (Barrer)	41.30	[50]
(2.5 wt%)—PVAm/PSf	MK (1.0 wt%)	Solution mixing and casting method[b]	FSC-FTM (Composite)	Flat-sheet (SL: 2.37 μm)	T = 25 °C P_{UPS} = 1 bar Humidified R.H = 100% Mixed gas CO_2/N_2 (20/80 vol%)	186.90 (GPU) 443.00 (Barrer)	2.27 (GPU) 5.37 (Barrer)	82.40	[51]
(89 wt%)—CMC/PES	(10 wt%)—PAMAM dendrimer/(1 wt%)—HT	Solution mixing and casting method[b]	FSC-FTM (Composite)	Flat-sheet (SL: 1 μm)	T = 90 °C P = 2 bar Humidified Mixed gas CO_2/N_2 (20/80 vol%)	123.00 (GPU) 123.00 (Barrer)	1.83 (GPU) 1.83 (Barrer)	67.00	[52]

(continued)

Table 7.2 (continued)

Materials	Carriers/fillers	Fabrication method	FTMs types	Module/thickness	Operating conditions	$P(CO_2)$	$P(N_2)$	$\alpha_{\frac{(CO_2)}{(N_2)}}$	References
PES	PDA-modified CNT (180 nm)/(30 (mg. mL^{-1})—PSS/(50 mg. mL^{-1})—TEPA	PDA- modified CNT substrate: vacuum filtration method[a] Solution mixing and casting method[b]	FSC-FTM (Composite)	Flat-sheet (SL: 0.22 μm) (CNT network: 0.18 μm)	T = 90 °C P = 1 bar P_p = 0.30 bar Humidified Mixed gas CO_2/N_2 (15/85 vol%)	3250.00 (GPU) 1300 (Barrer)	8.12 (GPU) 3.25 (Barrer)	400.00	[53]
PUU (10 wt%)	HDC (2.5 wt%)	PUU: Two-step polymerization method MMM: Solution mixing and casting method[b]	FSC-FTM (MMM)	Flat-sheet (120 μm)	T = 30 °C P = 10 bar Pure gas	140.00 (Barrer)	4.60 (Barrer)	30.40	[54]
PUU (10 wt%)	HDC (2.5 wt%)	PUU: Two-step polymerization method MMM: Solution mixing and casting method[b]	FSC-FTM (MMM)	Flat-sheet (120 μm)	T = 30 °C P = 10 bar Mixed gas CO_2/N_2 (50/50 vol%)	102.60 (Barrer)	4.20 (Barrer)	24.40	[54]
(15 wt%)—PVAm/PES	(65 wt%)—PZEA-Pro/(20 wt%)—PZEA-Sar	Deprotonating reaction[a] PES substrate: NIPS method, composite: Solution mixing and casting method[b]	MC-Hybrid FTM (Composite)	Flat-sheet (SL: 0.17 μm)	T = 57 °C P = 0.16 bar Humidified Water vapor = 17.20% Mixed gas CO_2/N_2 (20/80 vol%)	947.00 (GPU) 160.99 (Barrer)	4.50 (GPU) 0.76 (Barrer)	210.00	[55]

(continued)

Table 7.2 (continued)

Materials	Carriers/fillers	Fabrication method	FTMs types	Module/thickness	Operating conditions	$P(CO_2)$	$P(N_2)$	$\alpha_{\frac{(CO_2)}{(N_2)}}$	References
(35 wt%)—PVAm/PES	PZEA-Pro (65 wt%)	Deprotonating reaction[a] PES Substrate: NIPS method, Composite: Solution mixing and casting method[b]	MC-Hybrid FTM (Composite)	Flat-sheet (SL: 0.17 µm)	T = 57 °C P = 0.16 bar Humidified Water vapor = 17.20% Mixed gas CO_2/N_2 (20/80 vol%)	936.00 (GPU) 159.12 (Barrer)	4.84 (GPU) 0.82 (Barrer)	193.00	[55]
(35 wt%)—PVAm/PES	PZEA-Ala (65 wt%)	Deprotonating reaction[a] PES Substrate: NIPS method, Composite: Solution mixing and casting method[b]	MC-Hybrid FTM (Composite)	Flat-sheet (SL: 0.17 µm)	T = 57 °C P = 0.16 bar Humidified Water vapor = 17.20% Mixed gas CO_2/N_2 (20/80 vol%)	813.00 (GPU) 138.21 (Barrer)	4.23 (GPU) 0.71 (Barrer)	192.00	[55]
(35 wt%)—PVAm/PES	PZEA-Lys (65 wt%)	Deprotonating reaction[a] PES substrate: NIPS method, composite: Solution mixing and casting method[b]	MC-Hybrid FTM (Composite)	Flat-sheet (SL: 0.17 µm)	T = 57 °C P = 0.16 bar Humidified Water vapor = 17.20% Mixed gas CO_2/N_2 (20/80 vol%)	832.00 (GPU) 141.44 (Barrer)	4.42 (GPU) 0.75 (Barrer)	188.00	[55]

(continued)

Table 7.2 (continued)

Materials	Carriers/fillers	Fabrication method	FTMs types	Module/thickness	Operating conditions	$P(CO_2)$	$P(N_2)$	$\alpha \frac{(CO_2)}{(N_2)}$	References
(2.5 wt%)—PVAm/PSf	ZIF-8-d-MK (3 wt%)	Sonication, mixing and centrifugation[a] Solution mixing and casting method[b]	FSC-Hybrid FTM (Composite)	Flat-sheet (SL: 1.85 μm)	T = 25 °C P = 1 bar Humidified (R.H = 100%) Pure gas	169.00 (GPU) 312.65 (Barrer)	1.95 (GPU) 3.60 (Barrer)	86.70	[56]
(2.5 wt%)—PVAm/PSf	ZIF-8-d-MK (3 wt%)	Sonication, mixing and centrifugation[a] Solution mixing and casting method[b]	FSC-Hybrid FTM (Composite)	Flat-sheet (SL: 1.85 μm)	T = 25 °C P = 1 bar Humidified (R.H = 100%) Mixed gas CO_2/N_2 (20/80 vol%)	245.00 (GPU) 453.25 (Barrer)	2.50 (GPU) 4.62 (Barrer)	98.00	[56]
PVA (10 wt%)	[m-2HEA]-(Pr) (45.5 wt%)	Amine neutralization reaction[a] Solution mixing and casting method[b]	MC-Hybrid FTM (IL-MMM)	Flat-sheet (159 μm)	P = 6 barg Pure gas	25.60 (Barrer)	0.12 (Barrer)	213.00	[57]
(19.15 wt%)—PES/(15.32 wt%)—PEG/(0.38 wt%)—DA	PEI–TMC (molar ratio = 1:1)/(0.3 wt%)—SDS/(0.1 wt%)—mGO	mGO: Treating GO with NaOH and HCl[a] Spinning process and interfacial polymerization[b]	FSC-Hybrid FTM (Composite)	Hollow-Fiber (SL: 0.113 μm)	T = 25 °C P = 0.25 bar Pure gas	73.00 (GPU) 8.25 (Barrer)	1.21 (GPU) 0.13 (Barrer)	60.00	[58]
PVAm (5 wt%)	UiO-66-NH_2-PVA (24 wt%)	Hydrothermal synthesis method[a] Solution mixing and casting method[b]	FSC-FTM (MMM)	Flat-sheet (40 μm)	T = 25 °C P = 1.01 bar Humidified Mixed gas CO_2/N_2 (50/50 vol%)	76.15 (Barrer)	1.66 (Barrer)	45.76	[59]

(continued)

Table 7.2 (continued)

Materials	Carriers/fillers	Fabrication method	FTMs types	Module/thickness	Operating conditions	$P(CO_2)$	$P(N_2)$	$\alpha \frac{(CO_2)}{(N_2)}$	References
Pebax® MH 1657 (3.0 wt%)	Aniline (50 wt%)	Solution mixing and casting method[b]	FSC-MC Hybrid FTM (MMM)	Flat-sheet (63 μm)	T = 25 °C P = 7 bar Pure gas	151.00 (Barrer)	1.63 (Barrer)	92.50	[60]
Pebax® MH 1657 (3.0 wt%)	Aniline (50 wt%)	Solution mixing and casting method[b]	FSC-MC Hybrid FTM (MMM)	Flat-sheet (63 μm)	T = 25 °C P = 7 bar Mixed gas CO_2/N_2 (20/80 vol%)	123.12 (Barrer)	1.80 (Barrer)	68.34	[60]
(38 wt%)—PVA/(9 wt%)—PEG/(25 wt%)—PEI/(15 wt%)—TETA/PTFE	AFSNP (3.0 wt%)	Stöber process and amine functionalization[a] Solution mixing, crosslinking reaction and casting method[b]	FSC-Hybrid FTM (Composite)	Flat-sheet (SL: 27 μm)	T = 100 °C P = 2.5 bar Humidified Mixed gas CO_2/N_2 (20/80 vol%)	927.00 (Barrer)	2.85 (Barrer)	325.00	[61]
PAN	GO-PA-Mg^{2+} (0.08 g/m^2)	Post-immersion method (soaking process)[a] Vacuum filtration method[b]	FSC-FTM (Composite)	Flat-sheet (SL: 67 μm)	T = 25 °C P = 1 bar Humidified Mixed gas CO_2/N_2 (50/50 vol%)	23.00 (GPU) 1541 (Barrer)	0.47 (GPU) 31.49 (Barrer)	48.80	[62]

[a] Fabrication method for carrier
[b] Fabrication method for FTMs

to 20 wt%, the membrane showed a drastic decrement in CO_2/N_2 separation performance. This trend of gas separation performance attributed the excess amount of PAMAM hydrogen-bonding interactions with the PEO polymer, which increased the resistance of gas diffusion within the membrane matrix. At 2.5 wt% of filler loading, the PEO/PAMAM-based FTM showed the most optimum CO_2/N_2 selectivity of 42 with a CO_2 permeability of 32.30 Barrer [38].

In another study by Kim et al. [39] prepared different poly(vinyl alcohol)-g-poly(oxyethylene methacrylate) (PVA-g-POEM) based FSC composite FTMs by blending the said graft copolymer at concentration of 15 wt% with diethylenetriamine (DETA) as carrier at a filler loading range from 5 to 30 wt%. The active layer were casted on the 0.5 wt% poly(1-trimethylsilyl-1-propyne) (PTMSP)/PSF support substrate at a thickness of 0.20–0.35 μm. The permeation of pure gases such as CO_2 and N_2 through these composite FTMs was measured under a variable pressure range of 1–5 bar and at a constant temperature of 25 °C. According to the results, these membranes showed a higher CO_2/N_2 separation performance, which was caused by amine carriers facilitated transport of CO_2 and this response was occurred at 10 wt% of DETA loading. However, the CO_2 solubility on amine carriers was decreased by the increased DETA (above 10 wt%), which prevented the CO_2 permeance from increasing. The graft copolymer crystallinity was significantly enhanced with the addition of 10 wt% DETA, which reduced the permeance of N_2 gas. As a result, the CO_2/N_2 selectivity of the PVA-g-POEM/DETA-10 membrane was 52.10 and it had a high CO_2 permeance of 402.50 GPU (CO_2 permeability of 110.68 Barrer) [39].

Barooah and Mandal [40] studied the effect of silica nanoparticles (SNPs) loading on the CO_2/N_2 separation performance of crosslinked poly(vinyl alcohol) (PVA)/poly(ethylene glycol) PEG based composite FTMs. The silica nanoparticles (SNPs) were synthesized by the in-situ sol–gel technique and the composite membranes composed of a varied SNS content of 0–6.68 wt%, incorporated into the 10 wt% PVA/PEG blend were fabricated by using the solution mixing and casting method with crosslinking reaction. In these membranes, PTFE membrane was used as the support layer and has a selective layer thickness of around 27 μm. According to the mixed gas permeation experiment under humidified conditions, the temperature and sweep side water flow rate showed a significant influence on the CO_2 permeability and CO_2/N_2 selectivity. Argon (Ar) gas was used a sweep gas in this system. It was found that when the operating temperature was increased from 90 to 100 °C, the CO_2/N_2 separation performance was first increased up to 100 °C and then gradually dropped. This may be explained by the fact that when temperature increased, the membrane's water content decreased, which in turn reduced the solubility of CO_2 gas. When the sweep water content was raised to 0.04 mL/min, it was found that the CO_2/N_2 separation performance significantly improved. By increasing the silica concentration up to 3.34 wt% the composite membrane showed an excellent improvement of CO_2 permeability by 78% and CO_2/N_2 selectivity by 45% as compared to the pristine membrane at optimal operating conditions. The 3.34 wt% of SNPs based composite membrane showed the remarkable CO_2/N_2 selectivity of 210 with a high CO_2 permeability of 400 Barrer at a 2.5 bar pressure and temperature of 100 °C [40].

Another research study by Prasad and Mandal [41] claimed that FTMs produced via blending of poly(allylamine) (PAA) and chitosan (CS) showed excellent performance for CO_2/N_2 separation. The PAA concentration was varied from 15 to 45 wt% in the CS matrix to form different blend ratios of PAA-CS layers. Afterwards, these selective layers were coated on the surface of PES support membranes at a thickness of 4.50 μm for the preparation of composite based FSC-FTMs using the solution casting method. The CO_2 separation from binary gas (CO_2/N_2) was performed under humidified condition (R.H = < 80%) at a temperature of 90 °C and pressure of 2 bar. According to the gas permeation data, blended FTM demonstrated high CO_2/N_2 selectivity of 260 with a CO_2 permeance of 39 (CO_2 permeability = 175.50 Barrer) at an optimal concentration of 30 wt% PAA in chitosan matrix (PAA30), which outperformed the pristine CS membrane by ~ 3.81 fold and ~ three-fold, respectively. The amine groups in PAA/CSS blended composite FTM served as FSC for the facilitated transport of CO_2 molecules in combination with the solution diffusion mechanism. However, when the PAA concentration was raised over 30 wt%, the membrane's surface started to form cracks, which caused a decrement in the CO_2/N_2 selectivity [41].

Among the nanofillers, graphene oxide (GO) is the one of the most widely explored nanofillers due to its oxygen functionalities and easy modification of these functionalities. Ying et al. [42] prepared a GO-IL based support immobilized liquid membrane (SILM) using the vacuum filtration method for the application of CO_2/N_2 separation. For the fabrication of GO-IL SILM, GO was initially synthesized by using the Hummer's method and then 0.05 mg/ml of 1-butyl-3-methylimidazolium tetrafluoroborate—[BMIM][BF_4] IL was confined into the 2-dimensional (D) nano-channels of laminated GO for the rapid facilitated transport of CO_2 molecules. In this SILM, anodic alumina oxide (AAO) membrane was used as support substrate. The GO-[BMIM][BF_4]-SILM at a thickness of 1.05 μm showed the excellent transportation of CO_2 with a permeance of 68.5 GPU (CO_2 permeability = 71.93 Barrer) and CO_2/N_2 selectivity of 382 under a pure gas testing system at a pressure of 1 bar and a high temperature of 50 °C, which surpassed the revised upper bound limit of CO_2/N_2 separation [42].

In another study by Zhang et al. [43] a novel type of nano-capsules (NCs) containing sufficient carboxylic acid groups were synthesizes using the Stöber method. After that, the Pebax-based mixed matrix FTMs were fabricated at a thickness in range of 75–95 μm by loading NCs as nano-fillers at a variable concentration of 10–30 wt% by using the solution mixing and casting method. The membranes were then examined using the humidified pure gas permeation testing system at a constant pressure of 2 bar and a temperature of 25 °C. According to the gas permeation results, the CO_2 permselectivity of MMMs incorporated with 20 wt% of NCs showed the highest CO_2/N_2 selectivity of 69 with a CO_2 permeability of 1431 Barrer. Whereas 30 wt% NCs demonstrated the best CO_2 permeability of 1993 Barrer with a slightly lower CO_2/N_2 selectivity of 60 when compared to the Pebax FTM based on 20 wt% NCs. The gas separation performance of both these membranes exceeded the 2008 Robeson upper bound limit of CO_2/N_2 separation. The primary factor for the improved results of these FTMs was the hollow structure of the NCs, which offers the interconnected transport networks channels provided with sufficient carboxylic acid groups and absorbed water for the rapid facilitated transport of CO_2 molecules in combination with the solution diffusion mechanism [43].

Han et al. [44] prepared a novel copolymer wrapped MWCNTs based hybrid composite FTM for high separation of CO_2 from flue gas streams. This composite FTM was made up of PES as support membrane and a selective layer composed of 3.9 wt% of poly(1-vinylpyrrolidone-*co*-vinyl acetate) (PVP7-*co*-VAc3)/(3 wt%)—MWCNTs and 81.7 wt% of 2-(1-piperazinyl)ethylamine—Sarcosine (PZEA-Sar) based fillers, which were incorporated in the 14.4 wt% poly(N-vinylformamide-*co*-vinylamine)—PNVF-*co*-VAm polymeric matrix. The membrane was tested using a mixed gas testing system (CO_2/N_2 = 20/80 vol%) under humidified and dry conditions at a 57 °C temperature with the feed and vacuum pressure of 1–5 bar and 0.20 bar, respectively. Argon (Ar) was used as sweep gas. The mixed gas permeation analysis under humidified condition, revealed that the 3 wt% copolymer wrapped MWCNTs FTM having a selective layer thickness of 0.17 μm showed a high CO_2/N_2 selectivity of 162 with an CO_2 permeance of 970 GPU (CO_2 permeability = 164.90 Barrer) at a feed pressure of 2 bar. However, further increasing the pressure up to 5 bar reduces the CO_2/N_2 separation performance of the membrane due to the membrane compaction and carrier saturation phenomenon. While in the case of dry condition, the membrane demonstrated a stable performance of CO_2/N_2 selectivity value of 153 with a CO_2 permeance of 910 GPU (CO_2 permeability = 154.70 Barrer) over a testing period of 120 h at a feed pressure of 4 bar. This excellent gas separation performance was due to the PZEA-Sar and amino groups of PNVF-*co*-VAm, which act as MC and FSC in the facilitated transport process in combination with the solution diffusion mechanism within the membrane matrix. Moreover, this MWCNT-reinforced FTM have the capability to be used under crucial working conditions designed for 90% carbon capture from coal-derived flue gas streams [44].

Zhang et al. [45], studied the effect of zinc ions (Zn^{2+}) concentration in cellulose membrane for the CO_2/N_2 separation performance. The mixed matrix membranes (MMMs) were prepared at a thickness of 41 μm using the facile and green solvent system. The fabricated membranes were tested using the pure gas testing system at a pressure range of 2–4 bar and a constant temperature of 25 °C. The CO_2 permeability of the metal-free cellulose membrane was 23.50 Barrer, and the maximum CO_2/N_2 selectivity of 4.60 was achieved. The incorporation of Zn^{+2} in the FTMs leads to the improvement of CO_2/N_2 separation performance. The MMM having the Zn^{2+} concentration of 22.20% showed the most optimum CO_2/N_2 selectivity of 27.20 with an CO_2 permeability of 155 Barrer at a pressure of 2 bar. Whereas, by increasing the pressure up to 4 bar, the CO_2/N_2 separation performance of this membrane was further enhanced, which demonstrated the higher CO_2/N_2 selectivity of 30.20 with a corresponding CO_2 permeability of 166.20 Barrer. The π-complexation process between Zn^{2+} and CO_2 in the facilitated transport process is attributed to these good CO_2/N_2 separation performances [45].

In 2019, Zhang et al. [46] developed the aminosilane-functionalized graphene oxide (f-GO) based FTMs for the enhanced performance of CO_2 separation from the flue gas streams. The f-GO nano sheets were prepared by Hummer's procedure and functionalization reaction to graft 3-aminopropyltriethoxysilane (APTS) with the oxygen containing functional groups of GO nanosheets. Afterwards, by embedding these f-GO nanosheets at a concentration range of 0.3–0.9 wt% into 4.5 wt% Pebax MH-1657 membrane matrix, different f-GO based FTMs were fabricated using the solution mixing and casting method. Following that, different f-GO-based FTMs were prepared by using the solution mixing and casting method, in which the f-GO nanosheets at concentrations ranging from 0.3 to 0.9 wt% were embedded into a 4.5 wt% Pebax MH-1657 membrane matrix. These flat-sheet FTMs have a thickness in the range of 70–90 μm. The gas separation performance of these membranes was evaluated using a pure gas and a humidified mixed gas ($CO_2/N_2 = 20/80$ by vol.%) as the feed gas in the gas permeation testing system at a pressure range of 0.5–10 bar and a constant temperature of 35 °C. According to the gas permeation results, these membranes performed more effectively in a humidified condition than in a dry condition. The Pebax-FTM containing 0.9 wt% of f-GO exhibited the highest CO_2/N_2 selectivity of 71.1 and optimum CO_2 permeability of 934.30 Barrer under the humidified mixed gas testing system at a pressure of 2 bar, which surpassed the revised Robeson's upper bound limit of CO_2/N_2 separation. While at same operating conditions, under a dry condition pure gas permeation testing system, the Pebax-FTM having 0.7 wt% of f-GO showed a maximum CO_2/N_2 selectivity of 45.20 and a CO_2 permeability of 172 Barrer. In these membranes, the addition of f-GO reduced the crystallinity and improved polymeric chain mobility of the Pebax matrix. The amino groups attached to the GO provides the rapid facilitated transport channels for the CO_2 molecules at the polymer-filler interface, which significantly increased the CO_2/N_2 separation performance under a dry condition. While, under a humidified conditions due to the presence of water content, the facilitated transport process of CO_2 molecules within the FTM was more enhanced resulting in superior gas separation performance as compared to the dry condition [46].

Zhu et al. [47] fabricated the amine functionalized titanium (TiO_2) nanofiller-based composite FTMs for the enhanced CO_2/N_2 separation process. The TiO_2 nanoparticles were synthesized by the facile chelation method, followed by a modification process by grafting amine groups using the Schiff base reaction with dopamine (DA) and polyethyleneimine (PEI) to produce DA-PEI-TiO_2 nanoparticles. These nano-fillers were then loaded at a range of 1–4 wt% into the 5.0 wt% Pebax-1657 based selective layer solutions, which were then coated on the PVDF support membrane for the fabrication of different concentrations of DA-PEI-TiO_2 based Pebax/PVDF composite membranes. The gas permeation analysis showed that the pure Pebax membranes, CO_2 permeability and CO_2/N_2 selectivity were both improved by the addition of modified TiO_2 nanoparticles up to 3 wt% due to the enhanced CO_2 sorption capacity and rapid transport path provided by the grafted amine groups. The highest CO_2/N_2 selectivity of 101 and optimum CO_2 permeability of 67.7 Barrer were achieved by the Pebax composite membrane loaded with 3 wt% of DA-PEI-TiO_2 nanoparticles under a pure gas permeation system at a 3 bar pressure and temperature of 20 °C, which surpassed the revised Robeson upper bound limit. However, adding this nanofiller over 3 wt% led to defects in the membrane matrix, which reduced the composite membrane's ability to separate CO_2 from N_2 [47].

Zhang et al. [48] prepared the amine functionalized multi-wall carbon nanotubes (MWCNTs-NH_2)-based composite FTMs for the high-performance CO_2/N_2 separation process. In these FTMs, MWCNTs-NH_2 nanofillers at a concentration of 0.5 to 3.0 wt% were blended with the 3.0 wt% poly(vinylalcohol-*co*-vinylamine) (VAm-*co*-VOH) solutions, which were then coated on the PSF support membranes at a thickness of (3.0–3.3) μm via a using solution mixing and casting method to prepared composite FTMs. The CO_2/N_2 separation performance of these membranes was determined using the humidified mixed testing system under a pressure of 2–14 bar and a temperature of 25–50 °C, in which a flue gas mixture of (CO_2/N_2 = 20/80 vol%) and H_2 gas were used as feed gas and sweep gas, respectively. The gas permeability analysis results revealed that, when the feed gas pressure and temperature were increased, a decreasing trend was observed in both CO_2 permeance and CO_2/N_2 selectivity. The composite FTMs, which contained 2 wt% of MWCNTs-NH_2 nanofiller, showed the highest CO_2/N_2 of 74 with an optimum CO_2 permeance of 132 GPU (CO_2 permeability = 396 Barrer) at 2 bar pressure and a temperature of 25 °C. This excellent gas separation performance was due to the amine carriers attached on the MWCNTs, which provide a rapid facilitated transport mechanism for CO_2 molecules in the presence of water molecules with the combination of physical diffusion process. However, the transport of N_2 within the membrane matrix only depends on the solution diffusion mechanism. Furthermore, this composite membrane also demonstrated good stability when tested with CO_2/N_2 mixed gas for 110 h at a pressure of 5.4 bar [48].

Borgohain and Mandal [49] investigated the performance of Carboxymethyl Chitosan (CMC)-PAMAM dendrimer/PES based composite FTMs for the separation of mixed gases (CO_2/N_2 = 20/80 by vol%) at different temperatures (60–110 °C) and sweep/feed water flow ratios (0.33–3). These FTMs were fabricated by varying PAMAM dendrimer filler loading in the range of 10–30 wt% using the solution mixing and casting method. The gas permeation results revealed that by increasing the filler loading from 5 to 10 wt%, high CO_2/N_2 separation performances were observed. However, above this filler concentration, the FTM showed a decreasing trend in CO_2/N_2 selectivity with an improvement in CO_2 permeability. The composite membrane with 10 wt% PAMAM dendrimer having the selective layer thickness of 2.64 μm showed the highest CO_2 permeance of 98 GPU (CO_2 permeability = 258.72 Barrer) and CO_2/N_2 selectivity of 149 on the constant sweep/feed water flow ratios of 2.33 and 1.67, respectively, at temperature of 90 °C and pressure of 2 bar. This remarkable gas separation performance can be attributed to the salting-out effect provided by the CMC matrix and the PAMAM-based nanofiller molecular gate mechanism [49].

Janakiram et al. [50] prepared a nanocellulose-based hybrid composite FTM for high performance CO_2 capture from flue gas streams. The MFC-PEG4/PZEA-Sar nanofiller was synthesized using the carboxymethylation and 1-ethyl-3-(3-dimethylaminopropyl) carbodiimide (EC)/N-hydroxysulfosuccinimide (NHS) activation reaction. The blend of 2.5 wt% sterically hindered polyallylamine

(SHPAA)/PVA was used as a polymeric matrix, in which 30 wt% microfibrillated cellulose (MFC)-4arm-poly(ethylene glycol)-NH_2 (PEG4)/(40 wt%)—2-(1-piperazinyl) ethylamine sarcosinate (PZEA-Sar) filler was embedded to prepare the selective layer solution, which was then coated on the surface of PVDF support membrane at a thickness of 0.30 μm using solution casting method. The nanocellulose-based membrane was tested using the mixed gas testing system ($CO_2/N_2 = 90/10$ vol%) under humidified conditions (R.H = 100%) at an upstream pressure of 1.7 bar and a temperature of 35 °C. According to the gas permeation analysis, this composite membrane demonstrated the highest CO_2/N_2 selectivity of 41.30 with a CO_2 permeance of 652 GPU (CO_2 permeability = 195.60 Barrer). This hybrid fillers served as both MC and FSC due to the presence of PZEA-Sar and amine carriers, respectively. Additionally, it provided sites for water retention, which increased the CO_2 solubility, thus providing a rapid diffusion channels for the facilitated transport of CO_2 molecules coupled with the solution diffusion mechanism [50].

In 2020, Wang et al. [51] reported high performance mixed matrix composite based FSC-FTMs for CO_2/N_2 separation. These membranes were supported by PSF substrates and had a selective layer of metakaolin (MK) embedded within a 2.5 wt% PVAm matrix with filler loadings ranging from 0.5 to 2 wt% using the solution mixing and casting method. The prepared membranes were then evaluated using the humidified mixed gas testing system ($CO_2/N_2 = 20/80$ by vol%) at a constant temperature of 25 °C and a pressure range of 1–8 bar. According to gas permeation testing results, in comparison to the pure PVAm membrane, the MK-PVAm based composite FTM obtained by incorporating 1.0 wt% of MK having a selective thickness of 2.37 μm displayed an excellent CO_2/N_2 selectivity of 82.40 and a CO_2 permeance of 186.90 GPU (CO_2 permeability = 443 Barrer) at a pressure of 1 bar, which surpassed the revised Robeson's upper bound limit of CO_2/N_2 separation. This FTMs with 1.0 wt% MK also showed long-term performance stability of CO_2/N_2 separation process during more than 300 h of examination. However, by further increasing the pressure up to 8 bar, more moisture contents were started to vaporized from the FTMs matrix, which caused reduction in amine carrier mobility leading to the decrement in CO_2/N_2 separation performance. Moreover, above 1 wt% filler loading, these defects started to appeared in the FTMs matrix, which caused decreasing trend in CO_2/N_2 selectivity due to the agglomeration of nanofillers within the membrane matrix [51].

Borgohain and Mandal [52] investigated the synergistic effect of the hydrotalcite (HT) and poly (amidoamine) (PAMAM) dendrimer combination as a nano-filler in the membrane matrix for the application of CO_2/N_2 separation. In this study, a blend of 10% PAMAM dendrimer and 1% HT nanofiller was incorporated into an 89% carboxymethyl chitosan (CMC) polymeric solution to prepare a mixed matrix membrane (MMM) solution. Then this MMM solution was coated as a selective layer on the PES-based support membrane for the fabrication of the PAMAM-HT-CMC/PES composite FTM. The resulted membrane having a selective layer thickness of 1 μm was then tested for the CO_2/N_2 separation performance using a humidified mixed gas permeation system ($CO_2/N_2 = 20/80$ by vol%) at a pressure of 2 bar and an elevated temperature of 90 °C. In this system, pure Argon (Ar) was used as a sweep gas with a constant water flow ratio of 2.33 (sweep/feed). According to the gas permeation results, PAMAM-HT-CMC/PES composite FTM exhibited a good CO_2/N_2 separation selectivity of 67 with a high CO_2 permeability of 123 Barrer at a high temperature of 90 °C, which showed that this MMM can be considered as a viable option for the elevated temperature based flue gas separation processes [52].

Zhang et al. [53] developed a polymer modified carbon nanotube (CNT) network based FTMs for highly efficient and stable CO_2 separation from flue gas streams. For the preparation of this type of membrane, polydopamine (PDA)-modified CNT was first dispersed (180 nm) onto a PES porous support membrane via a vacuum filtration method. Afterwards, a blend of 30 mg/mL polystyrene sulfonate (PSS) and 50 mg/mL tetraethylenepentamine (TEPA) solution was coated at a thickness of 0.22 μm on the PDA-modified CNT-PES substrate using a solution casting method. This composite membrane was tested using a humidified mixed gas permeation system (CO_2/N_2 = 15/85 vol%) at a temperature of 90 °C and 1 bar feed pressure. The permeate pressure of 0.3 bar was controlled by a vacuum system. According to the gas permeation analysis, it was found that the very porous and connected CNT network encouraged the development of an ultrathin selective layer, which dramatically improved CO_2 transport. The reaction sites were provided by the amine carriers, which were stabilized by surface negative charge to give enhanced CO_2 transport and increased CO_2/N_2 selectivity. Thus, a remarkable CO_2/N_2 selectivity of 400 and a high CO_2 permeance of 3250 GPU (CO_2 permeability = 1300 Barrer) were obtained at 90 °C in vacuum control system, surpassing the upper limit of the state-of-the-art membranes [53].

Sadeghi et al. [54] improved the CO_2/N_2 separation performance of the polyurethane-urea (PUU) membranes by incorporating the Hexamethylenetetramine (HDC) as 2D metal organic frame work (MOF) based nanofillers. In this study, the PUU polymer was prepared using a two-step polymerization method. Whereas, the different PUU-HDC FTMs were synthesized at a thickness range of 110–180 μm by varying the concentration of HDC from 0.5 to 2.5 wt% in the 10 wt% PUU membrane matrix using the solution mixing and casting method. The PUU-HDC membranes were tested for CO_2/N_2 separation performance using both the pure gas and mixed gas permeation testing systems at a pressure of 10 bar and a temperature of 30 °C. The pure gas permeation results revealed that the PUU-HDC membrane containing 2.5 wt% of the MOF nanofiller and having a thickness of 120 μm demonstrated a improved gas separation performance by 89% of CO_2 permeability and 38% of CO_2/N_2 selectivity as compared to a pristine membrane. The CO_2/N_2 selectivity of the PUU-HDC-2.5 membrane was 30.40, and it had a CO_2 permeability of 140 Barrer, indicating the strong potential of nanocomposite membranes for the separation of CO_2 from flue gas streams. Whereas, in the case of a mixed gas permeation system, this membrane showed a lower CO_2/N_2 selectivity of 24.40 with a CO_2 permeability of 102.60 Barrer as compared to the pure gas permeation results due to the competitive transport between the gas molecules of the mixture within the membrane matrix. Furthermore, the effect of changing temperature from 25 to 35 °C on the CO_2/N_2 separation performance was also investigated in this research work, which showed that by increasing the temperature, an improvement in gas permeability was observed. However, the membrane's selectivity was compromised at high temperatures [54].

In 2021, Zhang et al. [55] studied the PVAm based hybrid FTMs, in which different types of amino acid salts based mobile carriers (MCs) were incorporated into the PVAm solutions and then coated on the surface of polyether sulfone (PES) support membranes at a thickness of 0.17 μm by using the solution casting method. The PES support membranes were prepared by the non-solvent induced phase separation (NIPS) method. Whereas, the MCs were prepared by a deprotonating reaction between the amino salts and 2-(1-piperazinyl)ethylamine (PZEA), which produced four different types of MCs: PZEA-Proline (Pro), PZEA-Alanine (Ala), PZEA-Lysine (Lys), and PZEA-Sarcosinate (Sar). In these membranes, amine groups present in PVAm polymer act as fixed-site carriers for the facilitated transport of CO_2 molecules. The CO_2/N_2 separation performance of these fabricated membranes was then evaluated using the humidified mixed gas testing system ($CO_2/N_2 = 20/80$ by vol%) at a feed pressure of 0.16 bar and a temperature of 57 °C. In this testing system, humidified Argon (Ar) was used as a sweep gas at a pressure of 0.06 bar. According to the results, PZEA-Pro was the most effective amino salt MC in facilitating transport of CO_2 molecules, followed by PZEA-Lys and PZEA-Ala. The FTMs containing 35 and 65 wt% PZEA-Pro demonstrated the promising CO_2 permeance of 936 GPU (CO_2 permeability = 159.12 Barrer) with a CO_2/N_2 selectivity of 193. This permeance was further improved to 947 GPU (CO_2 permeability = 160.99 Barrer) with an exceptional CO_2/N_2 selectivity of 210 when the PVAm was lowered to 15 wt%, by substituting its 20 wt% with PZEA-Sar. This membrane also demonstrated 110 h of stable CO_2 transport performance. Moreover, amino salts based PEZA-PVAm FTMs CO_2/N_2 separation results, surpassed both upper bound Robeson's limit and had superior thermal stability than the pure PZEA based membrane [55].

Wang et al. [56], prepared a ZIF-8-decorated metakaolin (ZIF-8-d-K) hybrid filler, which was then embedded into the 2.5 wt% PVAm polymeric matrix at a filler loading of 1–5 wt%. The selective layers were coated on the PSF support membrane using the solution casting method to produce mixed matrix composite hybrid FSC-FTMs. The selective layers produced in this study have variable thickness in the range of 1.85–2.00 μm. These membranes were examined using both pure and mixed gas permeation testing systems under humidified conditions (R.H = 100%) at a constant pressure and temperature of 1 bar and 25 °C, respectively. According to the results, these FTMs showed significant improvement in the CO_2 permeance with the increase in filler loading as compared to the pure PVAm membrane. In a pure gas testing system, at 3 wt% ZIF-8-d-K loading, the composite FTMs having the selective layer thickness of 1.85 μm, showed the highest CO_2/N_2 selectivity of 86.7 with an optimum CO_2 permeance of 169 GPU (CO_2 permeability = 312.65 Barrer), which resulted from the shape sieving effect and the high-aspect-ratio impact of the hybrids. Furthermore, the FTMs preloaded with 3 wt% ZIF-8-d-MK demonstrated long-term stability (over 360 h) in a CO_2/N_2 gas mixture (20/80 vol.%), with an maximum CO_2 permeance of 245 GPU (CO_2 permeability = 453.25 Barrer) and an CO_2/N_2 selectivity of 98.00, indicating that this membrane had good capability of CO_2 removal from flue gas streams [56].

Alcantara et al. [57], reported that the N-methyl-2-hydroxyethylammonium propionate IL [m-2HEA](Pr) based PVA FTMs also showed significantly enhanced CO_2/N_2 separation performance. The IL was prepared by the amine neutralization reaction with the carboxylic acid. Whereas, these IL [m-2HEA](Pr)-based FTMs were synthesized by incorporating 2.60–60.30 wt% of IL using solution mixing and casting method. According to the pure gas permeation experiments, increasing in the IL concentration in PVA-based membrane stimulated higher CO_2 permeability. The most promising results were achieved by FTM having 45.5 wt% of IL incorporated into the PVA matrix, which exhibited an excellent CO_2/N_2 selectivity of 213 with a CO_2 permeability of 25.60 Barrer. However, further increasing the IL concentration up to 60.3 wt% of [m-2HEA](Pr), the FTM decreased in CO_2/N_2 separation performance, which was attributed to the membrane water swelling properties that promote the polymeric chain mobility within the membrane matrix, which resulted in the sharp increase in N_2 permeability and a decrease in CO_2/N_2 selectivity. Furthermore, these membranes were able to sustain a high pressure differential of up to 6 barg [57].

Choi et al. [58] developed a modified graphene oxide (mGO) based thin-film composite hollow fiber FSC-FTM to achieve high performance in CO_2/N_2 separation process. For the preparation of this membrane, the GO nanosheets were first treated with NaOH and HCl to deplete the oxygen groups they contained in order to produce mGO nanosheets. The 19.15 wt% PES with 15.32 wt% PEG and 0.38 wt% were subjected to a spinning technique to prepare the substrate layer. Afterwards, by using the interfacial polymerization (IP) method, a thin layer of a blend composed of poly(ethylene imine) (PEI)—trimesoyl chloride (TMC) with 0.3 wt% sodium dodecyl sulfate (SDS)—surfactant and 0.1 wt% of mGO was formed on the support membrane at a thickness of 0.113 μm, which produce mGO based FSC-hybrid composite hollow fiber based FTM. The single-gas permeance was employed to evaluate the CO_2/N_2 separation performance of this membrane at a temperature of 25 °C and an optimum pressure of 0.25 bar. The membrane showed an enhanced CO_2/N_2 selectivity of 60 with a CO_2 permeance of 73 GPU (CO_2 permeability = 8.25 Barrer). The amino groups present in the coated layer acts as FSC and the presence of mGO increase the free volume of the membrane matrix, which provides a selective gas transport pathway for the CO_2 molecules during the flue gas separation process [58].

Ashtiani et al. [59] studied the UiO-66-NH_2-PVA functionalized (MOFs) based FSC-FTM for enhanced CO_2/N_2 separation performance. They synthesized a UiO-66-NH_2-PVA MOF-based nanofiller via a hydrothermal process. Afterwards, these nanofillers were embedded at a concentration of up to 55 wt% into a 5 wt% PVAm membrane matrix using the solution mixing and casting method for the preparation of different mixed matrix FTMs at a thickness of 40 μm. These fabricated membranes, when tested under a humidified mixed gas system (CO_2/N_2 = 50/50 by vol%), showed an improvement in CO_2/N_2 separation performance with the increase in nanofiller content up-to 24 wt% in the FTM at a 1.01 feed pressure and a temperature of 25 °C. The PVAm-FTM containing 24 wt% of UiO-66-NH_2-PVA (MOFs) showed the highest CO_2/N_2 selectivity of 45.76 with an optimum CO_2 permeability of 76.15

Barrer. The main factor for this improvement in the gas separation performance of these FTMs was due to the presence of UiO-66-NH_2-PVA MOF in the membrane matrix, which has an amino group-rich porous structure that offers faster transport channels for the CO_2 molecules and also hinders the passage of the N_2 molecules. However, above this filler loading concentration at 55 wt%, the FTM showed a drastic decrement in CO_2/N_2 selectivity with an increase in both CO_2 and N_2 permeabilities [59].

Kojabad et al. [60] were able to accomplish good results by incorporating aniline at a filler loading range of 25–75 wt% in the poly (ether-block-amide) (Pebax) matrix by using the solution mixing and casting method. These fabricated membranes have a thickness in the range of 60–71 μm. The mixed matrix FTMs were then evaluated using both pure mixed gas ($CO_2/N_2 = 20/80$ by vol%) at a constant pressure of 7 bar and a temperature of 25 °C. According to the results of the pure gas permeation test, by increasing the concentration of aniline in the Pebax matrix from 0 to 50 wt% the FTMs showed a significant improvement in CO_2 permselectivity. The aniline in these membranes acts as a semi-mobile carrier, which provides the facilitated transport of CO_2 molecules in addition to the solution diffusion mechanism within the membrane matrix. The FTM with 50 wt% of aniline had the most optimum gas separation results and exhibited a CO_2 permeability of 151 Barrer and a CO_2/N_2 selectivity of 92.50, which were both higher by 76% and 101%, respectively, than the pristine Pebax membrane. While in the case of the mixed gas permeation test, the lower CO_2/N_2 selectivity of 68.34 with CO_2 permeability of 123.12 Barrer due to the competitive interaction of the gas molecules in the membrane matrix. Moreover, above 75 wt% of aniline concentration in the Pebax matrix, the membrane displayed enhancement in both ideal gas permeability with a lower CO_2/N_2 selectivity [60].

Barooah et al. [61], used amine surface-functionalized silica nanoparticles (AFSNPs) in the composite FTMs to achieve enhanced CO_2/N_2 separation performance. In this study, the silica nanoparticles (SNPs) were prepared by Stober's method and followed by surface modification using the (3-aminopropyl) trimethoxysilane (3-APTMS) to produce AFSNPs. These nanofillers were embedded into the blended polymeric solution composed of 38 wt% PVA, 9 wt% PEG, 25 wt% PEI, 15 wt% TETA, and 10 wt% KOH, which was then coated on the polytetrafluoroethylene (PTFE) support membrane to produce a AFSNPs based composite membrane. For evaluation of the CO_2/N_2 separation performance of this membrane, a mixed gas ($CO_2/N_2 = 20/80$ by vol%) and Argon (Ar) were used as a feed gas and sweep gas under humidified conditions with a pressure of 2.5 and 1.2 bar, respectively, and at a constant temperature of 60–110 °C. According to the gas permeation analysis, the CO_2 permeance and CO_2/N_2 selectivity were significantly enhanced by the simultaneous positive effects of the amine carriers and surface functionalization of nanoparticles. The 3 wt% AFSNPs based composite membrane having the selective layer thickness of 27 μm showed a remarkable CO_2/N_2 selectivity of 325 with a CO_2 permeance of 36 GPU (CO_2 permeability = 927 Barrer) at a 2.5 bar pressure and optimal temperature of 100 °C, which exceeded the revised Robeson's upper bound limit of CO_2/N_2 separation. In comparison to the pristine membrane, these results revealed a substantial increase of 141% in CO_2 permeance (GPU) and a 55% rise in CO_2/N_2 selectivity [61].

Wang et al. [62], modified the GO channels by incorporating magnesium ion (Mg^{2+}) and phosphoric acid (PA) using the facile post-immersion method for high performance CO_2 separation from flue gas streams. In the membrane preparation process, GO and PA were initially blended with a mass ratio of 1:10, followed by GO-PA membrane synthesis on the PAN porous support membrane via a vacuum filtration method. Then this composite membrane was soaked in the magnesium chloride ($MgCl_2$) solution to produce the GO-PA-Mg^{2+}/PAN composite FTM. The GO amount was varied in the GO-PA-Mg^{2+} membrane in the range of 0.016–0.16 g/m^2 to study its effect on the CO_2/N_2 separation performance. The resultant membranes were tested using a humidified mixed gas testing system (CO_2/N_2 = 50/50 by vol%) with the sweep gas under a constant pressure of 1 bar and temperature of 25 °C. The gas permeation data reveals that the CO_2 permeance constantly decreases as GO deposition increases, whereas the CO_2/N_2 separation factor exhibits a pattern of initially rising and then remaining steady. When the GO deposition amount is 0.08 g/m^2, the GO-PA-Mg^{2+}/PAN composite membrane having a selective layer thickness of 67 μm demonstrated the best gas separation performance, with a CO_2/N_2 selectivity improved from 4 to 48.80 and a 1.5 times higher CO_2 permeance of 23 GPU (CO_2 permeability = 1541 Barrer) than the pristine GO membrane. This improvement in gas separation efficiency can be attributed to three main factors: (a) The amount of metal ions in the membrane was initially increased by introducing PA into GO channels due to its strong complexing capacity. (b) The Mg^{2+} metallic ions provide multiple roles, such as in the facilitating transport of CO_2 molecules with its salting-out effect to reduce the N_2 gas solubility and restrict its diffusion within the membrane matrix. (c) Water molecules also promoted the selective facilitated transport of CO_2 molecules during the CO_2/N_2 separation process. Additionally, the separation efficiency of GO-PA-Mg^{2+}/PAN-based composite membranes exhibits good long-term stability due to their structural durability [62]. All the discussed FTMs have improved the separation of CO_2 from flue gas with enhanced selectivity. The discussed results reflect the significant development in this field.

7.3 Trade-Off Robeson Plot of FTMs for Flue Gas (CO_2/N_2) Separation

Robeson explained that membranes for gas separation generally show a trade-off between gas selectivity and gas permeability by his upper bound curve in 1991 [63] that was further updated in 2008 [64]. The current membrane research is focused to develop such FTMs for separation of CO_2 from flue gas that can surpass the upper bound with their separation performance in the desirable commercial region. FTMs have gained importance because of their intrinsic transport properties, low cost and high processability. Figure 7.2 shows that most of the recently fabricated FTMs for separation of CO_2 from flue gas have crossed the "upper bound" limit in the trade-off between CO_2 permeability and CO_2/N_2 selectivity specifically for flue gas. The PDA modified CNT/PSS/PES based FSC-FTM stood out the best membrane in term of CO_2/N_2 selectivity among the other recently reported FTM, which showed the value of 400 with an high CO_2 permeability of 1300 Barrer. While, in term of CO_2 permeability, the 30 wt% NCs-Pebax based FTM showed a highest value of 1993 Barrer with a low CO_2/N_2 selectivity of 69.00. Furthermore, the other FSC composite and hybrid MMMs based FTMs also showed the most remarkable CO_2/N_2 separation performances. This signified that overall current research trend for the fabrication of CO_2 selective FTMs is in the right direction of producing economic and commercially feasible FTMs for the separation of CO_2 from flue gas.

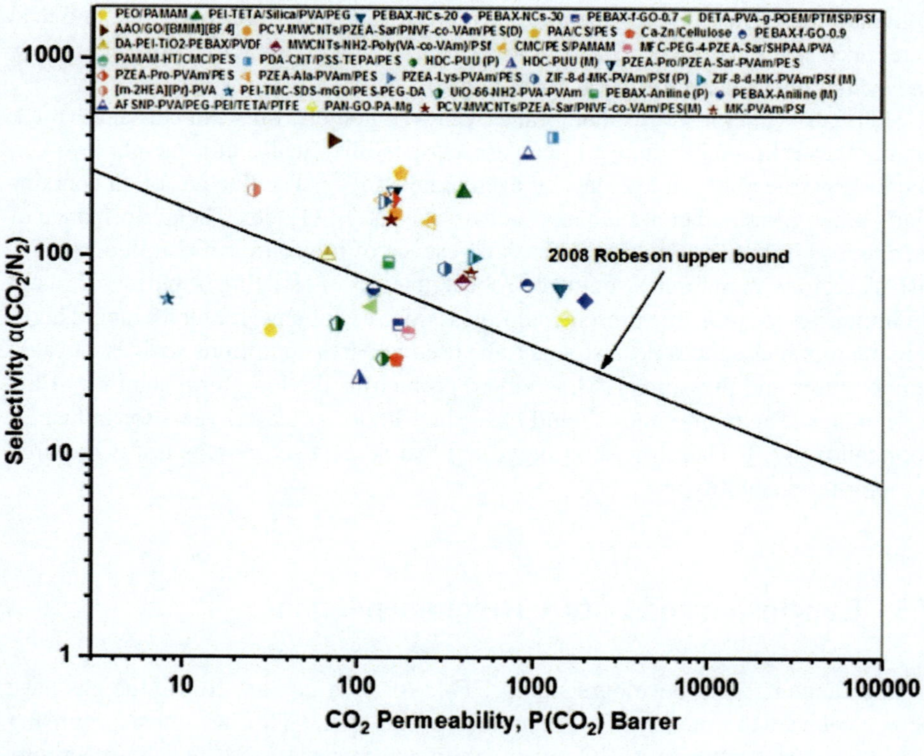

Fig. 7.2 Robeson's upper bound graph illustrating the flue gas separation (CO_2/N_2) selectivity for different FTMs versus CO_2 permeability from Table 7.2 data

7.4 Challenges of FTMs for Improved CO_2 Separation from Flue Gas

The most important problem that influences the performance of CO_2 membrane is plasticization [65–68]. It occurs due to the dissolution of penetrants into the polymer matrix at high pressure resulting in enhanced polymer free volume and chain mobility [69]. In CO_2-induced plasticization selectivity reduces; however, permeability of membrane increases. Thickness of membrane also influences this effect greatly [70]. In thinner membranes chain mobility is better due to smaller binding force of molecular chain. Therefore, it is important to optimize FTMs performance for gas separation at realistic conditions and improve the morphology of membranes for large scale applications. The second most common issue to be resolved is physical aging of membranes. The polymer chains keep rearranging themselves over time and lead to reduced specific volume. This improves the packing efficiency of polymer matrix and eventually reduces diffusion coefficients [71].

Here also thickness of membranes plays a major role. For instance thick membranes take longer time to decrease permeability while thin membranes can decrease permeability in few days or even in hours [72]. The flue gas often contains many other gases and contaminants such as SO_2, NO_x, O_2, etc. The performance of membranes decreases greatly due to the presence of impurities in real flue gas [73, 74]. Therefore, membranes should be robust in terms of resisting impurities. Except for impurities the moisture present in flue gas is also a challenge for membrane. Therefore, membranes should be designed to be used in harsh conditions such as elevated temperatures and pressures [75]. Another, problem is the long-term stability of the membranes. The membranes should have shelf life of at least 3 years for industrial applications [72]. Therefore, it is suggested that at least six months test is required for long-term stability test.

7.5 Conclusion and Future Recommendations

In this chapter recent developments in FTMs for CO_2 capture from Flue gas have been reviewed thoroughly. The literature reveals that FTMs are most promising membranes that offer high CO_2 permeability/permeance and CO_2/N_2 selectivity. The FTMs containing fixed-site, mobile-site and both are performing excellently in CO_2 separation processes. However, improvement is required to maintain miscibility of mobile carriers in the polymeric membrane. The FTMs based on interfacial polymerization, mixed matrix membranes and other hybrid nano-materials composite membranes showed remarkable CO_2/N_2 selectivity with high permeability for CO_2. Functionalized nano-materials may play a better role in improving strength and shelf life of membranes. This may allow the facile fabrication of prototype membranes and membrane modules. However, the technology is lacking behind because of the absence of field tests. Therefore, comprehensive field tests are required to understand and overcome the technical gap between material fabrication and their implementation in commercial-scale gas separation applications.

References

1. Dranga B-A, Lazar L, Koeser H (2012) Oxidation catalysts for elemental mercury in flue gases—a review. Catalysts 2(1):139–170
2. Song C, Pan W, Srimat ST, Zheng J, Li Y, Wang Y-H, Xu B-Q, Zhu Q-M (2004) Tri-reforming of methane over Ni catalysts for CO_2 conversion to syngas with desired H_2/CO ratios using flue gas of power plants without CO_2 separation. Stud Surf Sci Catal 153:315–322
3. Scala F, Lancia A, Nigro R, Volpicelli G (2005) Spray-dry desulfurization of flue gas from heavy oil combustion. J Air Waste Manag Assoc 55(1):20–29
4. Zevenhoven R, Kilpinen P (2001) Control of pollutants in flue gases and fuel gases. Helsinki University of Technology
5. Ismail AF, Khulbe KC, Matsuura T (2015) Gas separation membrane materials and structures. In: Gas separation membranes. Springer, Berlin, pp 37–192
6. Breck DW, Breck DW (1973) Zeolite molecular sieves: structure, chemistry, and use. Wiley
7. Matteucci S, Yampolskii Y, Freeman BD, Pinnau I (2006) Transport of gases and vapors in glassy and rubbery polymers. Mater Sci Membr Gas Vapor Sep 1:1–2
8. Ajibade FO, Adelodun B, Lasisi KH, Fadare OO, Ajibade TF, Nwogwu NA, Sulaymon ID, Ugya AY, Wang HC, Wang A (2021) Environmental pollution and their socioeconomic impacts. In: Microbe mediated remediation of environmental contaminants. Elsevier, pp 321–354
9. Chao C, Deng Y, Dewil R, Baeyens J, Fan X (2021) Post-combustion carbon capture. Renew Sustain Energy Rev 138:110490
10. Agarwal RK (2018) Carbon capture, utilization and sequestration. BoD–Books on Demand
11. Kárászová M, Zach B, Petrusová Z, Červenka V, Bobák M, Šyc M, Izák P (2020) Post-combustion carbon capture by membrane separation, review. Sep Purif Technol 238:116448
12. Karim SS, Farrukh S, Hussain A, Noor T, Younas M (2021) A comprehensive overview of dual-layer composite membrane for air (O_2/N_2) separation. Polym Polym Compos 29(9_suppl):S1630–S1640
13. Kusumocahyo SP, Kanamori T, Sumaru K, Aomatsu S, Matsuyama H, Teramoto M, Shinbo T (2004) Development of polymer inclusion membranes based on cellulose triacetate: carrier-mediated transport of cerium (III). J Membr Sci 244(1–2):251–257
14. Sugano K, Kansy M, Artursson P, Avdeef A, Bendels S, Di L, Ecker GF, Faller B, Fischer H, Gerebtzoff G (2010) Coexistence of passive and carrier-mediated processes in drug transport. Nat Rev Drug Discov 9(8):597–614
15. Hansen SP, Fyles TM (2011) Carrier-mediated electrodialysis. Chem Commun 47(22):6428–6430
16. Qiu Y-R, Mao L-J (2013) Removal of heavy metal ions from aqueous solution by ultrafiltration assisted with copolymer of maleic acid and acrylic acid. Desalination 329:78–85
17. Gao J, Qiu Y, Hou B, Zhang Q, Zhang X (2018) Treatment of wastewater containing nickel by complexation-ultrafiltration using sodium polyacrylate and the stability of PAA-Ni complex in the shear field. Chem Eng J 334:1878–1885
18. Ma J, Stevens GW, Mumford KA (2018) The performance of diphenyldichlorosilane coated ammonium exchange zeolite and its application in the combination of adsorption and biodegradation of hydrocarbon contaminated ground water. Chem Eng J 347:415–423
19. Li Y, Wang S, He G, Wu H, Pan F, Jiang Z (2015) Facilitated transport of small molecules and ions for energy-efficient membranes. Chem Soc Rev 44(1):103–118
20. St John AM, Cattrall RW, Kolev SD (2013) Determination of the initial flux of polymer inclusion membranes. Sep Purif Technol 116:41–45

21. Qiu Y-R, Matsuyama H, Gao G-Y, Ou Y-W, Miao C (2009) Effects of diluent molecular weight on the performance of hydrophilic poly (vinyl butyral)/pluronic F127 blend hollow fiber membrane via thermally induced phase separation. J Membr Sci 338(1–2):128–134
22. Wu Y, Alivand MS, Hu G, Stevens GW, Mumford KA (2020) Nucleation kinetics of glycine promoted concentrated potassium carbonate solvents for carbon dioxide absorption. Chem Eng J 381:122712
23. Fontàs C, Tayeb R, Dhahbi M, Gaudichet E, Thominette F, Roy P, Steenkeste K, Fontaine-Aupart M-P, Tingry S, Tronel-Peyroz E (2007) Polymer inclusion membranes: the concept of fixed sites membrane revised. J Membr Sci 290(1–2):62–72
24. Wang D, Hu J, Li Y, Fu M, Liu D, Chen Q (2016) Evidence on the 2-nitrophenyl octyl ether (NPOE) facilitating copper (II) transport through polymer inclusion membranes. J Membr Sci 501:228–235
25. Wang M, Wang Z, Zhao S, Wang J, Wang S (2017) Recent advances on mixed matrix membranes for CO_2 separation. Chin J Chem Eng 25(11):1581–1597
26. Deng L, Hägg M-B (2015) Fabrication and evaluation of a blend facilitated transport membrane for CO_2/CH_4 separation. Ind Eng Chem Res 54(44):11139–11150
27. Li S, Wang Z, Yu X, Wang J, Wang S (2012) High-performance membranes with multi-permselectivity for CO_2 separation. Adv Mater 24(24):3196–3200
28. Tong Z, Ho WW (2017) New sterically hindered polyvinylamine membranes for CO_2 separation and capture. J Membr Sci 543:202–211
29. Qiao Z, Wang Z, Zhao S, Yuan S, Wang J, Wang S (2013) High adsorption performance polymers modified by small molecules containing functional groups for CO_2 separation. RSC Adv 3(1):50–54
30. Qiao Z, Wang Z, Zhang C, Yuan S, Zhu Y, Wang J, Wang S (2013) PVAm–PIP/PS composite membrane with high performance for CO_2/N_2 separation. AIChE J 59(1):215–228
31. Zhao Y, Ho WW (2012) Steric hindrance effect on amine demonstrated in solid polymer membranes for CO_2 transport. J Membr Sci 415:132–138
32. Liu Y, Yu S, Wu H, Li Y, Wang S, Tian Z, Jiang Z (2014) High permeability hydrogel membranes of chitosan/poly ether-block-amide blends for CO_2 separation. J Membr Sci 469:198–208
33. Yuan S, Wang Z, Qiao Z, Wang M, Wang J, Wang S (2011) Improvement of CO_2/N_2 separation characteristics of polyvinylamine by modifying with ethylenediamine. J Membr Sci 378(1–2):425–437
34. Yuan F, Wang Z, Li S, Wang J, Wang S (2012) Formation–structure–performance correlation of thin film composite membranes prepared by interfacial polymerization for gas separation. J Membr Sci 421:327–341
35. Wang M, Wang Z, Wang J, Zhu Y, Wang S (2011) An antioxidative composite membrane with the carboxylate group as a fixed carrier for CO_2 separation from flue gas. Energy Environ Sci 4(10):3955–3959
36. Wang M, Wang Z, Li S, Zhang C, Wang J, Wang S (2013) A high performance antioxidative and acid resistant membrane prepared by interfacial polymerization for CO_2 separation from flue gas. Energy Environ Sci 6(2):539–551
37. Yao K, Wang Z, Wang J, Wang S (2012) Biomimetic material—poly (N-vinylimidazole)–zinc complex for CO_2 separation. Chem Commun 48(12):1766–1768
38. Lee JH, Lim JY, Park MS, Kim JH (2018) Improvement in the CO_2 permeation properties of high-molecular-weight poly (ethylene oxide): use of amine-branched poly (amidoamine) dendrimer. Macromolecules 51(21):8800–8807
39. Kim DH, Park MS, Kim NU, Ryu DY, Kim JH (2018) Multifunctional amine-containing PVA-g-POEM graft copolymer membranes for CO_2 capture. Macromolecules 51(15):5646–5655
40. Barooah M, Mandal B (2018) Enhanced CO_2 separation performance by PVA/PEG/silica mixed matrix membrane. J Appl Polym Sci 135(28):46481
41. Prasad B, Mandal B (2018) Preparation and characterization of CO_2-selective facilitated transport membrane composed of chitosan and poly (allylamine) blend for CO_2/N_2 separation. J Ind Eng Chem 66:419–429

42. Ying W, Cai J, Zhou K, Chen D, Ying Y, Guo Y, Kong X, Xu Z, Peng X (2018) Ionic liquid selectively facilitates CO_2 transport through graphene oxide membrane. ACS Nano 12(6):5385–5393
43. Zhang H, Guo R, Zhang J, Li X (2018) Facilitating CO_2 transport across mixed matrix membranes containing multifunctional nanocapsules. ACS Appl Mater Interfaces 10(49):43031–43039
44. Han Y, Wu D, Ho WW (2018) Nanotube-reinforced facilitated transport membrane for CO_2/N_2 separation with vacuum operation. J Membr Sci 567:261–271
45. Zhang X-F, Hou T, Chen J, Feng Y, Li B, Gu X, He M, Yao J (2018) Facilitated transport of CO_2 through the transparent and flexible cellulose membrane promoted by fixed-site carrier. ACS Appl Mater Interfaces 10(29):24930–24936
46. Zhang J, Xin Q, Li X, Yun M, Xu R, Wang S, Li Y, Lin L, Ding X, Ye H (2019) Mixed matrix membranes comprising aminosilane-functionalized graphene oxide for enhanced CO_2 separation. J Membr Sci 570:343–354
47. Zhu H, Yuan J, Zhao J, Liu G, Jin W (2019) Enhanced CO_2/N_2 separation performance by using dopamine/polyethyleneimine-grafted TiO_2 nanoparticles filled PEBA mixed-matrix membranes. Sep Purif Technol 214:78–86
48. Zhang B, Fu J, Zhang Q, Yi C, Yang B (2019) Study on CO_2 facilitated separation of mixed matrix membranes containing surface modified MWCNTs. J Appl Polym Sci 136(33):47848
49. Borgohain R, Mandal B (2019) pH responsive carboxymethyl chitosan/poly (amidoamine) molecular gate membrane for CO_2/N_2 separation. ACS Appl Mater Interfaces 11(45):42616–42628
50. Janakiram S, Yu X, Ansaloni L, Dai Z, Deng L (2019) Manipulation of fibril surfaces in nanocellulose-based facilitated transport membranes for enhanced CO_2 capture. ACS Appl Mater Interfaces 11(36):33302–33313
51. Wang Y, Li L, Zhang X, Li J, Wang J, Li N (2020) Polyvinylamine/amorphous metakaolin mixed-matrix composite membranes with facilitated transport carriers for highly efficient CO_2/N_2 separation. J Membr Sci 599:117828
52. Borgohain R, Mandal B (2020) Thermally stable and moisture responsive carboxymethyl chitosan/dendrimer/hydrotalcite membrane for CO_2 separation. J Membr Sci 608:118214
53. Zhang S, Li H, Li H, Sengupta B, Zha S, Li S, Yu M (2020) Negative charge confined amine carriers within the nanowire network for stable and efficient membrane carbon capture. Adv Func Mater 30(30):2002804
54. Sadeghi M, Isfahani AP, Shamsabadi AA, Favakeh S, Soroush M (2020) Improved gas transport properties of polyurethane–urea membranes through incorporating a cadmium-based metal organic framework. J Appl Polym Sci 137(20):48704
55. Zhang Z, Rao S, Han Y, Pang R, Ho WW (2021) CO_2-selective membranes containing amino acid salts for CO_2/N_2 separation. J Membr Sci 638:119696
56. Wang Y, Wang J, Zhang X, Li J, Li L (2021) Polyvinylamine/ZIF-8-decorated metakaolin composite membranes for CO_2/N_2 separation. Sep Purif Technol 270:118800
57. Alcantara ML, de Almeida OG, Liao LM, Borges CP, Mattedi S (2021) Amine/carboxylic acid ionic liquid composite membranes for CO_2 separation. Ind Eng Chem Res 60(11):4405–4419
58. Choi O, Hossain I, Jeong I, Park C-H, Kim Y, Kim T-H (2021) Modified graphene oxide-incorporated thin-film composite hollow fiber membranes through interface polymerization on hydrophilic substrate for CO_2 separation. Membranes 11(9):650
59. Ashtiani S, Khoshnamvand M, Regmi C, Friess K (2021) Interfacial design of mixed matrix membranes via grafting PVA on UiO-66-NH_2 to enhance the gas separation performance. Membranes 11(6):419
60. Kojabad ME, Babaluo A, Tavakoli A (2021) A novel semi-mobile carrier facilitated transport membrane containing aniline/poly (ether-block-amide) for CO_2/N_2 separation: molecular simulation and experimental study. Sep Purif Technol 266:118494
61. Barooah M, Mandal B, Su B (2021) Enhanced CO_2 separation performance of mixed matrix membrane by incorporating amine-functionalized silica filler. J Appl Polym Sci 138(47):51438

62. Wang H, Zheng J, Zhao J, Jin W (2021) Designing GO channels with high selectivity for CO_2/N_2 separation via incorporating metal ions. Chem Asian J 16(20):3141–3150
63. Robeson LM (1991) Correlation of separation factor versus permeability for polymeric membranes. J Membr Sci 62(2):165–185
64. Robeson LM (2008) The upper bound revisited. J Membr Sci 320(1–2):390–400
65. Suleman MS, Lau KK, Yeong YF (2016) Plasticization and swelling in polymeric membranes in CO_2 removal from natural gas. Chem Eng Technol 39(9):1604–1616
66. He Y, Benedetti FM, Lin S, Liu C, Zhao Y, Ye HZ, Van Voorhis T, De Angelis MG, Swager TM, Smith ZP (2019) Polymers with side chain porosity for ultrapermeable and plasticization resistant materials for gas separations. Adv Mater 31(21):1807871
67. Chi WS, Sundell BJ, Zhang K, Harrigan DJ, Hayden SC, Smith ZP (2019) Mixed-matrix membranes formed from multi-dimensional metal-organic frameworks for enhanced gas transport and plasticization resistance. ChemSusChem 12(11):2355–2360
68. Low Z-X, Budd PM, McKeown NB, Patterson DA (2018) Gas permeation properties, physical aging, and its mitigation in high free volume glassy polymers. Chem Rev 118(12):5871–5911
69. Hou Q, Wu Y, Zhou S, Wei Y, Caro J, Wang H (2019) Ultra-tuning of the aperture size in stiffened ZIF-8_Cm frameworks with mixed-linker strategy for enhanced CO_2/CH_4 separation. Angew Chem Int Ed 58(1):327–331
70. Kapantaidakis G, Koops G, Wessling M, Kaldis S, Sakellaropoulos G (2003) CO_2 Plasticization of polyethersulfone/polyimide gas-separation membranes. AIChE J 49(7):1702–1711
71. Rowe BW, Freeman BD, Paul DR (2009) Physical aging of ultrathin glassy polymer films tracked by gas permeability. Polymer 50(23):5565–5575
72. Baker RW, Low BT (2014) Gas separation membrane materials: a perspective. Macromolecules 47(20):6999–7013
73. Mubashir M, Fong YY, Leng CT, Keong LK (2018) Issues and current trends of hollow-fiber mixed-matrix membranes for CO_2 separation from N_2 and CH_4. Chem Eng Technol 41(2):235–252
74. Xu R, Wang Z, Wang M, Qiao Z, Wang J (2019) High nanoparticles loadings mixed matrix membranes via chemical bridging-crosslinking for CO_2 separation. J Membr Sci 573:455–464
75. Ogieglo W, Upadhyaya L, Wessling M, Nijmeijer A, Benes NE (2014) Effects of time, temperature, and pressure in the vicinity of the glass transition of a swollen polymer. J Membr Sci 464:80–85

Chapter 8
Carbon Dioxide (CO_2) Gas Storage and Utilization

Huma Shaikh, Shahnila Shah, Syed Shujaat Karim, Mohammad Younas, Syed Awais Ali, Sarah Farrukh, Mansoor Ul Hassan Shah, and Syed Nasir Shah

Abstract The disposal of carbon dioxide (CO_2) after its capture has become a limiting factor for its effective industrial applications. CO_2 is a major greenhouse gas as well as a valuable carbon resource. CO_2 utilization technology can bring a revival in the industrial applications of CO_2. The existing environmental problems due to CO_2 production and its swift increase in the atmosphere are discussed in this chapter. The efficient and CO_2-specific materials that can be utilized for CO_2 storage are discussed in detail along with their mechanisms. The possible geological storage pathways are also detailed here. CO_2 can be utilized for several industrial processes. Details of different ways to utilize CO_2 are also given in this chapter.

H. Shaikh · S. Shah
National Center of Excellence in Analytical Chemistry (NCEAC), University of Sindh, Jamshoro, Sindh, Pakistan
e-mail: huma.shaikh@usindh.edu.pk

S. Shah
e-mail: shahnila.shah@scholars.usindh.edu.pk

S. S. Karim · S. Farrukh (✉)
School of Chemical and Materials Engineering (SCME), National University of Sciences and Technology (NUST), Sector H-12, Islamabad, Pakistan
e-mail: sarah.farrukh@scme.nust.edu.pk

S. S. Karim
e-mail: sshujaat.che6scme@student.nust.edu.pk

M. Younas · M. U. Hassan Shah
Department of Chemical Engineering, University of Engineering and Technology, Peshawar, Pakistan
e-mail: m.younas@uetpeshawar.edu.pk

M. U. Hassan Shah
e-mail: mansoorshah@uetpeshawar.edu.pk

S. A. Ali · S. N. Shah
Department of Energy Engineering, Faculty of Mechanical and Aeronautical Engineering, University of Engineering and Technology, Taxila, Rawalpindi, Pakistan

S. Farrukh et al. (eds.), *Facilitated Transport Membranes (FTMs) for CO_2 Capture: Overview and Future Trends*, Green Energy and Technology,
https://doi.org/10.1007/978-3-031-21444-8_8

Keywords Carbon dioxide (CO_2) · CO_2 utilization · Gas storage · CO_2-specific materials · CO_2 storage and handling · Carbon capture

8.1 Introduction

The World Meteorological Organization (WMO) declared a 148% rise in the atmospheric concentration of CO_2 recently; i.e., the concentration of CO_2 in the atmosphere has risen from 278 ppm (1750) to 418 ppm (2022) based on the results of measurements performed at the Mauna Loa Observatory (HI, USA) (Fig. 8.1a, b) [1]. The rapid and unplanned industrialization has resulted in a drastic increase in this greenhouse gas. Ultimately, the whole world is suffering from Global Warming, which is an extinction threat to mankind and other living beings on the earth [2–6]. The overall increase in the concentrations of greenhouse gases has already resulted in approximately a 1° rise in average land and ocean surface temperatures globally [7].

In its latest report, the United Nations Intergovernmental Panel on Climate Change (IPCC) states that effective measures are required to control the global increase in temperature (by 1.5°) by 2050. Otherwise, living beings will suffer from severe consequences that include frequent and intense heavy rain falls, acidification, ocean warming, flooding, droughts, and rapid a decline in oxygen levels in the ocean [7]. Therefore, the international community has reached an agreement to control CO_2 emissions. The scientific community is trying its best to mitigate this problem by introducing CO_2 capture, utilization, and storage (CCUS) technology [8]. Although CO_2 is less potent than other greenhouse gases, it's extremely high emissions are a major contribution to the global warming effect.

Therefore, sequestration of CO_2 in oil/gas/coal/geological formations has been practiced as one of the options to mitigate the CO_2 emission problem. Initially, the CO_2 was captured from fossil fuel-based power plants [9]. During the process of

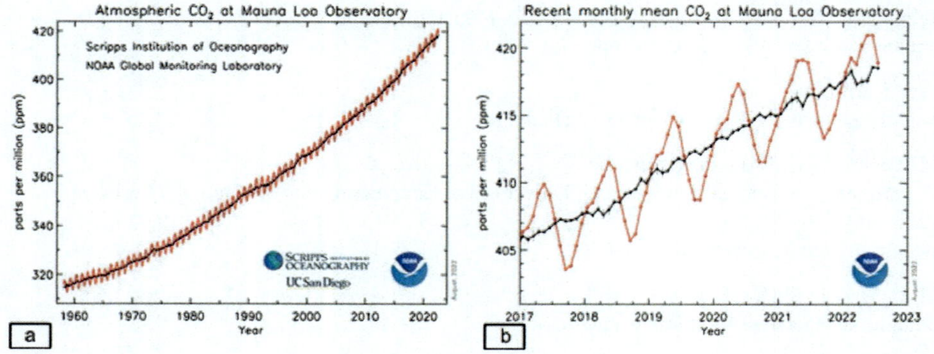

Fig. 8.1 **a** Increase in concentration of CO_2 during 1960–2020, **b** CO_2 levels during 2017–2022; measured at the Mauna Loa Observatory (Hawaii); monthly mean (red line); average seasonal cycle (black line). *Source* https://www.esrl.noaa.gov/gmd/ccgg/trends/

sequestration, the captured CO_2 is adsorbed in extremely abundant porous geological formations that possess porosity in the range of 10–200 m^2/g. The most common storage tanks explored for CO_2 include enhanced oil recovery (EOR), depleted oil and gas wells, salt formations, enhanced coal bed methane recovery (ECBM), enhanced gas recovery (EGR), and ocean storage [10]. Although these techniques are considered uncertain because it is thought that long-term CO_2 retention is not possible and there is leakage risk of stored CO_2 back into the atmosphere [11, 12], they are significant for the future development of long-term storage reservoirs for CO_2.

Besides, many small and intermediate-scale mitigation initiatives with optimal cost have also been taken into account [9]. Among those, porous materials that can store CO_2 at a temperature close to or above room temperature are an efficient solution. These porous reservoirs can be employed after CO_2 capture. In the next sections of this chapter, we will discuss in detail the geologic storage of CO_2 along with the storage efficiency of synthetic porous solids. More recently, the utilization of CO_2 for producing value-added goods has gained the attention of the scientific community. They are on the verge of developing industrial procedures that utilize CO_2 as one of the main production ingredients. This strategy will lead to an economical CO_2 capture process and will prove to be a valid alternative to CO_2 geological storage. The possible pathways for the utilization of CO_2 gas will also be highlighted in this chapter.

8.2 Storage of Carbon Dioxide (CO_2) Gas

8.2.1 Synthetic Porous Solids for Storage of CO_2

In this part of the chapter, we will typically discuss different porous materials that have been proposed for storage of CO_2 in the recent literature. The porous materials that possess a high surface area and a high volume of pores are generally known as solid-state porous media for the storage of gas. High porosity is an important feature of efficient storage material. The physico-chemical properties of the gas to be stored have a great influence on the structural properties of the storage material at a specific temperature and pressure.

The ideal gas storage material stores gas at ambient temperature (298 K) or above (318 K maximum) without having any effect on its working capacity, even if gas condenses. Generally, CO_2 appears to be a highly absorbable gas even at low pressure conditions due to its specific subcritical-supercritical borderline at a temperature of ~ 303 K [13]. Some of the most suitable porous materials for CO_2 storage are discussed below.

(a) **Carbons**

Carbon-based porous materials have always gained popularity among other porous materials for CO_2 storage due to their cost effectiveness, huge surface area, excellent thermal and chemical stability, hydrophobicity, and low cost regeneration. The presence of a high surface area and a high porosity with a homogeneous structure are key elements for improved CO_2 storage. Accessibility to pore networks via proper activation of storage material is also another important milestone to be reached during synthesis [14]. Wei et al. demonstrated the effect of the mentioned parameters in their study reporting production of bamboo-derived activated carbons (specific surface area of 3500 m^2/g) with excellent adsorption capacity for CO_2 [15]. The study revealed that activation temperature, KOH/C mass ratio, and precursors used have a significant impact on CO_2 adsorption capacity.

The hydrothermal carbonization (HTC) is also one of the popular methods to produce porous carbon with a homogenous composition. In a study based on hydrochars derived from biomass precursors Jujun grass and Camellia japonica; a series of microporous activated carbons (surface area up to 3537 m^2/g and pore volume up to 1.85 cm^3/g) were reported to exhibit high CO_2 uptake [16]. The homogeneous carbon prepared via this approach showed excellent storage capacity for CO_2 in other studies as well [17, 18]. Porous carbon is also prepared using biomass-based precursors and polymers for remarkable storage of CO_2. For instance, nitrogen-doped microporous carbons are reported with a surface area of ~ 3000 m^2/g from polypyrrole [19]. The nanocarbons such as polyaniline decorated graphene sheets (PANI-f-HEG) nanocomposites were reported to have a CO_2 uptake capacity of 75 mmol/g at 25 °C and 11 bar [20]. The ideal carbon-based materials need to have high adsorption capacity for CO_2 at pressures in the range of 10–50 bar with surface area and pore volume of greater than 3000 m^2/g and 2.0 cm^3/g, respectively, and high mesoporosity.

(b) **Metal Organic Frameworks (MOFs)**

Metal organic frameworks (MOFs) are considered as one of the efficient materials for gas storage because of their distinctive structural arrangements, for instance, extraordinary surface areas and porosity, and their ability to modify internal pore structure [21, 22]. Their structures are based on metal ion clusters and organic liquids; based on their design and synthesis strategy; they are capable of taking up record amounts of guest molecules in their pores. Furthermore, the flexibility in their structural design entertains incorporation of various functionalities that further improve their capability to adsorb guest molecules. Therefore, MOFs that possess an extraordinary surface area (> 4000 m^2/g) and excellent adsorption capacity at pressures of 10–50 bar are the most eligible candidates for CO_2 storage.

Currently, material scientists are focused on the fabrication of MOFs having enhanced porosities and CO_2 storage capability. However, there is a lot of research to be done in terms of stability, capacity, and cost of MOFs [23–26]. The porous materials are required to adsorb CO_2 at elevated pressures because porous surfaces are more efficient at packing CO_2 molecules into their pores at high pressures. Several MOFs such as MOF-2, IRMOF-1, IRMOF-3, IRMOF-6, IRMOF-11, MOF-74, MOF-505, HKUST-1, and MOF-177 have already been evaluated and proved themselves as better candidates for the storage of CO_2. The promising characteristics of MOFs motivated researchers to design MOFs with extremely high pore volumes and surface areas. Furukawa et al. designed $Zn_4O(CO_2)_6$ based MOF-180, MOF-200, MOF-205 and MOF-210 with one or two organic linkers [23]. An extraordinary BET surface area of ~ 6240 m^2/g and a pore volume of ~ 3.6 cm^3/g was revealed by MOF-210. The MOF-200 and MOF-210 showed CO_2 uptake values of 74.2 wt% and 73.9 wt%, respectively, at 298 K and 50 bars. These MOFs are among the porous solid materials with excellent CO_2 storage capability (Fig. 8.2).

Here, it is important to mention that surface area is not the only factor that affects the CO_2 storage capacity. The interactions between adsorbent and CO_2 also play a significant role. For instance, Ni-MOF-74 and Mg-MOF-74 showed CO_2 adsorptions of 54.2 wt% (at 22 bar and 298 K) and 68.9 wt% (at 36 bar and 278 K), respectively [28]. Several MOFs were studied for CO_2 storage at elevated pressures [27, 29]. The

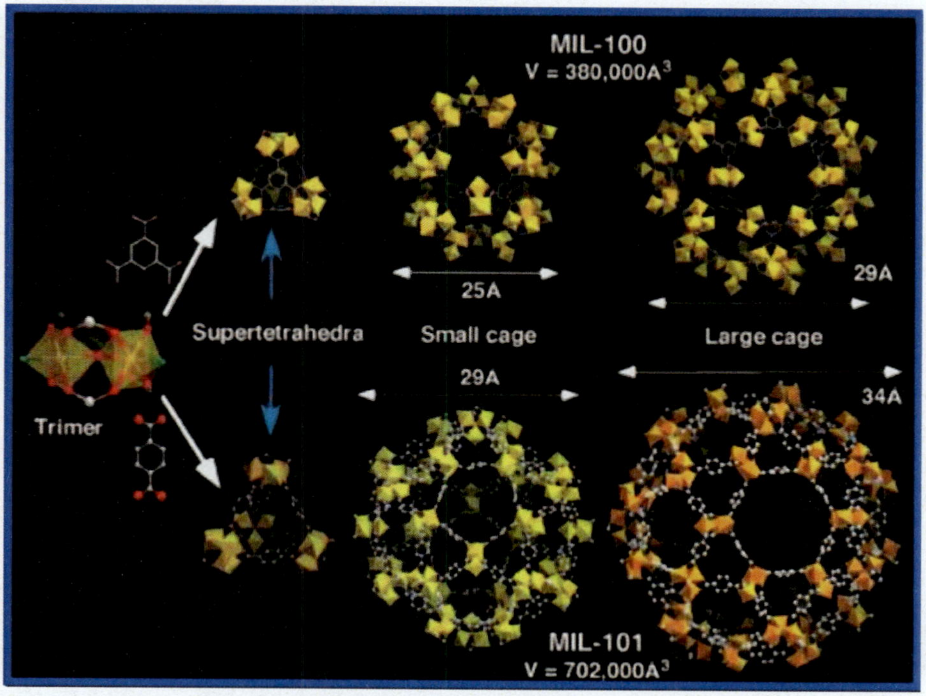

Fig. 8.2 Structures of MIL-100 and MIL-101. (Left) A trimer of Cr octahedra and linkers (1,3,5-BTC for MIL-100 and 1,4-BDC for MIL-101). (Center) Hybrid supertetrahedra. (Right) Cages of MIL-100 and MIL-101. Reprinted with permission from Ref. [27] copyright © 2022 American Chemical Society

Table 8.1 List of MOFs with extra-ordinary adsorption uptake for CO_2

MOF	BET SA (m^2/g)	Vp (cm^3/g)	Uptake (wt%)	Pressure (bar)	Temperature (K)	References
MOF-210	6240	3.6	74.2	50	298	[23]
MOF-200	4530	3.59	73.9	50	298	[23]
MOF-205	4460	2.16	62.6	50	298	[23]
MOF-177	4500	1.89	60.8	50	298	[23]
Be-BTB	4030	1.70	58.5	40	313	[30]
NU-100	6143	2.82	69.8	40	298	[25]
Al-soc-MOF-1	5585	2.3	66.7	40	298	[26]
PCN-68	5109	2.13	57.2	35	298	[24]
PCN-66	4000	1.63	53.6	35	298	[24]
PCN-61	3000	1.36	50.8	35	298	[24]
MIL-101(Cr)	4230	2.15	56.9	50	304	[27]
UMCM-1	4100	2.14	52.7	24	298	[31]

MOFs that possess better CO_2 storage ability are summarized in Table 8.1. The MOFs that possess unsaturated coordinative metal sites and heteroatomic frameworks are favourable materials for CO_2 storage. MOFs can be good alternatives for short-term storage of CO_2. However, their use for long-term storage of CO_2 is difficult due to their cost. So, there is a need to fabricate cost-effective MOFs for their commercial use as CO_2 storage materials.

(c) **Zeolites**

The zeolites were first introduced in 1756 by Axel Fredrik Cronstedt, a Swedish mineralogist. They are extremely ordered crystalline aluminosilicate materials with microporous structures. Zeolites possess homogeneous structures with uniform micro-pores and hydrophilic surfaces [32]. Their unique structural features allow them to have strong electrostatic interactions with CO_2. They are considered as promising molecular sieves broadly for carbon capture and separation applications [33, 34]. With respect to CO_2 storage, zeolite 13X is the most studied adsorbent material. Thakkar et al. [35] first reported 3D-printed monoliths using zeolites 13X and 5A. They used bentonite clay, methyl cellulose, and PVA as binders. The monolith was strengthened because of the presence of hydroxyl groups of methyl cellulose that provided increased cohesion between particles. Figure 8.3 shows zeolite monoliths that are cylindrical with a smooth surface and square channels, obtained via 3D-printing. The resulted monoliths were comprised of interconnected micro-, meso-, and macro-pore networks. The study showed that monoliths obtained using a high amount of 13X and 5A (90% wt) showed excellent adsorption capacity for CO_2.

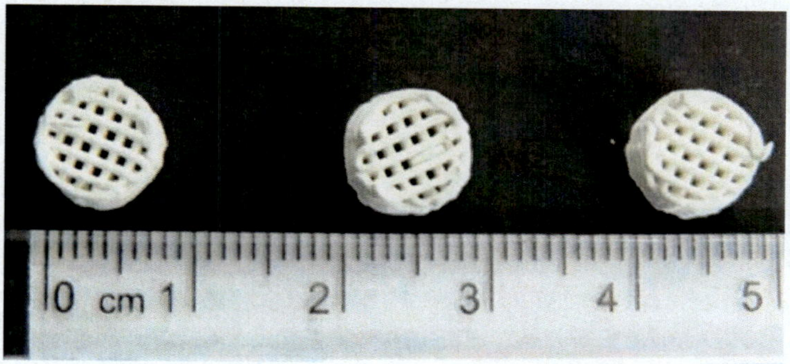

Fig. 8.3 Zeolite 13X monolith obtained via robocasting 3D printer. Reprinted with permission from Ref. [35] copyright © 2022 American Chemical Society

Although the storage capacity of 13X zeolites for CO_2 is more than activated carbons at low pressures, they offer significantly lower storage at elevated pressures. Therefore, they have limited commercial application as current CO_2 storage processes mainly operate at high pressures.

(d) **Other Silica Based Materials**

In addition to zeolites, silicoaluminophosphates (SAOPs) have also been explored as porous materials for the storage of CO_2. The SAPOs are crystalline and porous phosphates composed of oxides of Al, P, and Si. Their pore openings are narrow with eight-ring windows [36]. SAPOs are broadly studied for adsorptive separations of CO_2 at low pressures because they get saturated at 1 bar. SAPO-34 is one of the most widely studied materials for CO_2/N_2 separation [37]. The other silica materials that have found their application in CO_2 storage are MCM-41, SBA-15, KIT-6, etc. [38, 39]. The adsorption capacity of MCM-41 and its counterparts such as pore-expanded MCM-41 (PE-MCM-41) and triamine surface-modified counterpart (TRI-PE-MCM-41) were evaluated for CO_2. Increased CO_2 uptake was obtained for PE-MCM-41 as compared to MCM-41 at high pressures (Fig. 8.4). However, TRI-PE-MCM-41 showed excellent improvement in adsorption capacity at low pressures to moderate [39]. The meso-porous silica having amine functionalities showed better storage performance for CO_2. However, to utilize silica-based storage materials commercially, material scientists need to address the cost-effectiveness, long-term stability, and storage capacity of silica-based materials. Lashaki et al. [40] have discussed in detail the important stability issues related to different functionalized silicas, MOFs, and zeolites in their review article.

8.2.2 Geological Storage of CO_2

The captured CO_2 is injected into the deep rocks by soring it in upper crest of earth for many years [42, 43]. The CO_2 is extracted from biological and ignition activities

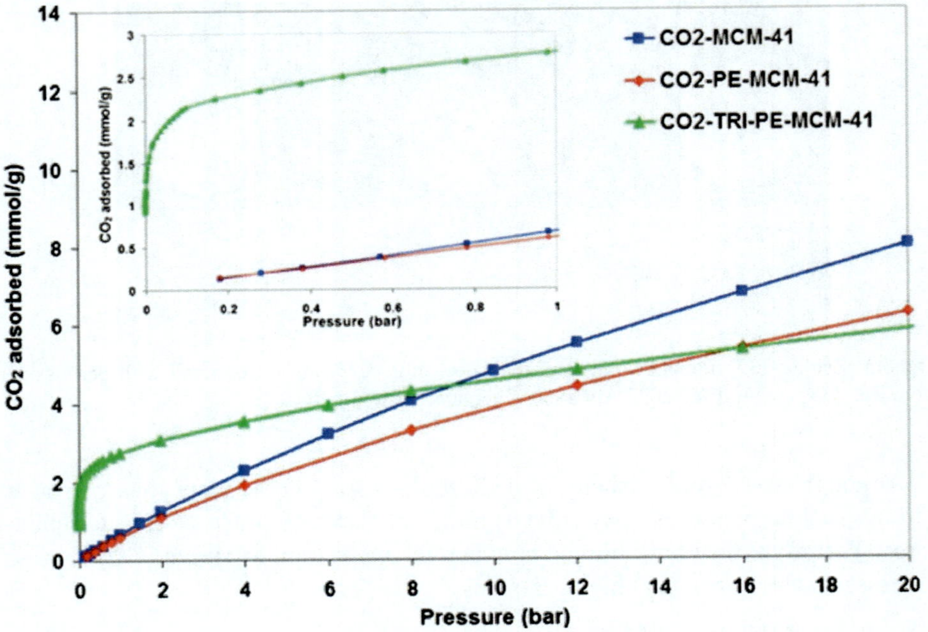

Fig. 8.4 Gravimetric CO_2 uptake for the MCM-41, PE-MCM-41 and TRI-MCM-41 at 298 K. Reprinted by permission from Springer Science Business Media, LLC, Springer Nature, Adsorption (Journal), Ref. [41] Copyright © 2022)

due to chemical reactions. Then it is collected in the natural subsurface in either carbonate mineral form or in solution and gaseous form. As the temperature rises with the earth's depth, the density of CO_2 also increases. Therefore, CO_2 must be in a dense, supercritical state while injected into the deep rocks [44]. Geological storage of CO_2 occurs in various sedimentary basin formations such as oil fields, coal seams, depleted gas fields, and saline formations [45, 46].

(a) Mechanism of CO_2 Storage in Geological Formation

The efficiency of geological storage of CO_2 basically depends on the trapping mechanism, either geochemical or physical, where CO_2 is immovable, trapped under a low permeable seal, transformed into solid mineral and adsorbed in coal micropores surfaces. Figure 8.5 describes various mechanisms available for CO_2 storage in geological formations. The potency of storage phenomena is governed by the mingling of both mechanisms to ensure storage for a long period [47].

(b) Physical Trapping (Stratigraphic, Structural and Hydrodynamic Trapping)

Physical trapping of CO_2 below low permeable seals (low permeable salt beds or shale) is the way of storing CO_2 in the deep geological formations. Shallow gas hydrates act as a seal in some high latitude areas [48, 49]. Sedimentary basins contain such physically bound traps that are occupied by gas, oil, and saline water. Figure 8.6

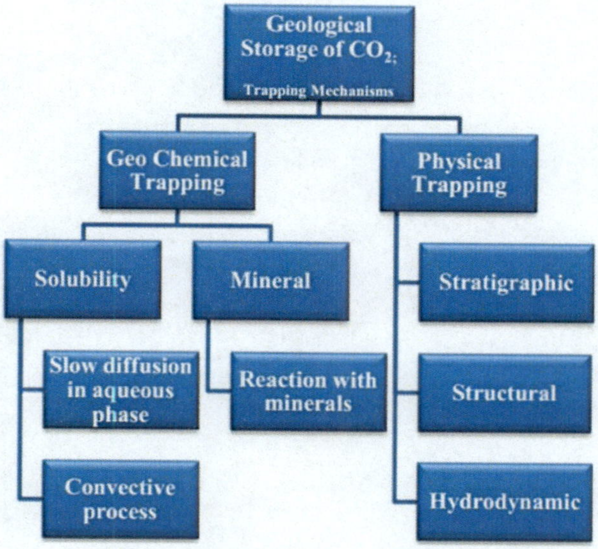

Fig. 8.5 Various mechanisms available for CO_2 storage in geological storage

shows the example of structural and stratigraphic traps for sequestration of CO_2 due to structure formation. Structural traps arise due to fracture rocks, which act as a preferential pathway for fluid flow and permeability barrier [50]. Figure 8.7 shows the CCS infrastructure using geological formations.

Stratigraphic traps are formed due to changes in rock formation because of variations in the sequence where the rocks were settled. Both types of traps (structural and stratigraphic) are considerable for CO_2 storage [43]. Hydrodynamic trapping occurs in saline formations where there are no closed traps but fluid travels over lengthy distances slowly. Once CO_2 is inserted into the formation, it displaces the water inside the saline formation and moves upward as it is less dense than water [52]. On reaching the formation's top, it migrates as a separate phase until stuck in structural and stratigraphic traps in the sealing formation or as a residual CO_2 saturation [48, 50]. In this way, the higher amount of CO_2 is dissolved in water and

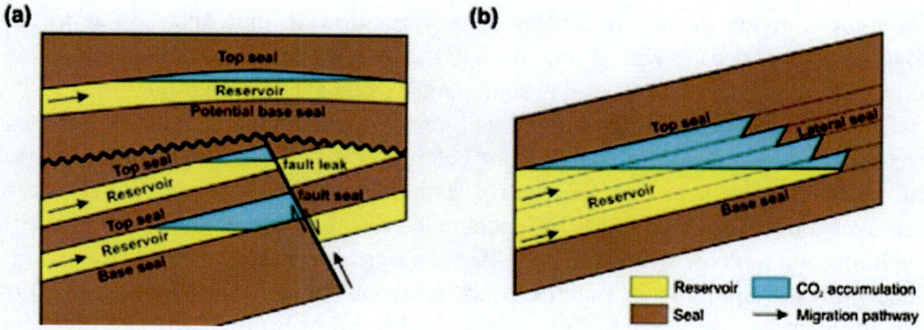

Fig. 8.6 Example of **a** structural and **b** stratigraphic trapsfor CO_2 sequestration. Reprinted with permission from Ref. [51] copyright © 2022 Elsevier

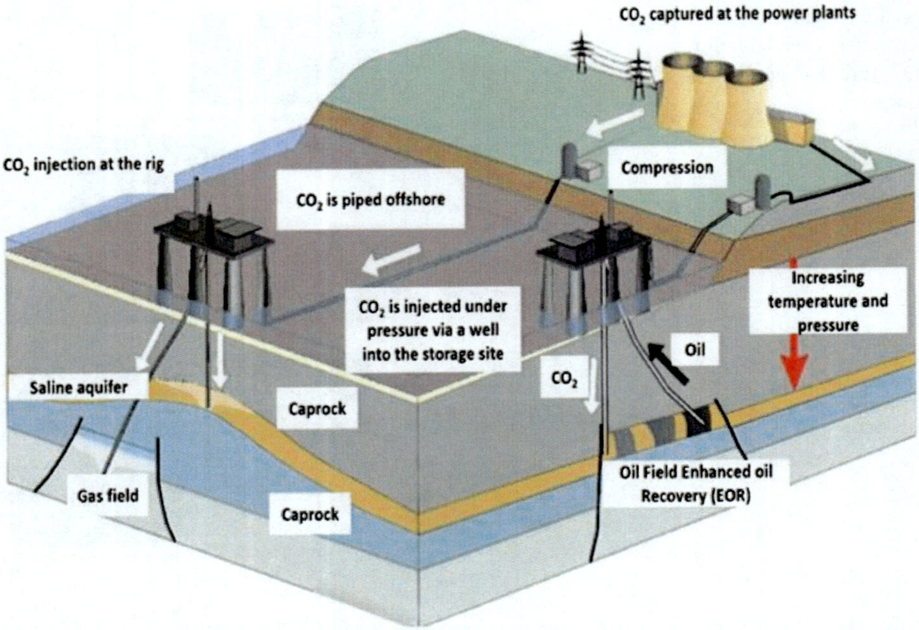

Fig. 8.7 Carbon capture and storage infrastructure using geological formations. Reprinted with permission from Ref. [54], copyright © 2022 Elsevier

migrates with underground water. The fluid takes a timescale of about a million years to reach the surface from the bottomless basin where the distance between the deep site of injection and the end of the overlying impermeable formation is more than a hundred kilometers [53].

(c) **Geochemical Trapping**

After trapping in the subsurface, CO_2 undergoes the sequence of geometrical trapping. Firstly, it dissolves in water to acidify it. The phenomenon is termed as solubility trapping. Figure 8.8 shows the pictorial representation of solubility trapping via convective mixing process (mechanisms for the dissolution of CO_2 into aquifers). The ionic trapping occurs by achieving the formation of ionic species due to the dissolving of host rock with a rise in its pH. At the end, some fraction of CO_2 is converted into mineral phase as dawsonite and carbonates in mineral trapping [55].

The formation of carbonates arises as a result of the reaction of bio-carbonates ions with iron (Fe), magnesium (Mg), and calcium (Ca) from feldspars, micas, clays, and chlorites present in the rock matrix. Mineral trapping is comparatively slow and takes thousands of years [52]. Geochemical reactions are neither the necessary conditions nor a secure storage for CO_2 in a geological setting with closed traps [57]. Geochemical trapping is also a crucial phenomenon for the secure storage of CO_2 in hydrodynamic traps. When CO_2 is dissolved in oil or brine, it will not travel as a separate phase upward but travel as in-situ fluids at the same rate as it remains underground for millions of years before its discharge at the surface [58].

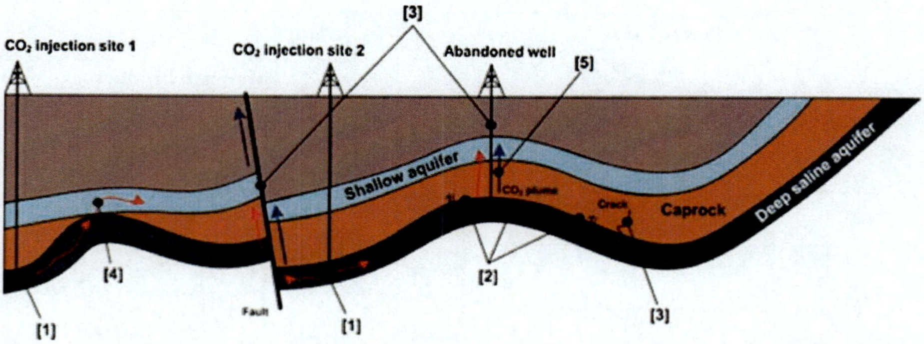

Fig. 8.8 Pictorial representation of solubility trapping process showing CO_2-rock interactions in subsurface. Reprinted with permission from Ref. [56] copyright © 2022 Elsevier

(d) Saline Aquifers

Saline aquifers are the permeable and porous rocks reservoirs containing saline fluids between the pores of rocks grains that occur at a depth larger than the aquifer that contains portable water [59]. Due to high quality of saline formation and higher depth, contained water cannot be economically subjugated for surface use [60]. Similarly, saline aquifers are located in the Northern sea at Sleipner, where CO_2 is inoculated into the Utsira Sandstone Formation [61]. Saline aquifers have the ability to a store great amount of CO_2 for a long time. Storage sites must be positioned in those areas which are geologically stable because the tectonic action could form a way for CO_2 to escape out from the reservoir through caprock into overburden and then escape to the earth's surface [62]. Saline aquifers could be of limestones or sandstones but an efficient CO_2 storage reservoir must have the size large enough to store excessive capacity of CO_2 for a longer time; and permeability and porosity parameters of reservoirs should be higher enough to attain the sufficient volume of CO_2 and to allow its injection. As CO_2 is inserted into the pores spaces inside the reservoir rocks, it displaces in-situ pore fluid. If the barriers exist in the flow of fluids and rock is less permeable, then injection causes a higher pressure of fluid centered at the injection point, which limits the rate at which CO_2 is injected as well as the stored CO_2 amount [63]. Secondly, the depths of the aquifers at about 800 m below the sea level are considered as efficient for CO_2 storage. CO_2 exists as a liquid in its dense phase and in its gaseous phase occupy less pore volume at subsurface pressure and temperature around 600–800 m [63, 64].

(e) Coal Bed Reservoir

A coal bed is another sink to store CO_2, which can be used to produce coal and coal bed methane at a shallower depth. A coal bed consists of dual primary and secondary porosity systems. A primary porosity system arises due to pores in the matrix, and a secondary porosity system is generated due to fractures in the porous network [65]. The primary target of coal bed drilling is methane, which is by adsorption stored in the coal matrix. CO_2 has a good affinity for coal as compared to methane. Thus,

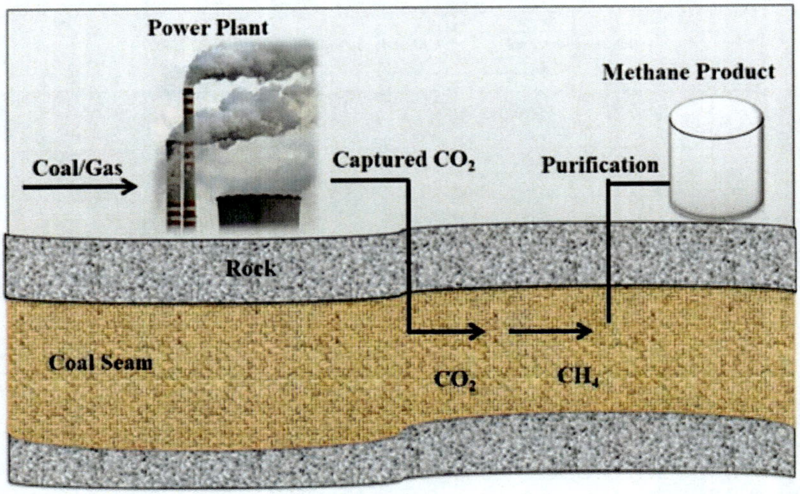

Fig. 8.9 Schematic representation of CO_2-enhanced methane (CH_4) recovery

CO_2 is a good candidate for enhancing methane recovery, and coal beds are also considered a suitable place for CO_2 storage [66, 67].

In terms of production and mechanism of storage, coal bed reservoirs are typically different in nature as compared to hydrocarbon reservoirs. CO_2 is dissolved in oil in conventional oil reservoirs, which causes a decrease in the viscosity of oil, resulting in a great amount of CO_2 to be recovered with the produced oil at the surface [67–69]. In the coal bed case, the majority of CO_2 is absorbed directly on the surface of the coal bed due to an efficient sequestration mechanism, forcing methane off the coal bed surface and aiding in the release of produced residual gas. Figure 8.9 shows the schematic process of methane enhanced recovery. Coal beds act as an effective CCS contributor due to their efficient CO_2 based-enhanced oil recovery (EOR) economics mechanism and their sequestration [68].

(f) Oil and Gas Fields

Depleted reservoirs of oil and gas are considered as a primary candidate for CO_2 storage. The oil and gas accumulate naturally in the structural and stratigraphic traps and do not outflow even for millions of years, signifying their safety and integrity [43]. In the CO_2 storage handling mechanism, the depleted fields cannot be affected harmfully by CO_2 (it already contains hydrocarbons). The schemes for CO_2 storage can be optimized to enhance the production of gas and oil during the formation of hydrocarbon fields [70]. Primary production of gas is followed by CO_2 injection then an afterward depletion strategy is adopted as a secondary practice for recovery and storage of gas, which basically depends on the injection strategy, characteristics of the reservoir, and operational parameters. For the storage purpose of CO_2, the depleted oil and gas reservoirs are regarded as an efficient candidate due to their surface integrity and sub-surface condition [71, 72]. These reservoirs have considerably limited operational costs and have the ability to confine gas or liquid for millions of years [43]. However, after depletion, a higher fraction of natural gas was left inside the reservoir, referred to as trapped gas, which includes both swept and residual gas. During enhanced gas recovery (EGR), once injected CO_2 is mixed with residual gas, the produced gas quality is significantly decreased as the mixing is not extensive [73].

(g) Salt Caverns

Salt caverns are mostly formed in underground deposits of salt. These are basically the cavities that can be utilized for the storage of hydrocarbons, as well as waste disposal [74, 75]. Salt caverns for the purpose of gas storage are formed due to the mechanism of leaching in which water at a high temperature is pumped to dissolve salt. The brine is removed by a single well, which facilitates the extraction and gas injection. Large cavities up to a volume of 5×10^5 m^3 can be formed by solution mining where water is inoculated inside the dome or salt beds and brine solution is pumped out [76]. Salt caverns offer long-term storage of waste disposal, gases and petroleum products. They can also be used for CO_2 sequestration, but this technique is relatively expensive as compared to geological sequestration of CO_2 and its implementation on a large scale is also difficult [77, 78]. Nevertheless, in northeastern Alberta, Canada, the deposits of oil sand and a large number of salt beds are sited in areas where the options for CO_2 geological storage such as deep saline aquifers and coal beds are limited. Therefore, CO_2 sequestration by salt caverns becomes effective option for reducing the emission of CO_2 for oil and gas industry [45].

(h) Shale Reservoirs

The development of horizontal drilling and shale formation technology using hydraulic fracturing further explores the storage of oil and gas reservoirs. Shale formation has low permeability as compared to primary production and is not able to produce the high quantity of oil [79]. In some cases, tertiary production is initiated with gas injection instead of water flooding due to the low permeability of shale and the reaction risk between water and clay [80]. During this process, some of the injected CO_2 is stored permanently in the formation of different storage mechanisms. Similarly, injected CO_2 may also be produced with the oil stream and is recycled back into the formation. CCS in shale reservoirs is more difficult due to the lack of information about geology and long-term behaviour [81]. In the future, shale reservoirs might play a pivotal role in CCS because of scaling up of the various shale reservoirs, maintaining their quality of efficient sealing, and the improvement of EOR techniques on the existing ones [82].

(i) CO_2 Storage in Oceans

World's 70% of the earth surface is covered with oceans, having an average depth of about 3800 m. An efficient way to store CO_2 is to inject the captured CO_2 into the deep oceans where most of its quantity is isolated from the atmosphere and stored for longer periods. This action can be achieved by the transportation of CO_2 via ships or pipelines to the sites of ocean storage where it is injected into the ocean water column [83]. Figure 8.10 shows the schematic of CO_2 ocean storage strategies. As

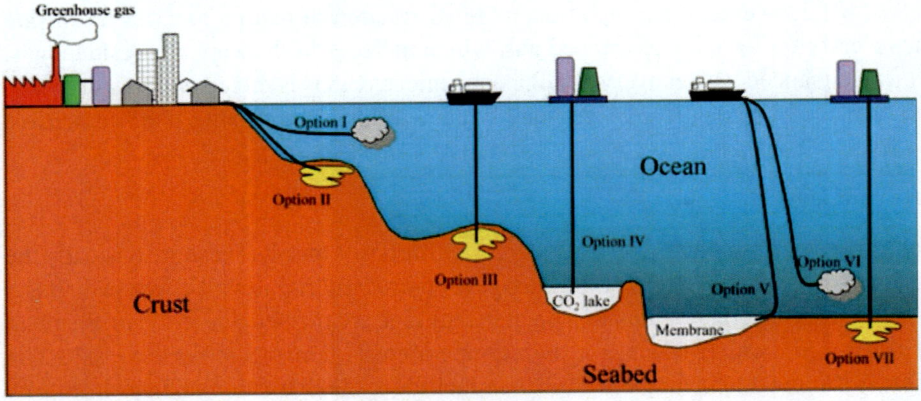

Fig. 8.10 Schematic illustration of some ocean storage strategies. Reprinted with permission from Ref. [85] copyright © 2022 Elsevier)

CO_2 is insoluble in water, there exists a natural exchange of CO_2 between water and atmosphere at the surface of the ocean, which occurs on reaching equilibrium. If CO_2 atmospheric concentration increases, the ocean takes up further CO_2 gradually. In this way, the oceans have occupied about 500 Gt CO_2 of the total 1300 Gt CO_2 of anthropogenic emissions released to the atmosphere over the past 200 years [84]. Due to increased atmospheric CO_2 concentration above the pre-industrial level, currently oceans are taking up CO_2 at a rate of about 7 Gt CO_2 year^{-1}. Most amount of this CO_2 resides in the ocean (upper surface), causing a decrease in pH at the ocean surface of about 0.1 pH due to the acidic nature of CO_2 in water.

There is no physical change observed in pH. However, simulation models predict that the oceans will cover more amount of CO_2 due to its dissolution with surface water at the ocean surface. However, the amount of CO_2 stored in the ocean is not yet quantified [86]. If the CO_2 atmospheric concentration is stabilized between 350 and 1000 ppmv, it indicates that the ocean can ingest 2000–12,000 Gt of CO_2 through active injection. However, environmental factors like pH may affect the storage capacity [87, 88].

8.3 CO_2 Gas Utilization and Its Products

The term carbon dioxide utilization (CDU), which is also referred to as carbon capture and reuse (CCR), utilization and storage (CCUS), carbon capture and utilization (CCU), and other associated terms, is a critical factor in addressing the global warming by controlling the emission of CO_2 in the atmosphere and also the best choice to use it as a renewable source of energy for the production of numerous valuable products. CDU is all about the conversion of carbon dioxide into useful organic, inorganic, and polymeric chemical products. This process is characterized by the formation and linking of new bonds to the CO_2 and transforming it into other

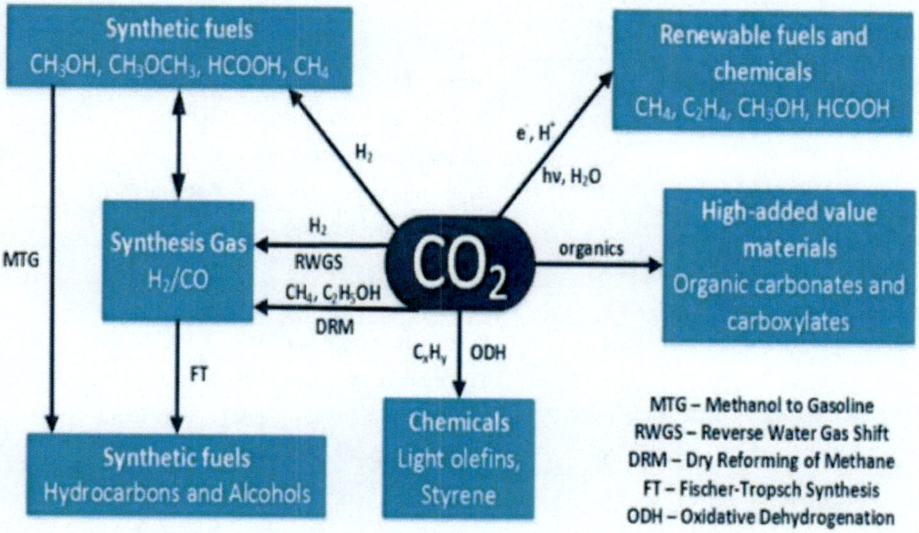

Fig. 8.11 Different options for CO_2 utilization reported in literature Ref. [94]

chemicals [89]. The choice of technique must be economically safe, practicable, and eco-friendly [90]. Prime routes of CO_2 utilization are chemical conversion, desalination, enhanced gas/oil recovery, and mineralization. It also entails the conversion of CO_2 into mineral carbonates, its use in food and beverage processing, and in the synthesis of urea [91].

CO_2 can also be used in water treatment and polymer synthesis, etc. [92, 93]. Figure 8.11 depicts the various pathways for CO_2 utilization in different applications [94]. However, further efforts are required to utilize the already captured and stored carbon dioxide in to valued possessions such as transportation of chemicals or fuels. In 2015, during Faraday's discussion on CO_2 utilization, twelve (12) principles of CO_2 chemistry analogues to the principles of green chemistry were proposed by Poliakoff (Table 8.2) [95]. These approaches help us to estimate the possible products that can be produced.

8.3.1 CO_2 as Feed Stock for Production of Chemicals

Utilization of non-fossil sources of carbon is one of many ways to reduce the emission of CO_2 and to address the problems related to this [96]. CO_2 can be used as a feedstock for the production of a large range of fine chemicals. Some of the major applications include that CO_2 gas is utilize in the production of urea, carbonates (inorganic), polyurethane, acrylates, acrylic acid, and alkylene carbonates [93, 97, 98]. Organic carbonates such as acrylic carbonates (Dimethyl carbonate, diallyl carbonates, diphenyl carbonates, diethyl carbonate, etc.) and cyclic carbonates (propylene carbonate, ethylene carbonate, cyclohexane carbonate, polycarbonates, bisphenol

Table 8.2 Principles of CO_2 chemistry

C	Catalysis is crucial
O	Origin of the CO_2?
2	Tomorrow's world may be different
C	Cleaner than existing process?
H	High volume of high-value chemicals?
E	E-factor must be low
M	Maximize integration
I	Innovative process technology
S	Sustainability is essential
T	Thermodynamics cannot be beaten
R	Renewable (and reasonable) energy input
Y	Your enthusiasm is not enough

polycarbonate, and styrene carbonate), that have major applications in lubricants, catalytic reactions, agrochemicals, polymers, coatings, and pharmaceuticals are also produced from CO_2. Polyurethane can also be produced by reaction of azetidines, n-analogues & aziridines of epoxides and CO_2 [92, 99].

Formic acid is another chemical that can be produced by CO_2 utilization. CO_2 hydrogenation to formic acid has great interest due to multiple reasons including; (a) mild reaction condition (b) no by-product formation (c) ability of hydrogen storage in liquid form and (d) easy decomposition of formic acid in CO_2 and hydrogen [97, 100]. Figure 8.12 illustrates the different possible routes for the transformation of CO_2 into useful chemicals [101]. High pressure and temperature of feed stock and need for large catalytic inventory are the main obstacles to adopt this route in the production of chemicals.

Furthermore, catalytic separation from the product is the other challenge of this route. For example, in polycarbonates production from the reaction of CO_2 with epoxy, the most commonly used catalysts are aluminum (Al)-based catalysts which are not environmentally friendly. Another route is oxidative carboxylation, which is a substitute for polycarbonate synthesis from olefins and CO_2. Biological utilization of CO_2 promotes another trail for the production of biodiesel and other chemicals used as a fertilizer, biogas, and for silage purposes. The benefits of this approach include a short growth cycle, a higher growth rate, and the production of diverse valuable by-products. However, CO_2 must be purified before its feeding into a photo-bioreactor for the removal of heavy metals, NO_X, and SO_X, which are toxic to microalgae evolution [102, 103].

Overall, there is an excessive market for CO_2 adaptation into fuel and chemicals, but the technology at lab stage is far away from being implemented on an industrial scale. The main reasons are that the investigated materials are expensive and chemically not much stable. Similarly, the conversion rate of CO_2 and the production of key products are comparatively low, which means they are not feasible to meet the requirements of large scale up of the technology [51].

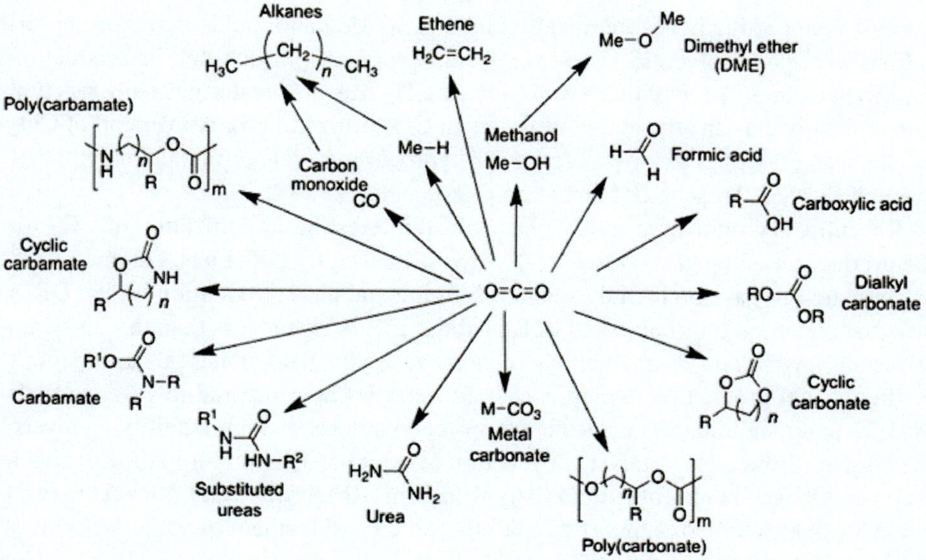

Fig. 8.12 Various possible chemical production via CO_2 utilization. Reprinted with permission from Ref. [104] copyright © 2022 Elsevier

8.3.2 CO_2 *Utilization for Fuel Production*

Utilization of CO_2 as a feedstock for fuels is also considered as an important aspect that could not only help to retain the climate change but also reduce the reliance on fossil fuels. Fuels such as methanol, syn gas, methane and alkanes can be produced with captured CO_2 as a feed stock which can be further used in several sectors like power plant, fuel cell, transportation etc. [105]. CO_2 has a stable thermodynamic nature and its utilization involves the existence of a catalyst and excess heat to obtain high yields of fuel [106]. Dry methane reforming and hydrogenation are important pathways for the production of fuel from CO_2 [103]. CO_2 hydrogenation has an edge in the recycling possibility of CO_2, H_2 storage ability, solving the electrical energy storage issue and producing fuel, whereas dry methane reforming (DMR) is also considered as one of the advantageous phenomena for methanol and fuels (liquid) production by the Fischer–Tropsch (FT) process [92, 107, 108].

Hydrogenation of CO_2 into formic acid (CH_2O_2) [109], methanol (CH_3OH) [110], methane (CH_4) [111] and carbon monoxide (CO) by using hydrogen (H_2) as a source from fossil fuel seems to be challenging as it leads to CO_2 emission into the atmosphere itself. However, renewable sources such as biomass, solar, and wind are alternatives to fossil sources to alleviate the additional emission of CO_2 during hydrogenation [112]. CH_4 has a low volumetric gas density, so producing more CH_4 will not be worthwhile to capture CO_2 due to its high availability. Rather, CO_2 hydrogenation into methanol is regarded as an efficient pathway. The activation of C–H bonds over copper (Cu) based catalysts for CH_4 production is extremely challenging and the catalyst is not attractive economically [106, 113]. Methanol holds numerous applications in organic solvents, paints, combustion engines, plastics, and its production contributes about 0.1% reduced emission of CO_2. Reverse water gas shift reaction is also regarded as an advantageous routes of CO_2 utilization for conversion of CO_2 to CO, as CO is used as raw material for hydrocarbon fuels & methanol synthesis through Fischer–Tropsch (FT) reaction [106].

Recently, dry reforming (DRM) has gained interest in the utilization of CO_2 for the production of syngas [114–116]. Syngas produced by DRM has a higher purity as compared to gas obtained by steam reforming and partial oxidation. In the DRM process, the amount of unreacted CH_4 is about 2% (which is less than that in steam reforming), which makes it an efficient choice to apply DRM at natural gas locations for liquid fuel production, which is easy to transport as compared to gaseous fuels [92]. Despite significant advancements in the synthesis of high stability catalysts, the choice of the right catalyst for reaction at a higher temperature still remains a major challenge [117]. Ruthenium (Ru), Rhodium (Rh), Iridium (Ir), Nickel (Ni) and Nickel–Cobalt (Ni–Co) support on alumina, silica and lanthanum oxide are widely used catalysts in reaction of DRM.

Oxidative dehydration of lighter alkanes to alkenes with CO_2 is another approach which lessens the amount of coke formation as well as maintains the catalyst stability at elevated temperatures. Furthermore, CO_2 boosts the dehydrogenation of light alkanes by eliminating H_2 through a reverse water gas shift (RWGS) reaction. But attention must be needed in temperature monitoring as the extreme heat could cause the olefins to suffer from over-oxidation that produces CO_2 by ensuing low selectivity [118–120]. The core barrier to using the stored CO_2 for manufacturing of syn fuel relies on the design and implementation of an optimized catalyst having higher catalytic action under different conditions as well as the ability to resist coke formation and prolong structural and thermal stability [92, 121].

8.3.3 CO_2 *Employment for Biofuel/renewable Energy Sources*

One of the sustainable methods to utilize CO_2 is its conversion to a valuable, renewable, and unlimited carbon source. Considering the limited sources of fossil fuels, looking for some alternative cheap fuels (i.e. biofuels) has become the foremost concern worldwide. The CO_2 utilization for biofuel production will not only helps to deal with issues associated to global warming but also provides an endless and easily accessible carbon source for generations.

The electro-catalytic or electrochemical reduction of CO_2 to other C_1 (C_2H_5OH, CO, CH_4, CH_3OH, C_2H_4) and Cn molecules such as (Cn-hydrocarbons, olefins, alcohols) has long been examined as a method of energy storing [122–124]. Such reactions help to reduce the anthropogenic emission of CO_2 into atmosphere. Conversion of CO_2 into energy chemicals can be considered as the storage of excess electric energy or "intermittent energies", which are not usually subjugated because they do not provide continuous energy generation as demanded by most process uses [125].

For some CO_2 conversion reactions, generally in carbonation reactions, modest external energy supply is required to either keep the CO_2 moiety intact, such as in carbonates and polycarbonates, or change one or two C–O bonds, such as in urea. In both cases, the oxidation state of carbon is kept at + IV. Some other reactions involve the reduction of oxidation state as 0 in HCOH, − II in H_3COH, + II in RCOOH, and − IV in CH_4, and can also imply C-bond functionalization, as summarized in Fig. 8.13.

(a) Catalyzed CO_2 Hydrogenation Reactions

Hydrogenation of CO_2 has widely been accepted as a promising reaction method for the production of hydrocarbons i.e., CH_4, CO, C_{2+} products etc., and clean fuels. Generally, there are three chief routes for this purpose, including photo-catalysis, thermal catalysis, and photo-thermal catalysis. Among them, thermal catalysis is extensively studied due to its high yield and greater efficiency by means of production rate. The process is initiates by the hydrogenation of CO_2 to produce CO via a ***Reverse Water Gas Shift*** (RWGS) reaction, as described in Eq. (8.1). Hydrogenation of CO then continues to produce CH_4 Eq. (8.2). This could also be achieved directly by the ***Sabatier reaction***, which proceeds through direct methanation of CO_2 as mentioned in Eq. (8.3). Methanation of CO_2 and CO are both exothermic reactions and occur at fairly high temperatures, i.e. 200–500 °C. Paraffins (alkanes) and ***olefins*** (alkenes)

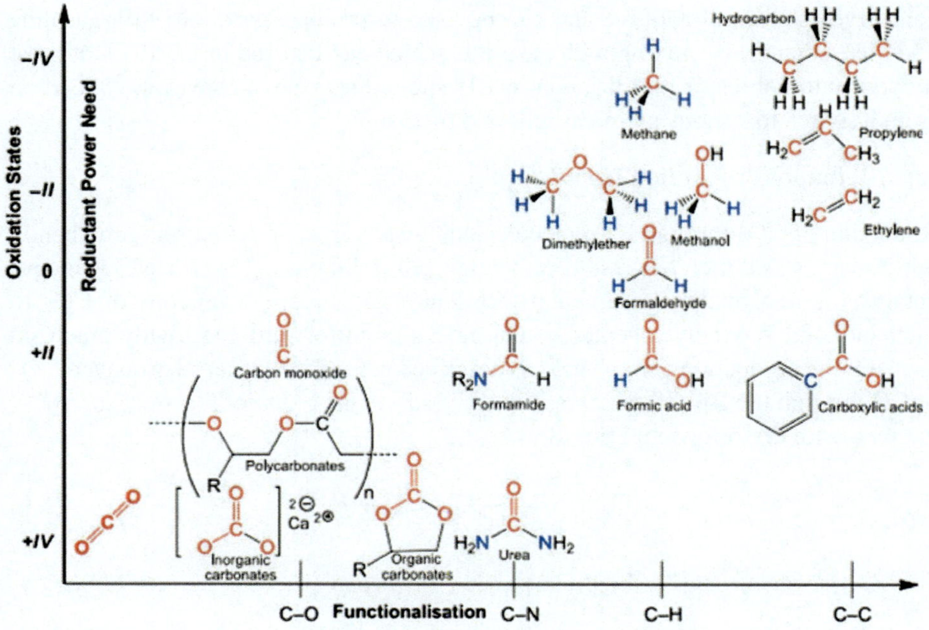

Fig. 8.13 Some representative molecules issued from CO_2 utilization reactions. Reprinted with permission from Ref. [126] copyright © 2022 Elsevier

are other essential and expensive chemical feedstocks that can be produced via the Fischer–Tropsch reaction, as described in Eqs. (8.4) and (8.5), respectively [127].

$$CO_2 + H_2 \leftrightarrow CO + H_2O \quad (8.1)$$

$$CO_2 + 3H_2 \leftrightarrow CH_4 + H_2O \quad (8.2)$$

$$CO_2 + 4H_2 \leftrightarrow CH_4 + 2H_2O \quad (8.3)$$

$$nCO + (2n + 1)H_2 \leftrightarrow CnH_{2n+2} + nH_2O \quad (8.4)$$

$$nCO + 2nH_2 \leftrightarrow CnH_{2n} + nH_2O \quad (8.5)$$

(b) **Bioalcohols as Another Potential Biofuel Product from CO_2 Emissions**

Alcohol-based fuels frequently originate from biological rather than petroleum sources and are generally identified as bioalcohols. Methanol (CH_3OH), ethanol (C_2H_5OH), propanol (C_3H_7OH) and butanol (C_4H_9OH) are commonly used bioalcohols for motor fuels. Even though only methanol and ethanol are economically and technically appropriate for usage as fuels. Methanol has a higher octane value and only half the energy density in comparison to gasoline, which helps the fuel to endure higher compression pressure before ignition [128]. On the contrary, ethanol can serve as a petrol additive and can be used to produce gasohol (90% gasoline and 10% ethanol) by mixing with gasoline, which can be used in the ICEs of most modern automobiles [129, 130]. Figure 8.14 shows the cycle of conversion of carbon from a source to various hydrocarbons and methanol.

(c) **Methanol Production from CO_2**

Reduction of CO_2 through H_2 molecules that results in the production of methanol, which serves as a fuel, has also been recognized in literature [131]. Heterogeneous catalysis is one of the fundamental techniques for direct conversions of CO_2 to methanol and is widely accepted as the most straightforward and highly practiced route for the production of methanol, as described in Eq. (8.6). It can also convert CO_2 to CO through the RWGS reaction, Eq. (8.7), or via the Sabatier reaction, Eq. (8.8), coupled with dry reforming, Eq. (8.9).

$$CO_2 + 3H_2 \rightarrow CH_3OH + H_2O \quad (8.6)$$

$$CO_2 + H_2 \rightarrow CO + H_2O \quad (8.7)$$

$$CO_2 + 4H_2 \rightarrow CH_4 + 2H_2O \quad (8.8)$$

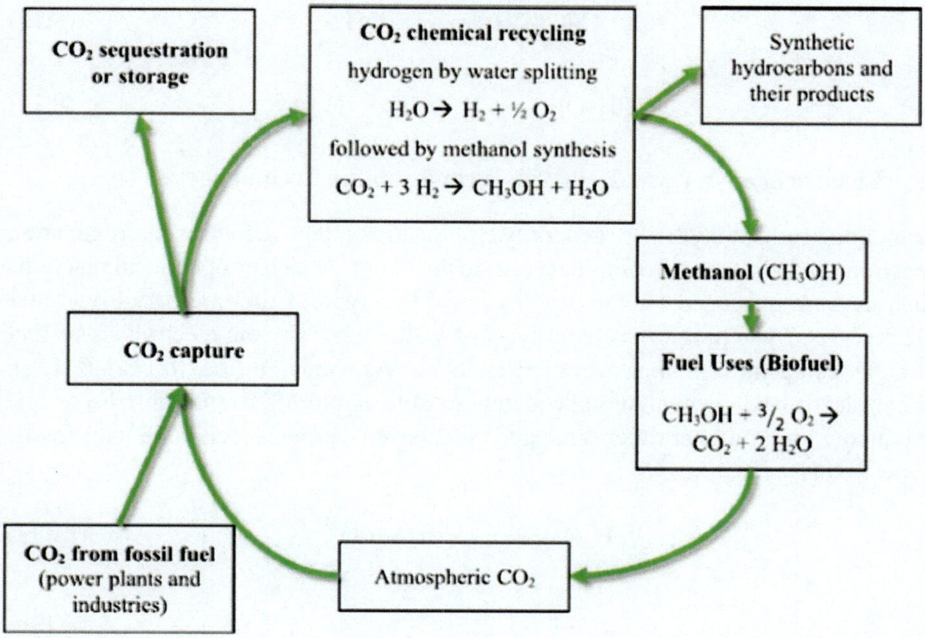

Fig. 8.14 Carbon conversion cycle from source to methanol and other hydrocarbon products [128]

$$CO_2 + CH_4 \rightarrow 2CO + 2H_2 \tag{8.9}$$

These reactions of methanol generation have been known by chemists for more than 80 years. Actually, in the 1920s and 1930s, the methanol production plants established in the USA were using CO_2, which was generally obtained as a by-product of other processes such as fermentation [128]. Additionally, a metal oxide-based proficient catalyst, particularly the combination of zinc oxide (ZnO) and copper (Cu), has been developed for this conversion [132].

(d) Dimethyl Ether (DME) Production from CO_2

Dimethyl ether (DME), produced from CO_2 and H_2, has potentially been considered efficient to be used as a clean alternative fuel for diesel engines. As it is well recognized as a highly effective ignition fuel with low SO_x, NO_x and particulate matter, it has the capability to be reformed to hydrogen at low temperatures [133]. Usually, two consecutive reactions are involved in the generation of DME; methanol synthesis followed by the dehydrogenation of methanol, as described in Eqs. (8.10) and (8.11) [134]. Furthermore, production of DME could also be achieved by direct conversion of syngas using an appropriate catalyst. In the recent past, researchers have been more in search of a catalyst that can help to achieve direct hydrogenation of CO_2 to methanol. Among the several, chromium (Cr), copper (Cu), palladium (Pd), and zinc (Zn) are the most widely used for this purpose as they help to maximize the yield and selectivity of methanol by minimizing the formation of other byproducts [135].

$$CO + 2H_2 \rightarrow CH_3OH \tag{8.10}$$

$$2CH_3OH \rightarrow CH_3OCH_3 + H_2O \tag{8.11}$$

(e) **Electrochemistry as a Key Mid-Term Enabling Technology**

Dihydrogen and dioxygen are generally employed as effective fuels in hydrogenation reactions and also in hydrogen fuel cells downstream. The electrochemical reactions that are currently used for the production of this type of fuel are based on water electrolysis. Switching to coelectrolysis of water and CO_2, the electrolysis-to-fuel strategy can achieve higher overall efficiencies. As shown in Eqs. (8.11a)–(8.11c), water electrolysers, which are adequately established to offer renewable H_2 (r-H_2) production in a rational time frame and cost, have become more viable with fossil-generated H_2.

$$H_2O + 2e^- \rightarrow H_2 + O^{2-} \tag{8.11a}$$

$$O^{2-} \rightarrow 2e^- + \frac{1}{2}O^{2-} \tag{8.11b}$$

$$H_2O \rightarrow H_2 + \frac{1}{2}O^{2-} \tag{8.11c}$$

Another way to temporarily store electrical energy in the bond energy of chemicals is the coupling of an electrolyser with its corresponding fuel cell, which could be used as fuel for the downstream cell. When the circuit is switched from the electricity-consuming to the electricity-producing mode of the device, the stored energy will be released back in parts.

(f) **The Long-Term Goal of Solar Fuels Through Photochemistry**

Carbon dioxide is a well-considered bridge between the complexity of life forms and almost unlimited solar energy, including the molecules that serve as their energy sources. There are currently three main research aspects enthused by this, which include; mimicking photosynthesis on solid photo-catalysts, harvesting photosynthesis products by developing adequate photo-bioreactors and altering microorganisms to steer photosynthesis towards the desired non-natural molecule to be used as fuel. These CO_2 utilization technologies are attracting substantial interest. At the same time, their timelines place these approaches among the long-term strategies [136, 137].

8.3.4 Dual-Function Materials (DFMs) for In-Situ CO_2 Conversion and Utilization

Materials that are composed of both catalytic active components and sorbents are called dual-functional materials (DFMs). On account of redox catalytic sites and basicity, these DFMs offer in situ CO_2 capture and utilization. These systems are operated by sequential feeding and alternate switching between different gas streams. Initially, the diluted CO_2 stream is fed up to complete saturation. Then the conversion of CO_2 is brought by the flowing of the reactive components such as CH_4/C_2H_6 or H_2. This method helps to address the problems associated with CO_2 storage and transportation.

Moreover, it also helps to overcome the challenges related to the high energy cost of thermal swing regeneration associated with other similar schemes, such as post combustion in amine-scrubbing [138]. It is also worth noticing that the working capacity of DFMs could be affected by both the physisorption and chemisorption properties of sorbent, as they are both associated with the occurrence of van de Waals and covalent chemical reactions, respectively. In the designing of DFMs, parameters such as feed gas composition, temperature of adsorption and reaction, sorbent to catalyst ratio, and weight loading of sorbent are of great importance as they help to optimize the stability, reaction potential, and sorption capacity. A few of the DFMs developed for CO_2 utilization are discussed below.

(a) **Nickle (Ni)-Based DFMs**

Owing to its high activity, cost effectiveness and abundancy, nickel has been greatly used in conversion reactions of CO_2, such as reforming of light alkanes, methanation and RWGS. Using a bifunctional CaO–Ni sorbent prepared through the Pechini sol–gel method, a calcium-looping methane reforming process is reported by Tian et al. [139]. In this process, nickel and calcium nitrates were mixed with anhydrous citric acid in water at 75 °C with continuous stirring. Then ethylene glycol was added to induce the gelation. The obtained gel was dried overnight and calcinated at 800 °C for 2 h to obtain the CaO−Ni sorbent catalyst. Another study investigated by Kim et al. reports the use of $Ni/MgO-Al_2O_3$ (a Ni-based catalyst) containing CaO for the dry reforming of methane [140]. Ni-impregnated MgO and CaO-based double salts supported on γ-Al_2O_3 has also been reported for conversion of CO_2 to syngas via dry ethane reforming, as described in Eq. (8.12) by Al-Mamoori et al. [141].

$$2CO_2 + C_2H_6 \rightarrow 4CO + 3H_2 \quad \Delta H_r^o = +428\,\text{kJ mol}^{-1} \tag{8.12}$$

(b) **Ruthenium (Ru) and Rhodium (Rh)-Based DFMs**

Ruthenium (Ru), owing to high selectivity and activity towards methanation of CO_2, has been extensively studied for DFMs. A lot of attention has been paid to Ru due to its fast reversible redox reaction that offers an easy transition between its oxide and metallic states [142]. Thus, due to these attributes, it is considered a strong candidate for the development of DFMs despite its high cost. One of the earliest studies on methanation of CO_2 using Ru-based DFMs was published by Duyar et al. [143], in which both Ru as a catalyst and CaO as a sorbent were dispersed into γ-Al_2O_3. Impregnation of $Ca(NO_3)_2$ on 10 wt% Ru/γ-Al_2O_3 was also used in another study to develop a DFM, the process was followed by in situ reduction to decompose $Ca(NO_3)_2$ into CaO at a temperature of 320 °C. These authors also compared the performance of Rhodium (Rh) and Ru-based DFMs using $Rh(NO_3)_3$ and $Ru(NO)(NO_3)_3$ as precursors. Results revealed that Rh-based DFMs with only 0.1% Rh loading were more active than those containing Ru, obtaining 0.40 g-mol CH_4/kg DFM [144].

(c) **Other Novel DFMs**

Recently, research has been focused on using earth-abundant metals such as Cr, Cu, and Fe, which are economically feasible and can enormously promote the development of DFM technology [112, 145]. Hyakutake et al. [146] investigated the fundamental active sites and chemical processes involved in CO_2 capture and conversion using a Cu/Al_2O_3 catalyst promoted with Ba and K. Among three different samples of Cu/Al_2O_3, K-promoted (Cu−K/Al_2O_3) and Ba-promoted (Cu −Ba/Al_2O_3), Cu−K/Al_2O_3 DFM showed the best behaviour by means of effectively capturing CO_2 and converting it to CO upon hydrogenation.

8.3.5 CO_2 *Utilization with Fly Ash*

60–88% of total combustion residues from the byproduct of a coal-fired power plant are comprised of fly ash. The annual production of fly ash is about 0.75–1 billion tonnes, worldwide, which is estimated to increase in the near future as the demand for cheap power supplies rises in developing countries [147]. With this increase, the disposal of fly ash will lead to many environmental threats due to the engagement of particulates in the air and oozing the toxic components into the water bodies and soil [148]. Among the various proposed methods to overcome this problem, carbon capture and its utilization is one of the most favoured methods which not only contributes to controlling the global CO_2 emission problems but also deals with fly ash disposal. The use of this inexpensive fly ash in carbon utilization could appear as an effective way to address the reduction in cost and decrease the environmental risks associated with fly ash disposal [149]. Among the range of opportunities for fly ash utilization with CCUS, a few of them are discussed below.

(a) **CO_2 Capture with Fly Ash Derived Sorbents**

Though fly ash can be directly used as a sorbent, but it is more effective and offers better adsorption characteristics in the form of porous materials. For CO_2 capture, the porous materials derived from fly ash can be categorized into four classes, including activated carbons, zeolites, mesoporous silica, and alkali silicate. Several pollution control applications, such as the removal of heavy metals in water treatment or the removal of volatile organic compounds (VOCs) from effluent gases takes advantage of activated carbon as an effective adsorbent. Generally, the synthesis of activated carbons is achieved through different raw materials such as coal, wood, petroleum pitch, and coke pitch. However, certain fly ashes that possess high levels of unburned carbon can also be used for the preparation of activated carbon. This synthesis could be achieved by a number of methods. Usually, the content is first isolated from other fly ash components, which is then thermally activated by heating it up to about 850–900 °C in an N_2, steam or CO_2 atmosphere for pore network development [150].

The application of fly ash as a solid sorbent for carbon capture and then as an admixture in cement mixtures has recently been investigated by Siriruang et al. [151]. The process was initiated by bonding CO_2 to the fly ash in the form of $CaCO_3$. In another study reported by Wei et al. [152] carbonated fly ash was employed as an alternative to Ordinary Portland Cement (OPC). From their work, it appears that fly ashes (rich in Mg and Ca) upon carbonation can lead to the formation of cementitious materials with properties good enough for structural construction. The production of industrial chemicals such as cyclic carbonates is another promising way to utilize CO_2. Chances are there to utilize fly ash derived from mesoporous silica as a catalyst for such kind of reactions. The synthesis of cyclic carbonates using silica-supported and non-supported amines has been reported recently by Yu et al. [153].

8.3.6 Utilization of CO_2 *for Enhanced Gas and Oil Recovery*

CO_2-enhanced oil recovery (CO_2-EOR) has emerged as a major option for productively utilizing CO_2 emissions captured from electric power and other industrial plants. The production of unconventional natural gas reservoirs through the recovery of enhanced gas by injecting CO_2 is assumed as a promising method for CO_2 sequestration and to improve the recovery of gas as well [154]. In the process of enhanced oil recovery (EOR), a constituent is inoculated into the reservoir that re-pressurizes the rock formation and releases gas or oil trapped in spaces. During this process, the injected CO_2 is mixed with oil to recover the rock formation. The stream is then pumped to the surface where the CO_2 that was mixed with oil is separated out and recycled to repeat the process. As compared to the traditional methods of oil recovery, this process produces more barrels per reservoir [155]. In EOR, CO_2 flooding is also considered as one of the most advantageous method as it mixes with oil and makes it easier and lighter for recovery [156].

Continuous gas injection and water alternating gas are two commonly used methods for CO_2-enhanced oil recovery (EOR). Furthermore, the addition of hydrocarbons (for instance, propane) can also improve the diffusion coefficient, displacement efficiency, as well as the separation efficiency [157]. Most systems of CO_2-EOR use naturally occurring CO_2, but research is being carrying out on capturing and storing the CO_2 from industrial effluents and flue gases to reduce the anthropogenic emission of CO_2 [158]. Even with a lot of applications of CO_2-/EOR in reservoirs, this approach still faces challenges due to rock formation heterogeneity between wells, properties of fluids, and capillary pressure, which reduce the CO_2 flooding effectiveness. Currently, the EOR contribution in CO_2 utilization and advancement in this approach is low i.e., 3%, due to the rising cost of CO_2. However, it is progressively rising by implementing it with several facilities [155, 158].

8.3.7 CO_2 *Employment for Mineral Production*

The process in which carbonates such as calcium carbonate ($CaCO_3$) and magnesium carbonate ($MgCO_3$) are produced by the reaction between dissolved CO_2 and metal oxides, i.e., calcium, magnesium, and iron oxide, is known as carbonation or mineral sequestration. The silica and carbonates produced via mineral carbonation are stable over long periods of time. That is why they can be reused in construction processes or can be disposed of in areas of silicate mines. However, high energy requirements and slow dissolution kinetics are the main challenges that are associated with mineral processing, which need to be addressed [159]. Artificial improvements in the kinetics of carbonation can be attained by the injection of a considerable CO_2 concentration and increasing the temperature. Significant efforts are being made to accelerate this reaction, but the main disadvantage of scaling up the mineralization process is slow kinetics. Furthermore, this method is energy-intensive as it requires high temperature (150–600 °C), high pressure (100–150 bar), extraction, processing, and transportation of rocks to achieve the high carbonation efficiency of about 80% [93, 160]. The requirement of additives for extraction of reactive species and large plant size makes the process cost forfeit.

Along with the storage of CO_2 for long time periods, mineral carbonates offer several other applications, particularly as construction materials in the construction industry, and thus this sort of storage can be considered as a better utilization option [161]. Specifically, igneous silicate rocks restrain essential silicate minerals such as wollastonite ($CaSiO_3$), olivine ($(Fe_x, Mg_{1-x})_2SiO_4$), and serpentine ($Mg_3Si_2O_5(OH)_4$), which are potential feedstock for mineralization and are abundant worldwide. Besides these kind of mining-based applications, alkaline waste from several industrial processes is also considered as an important feedstock for mineral carbonation. Apart from this, the other potential feedstocks include waste such as ash from cement production, coal combustion, and municipal waste incineration. Slag from asbestos mine tailings and steelmaking is also considered as a significant feedstock. Where there is limited availability of local geological storage sites or CO_2 transportation or for more dispersed and smaller industrial sources, mineral carbonation is considered as a preferred option for sequestration [162].

8.3.8 CO_2 *Employment for Desalination and Water Treatment*

Captured CO_2 can be used for removal of dissolved solids as well as transformation of brine into water, and the resultant portable water can be utilized where there is a deficiency [163, 164]. Many plants for desalination do not pay attention to the employment of CO_2 in their processes owing to economic obligation. However, the development of innovative technologies for the economical and effectual operation of CO_2 is in progress. During the mixing of sea water with ammonia molecules, the mixture is exposed to CO_2 that weakens the bonds, leading to the formation of ions for their removal. As a result, ammonium chloride (NH_4Cl) and sodium carbonate (Na_2CO_3) are formed as product which settle down at the lowermost of the tank and can be recycled later by thermal operation with calcium oxide (CaO) and can then be utilized as a feedstock for chlorine and ammonia synthesis [164, 165]. Hydrate forming is an alternative process used for desalination involving hydrate formation by using CO_2 for salt separation from water. In the present technique, CO_2 is in the form of liquid or gaseous, and CO_2 hydrates are transported elsewhere or dumped into oceans [166].

Another technique of desalination is forward osmosis (FO). In FO, instead of applying hydraulic pressure, the driving pressure is osmotic pressure, which utilizes brine and fresh water in very large quantities [167]. Additionally, metal corrosion, high concentrations of salt, solvents, and chemical residue have the capacity to destabilize the local eco-system. To overcome the problem, three main units are used, which include carbonation, recovery of amine and chlorine compounds, and filtration. The estimated cost of desalination is relatively higher than the municipal and agricultural water costs, which makes CO_2 desalination technology not as attractive to address the water market and it is highly unlikely to use CO_2 based desalination technology for supplying agricultural quality water due to its higher cost. However, for desalination, the use of CO_2 remains an efficient choice, and the current portable filtered water cost by CO_2-based purification technologies is far from being competitive [168, 169].

8.4 Combined CO_2 Capture and Storage

Combining CO_2 capture and storage is an efficient strategy to minimize the requirement for high energy for direct fuel or chemical production from waste streams exclusively in cases when capturing and utilization are attained at the same temperature. The augmentation of such processes leads to high energy-efficient and clean processes. Figure 8.15 shows the schematic way of combined CO_2 storage and utilization. The perception of hybrid processes in the separation and reaction of gas has been previously applied to numerous applications. However, membrane reactors combine the process of membrane separation with chemical reaction in a single unit as well

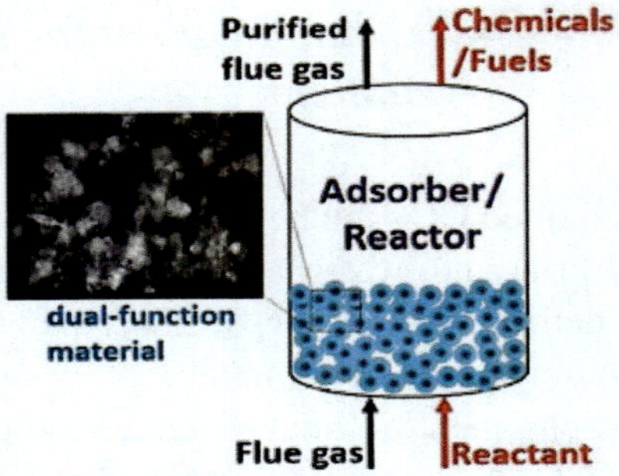

Fig. 8.15 Schematic representation of combine CO_2 capture and utilization. Reprinted with permission from Ref. [177] copyright © 2022 John Wiley & Sons, Inc.

as have an ability to encourage the reaction by the removal of product from zero reaction through the membrane by making the shift of equilibrium reaction to the product side [170–172]. Similarly, sorption enhanced reaction is another way that combine the adsorption and reaction in a single unit [173–176].

These innovative perceptions can be implemented for instantaneous CO_2 capture and utilization in numerous industries. The concept of direct fuel and chemical production from flue gases of industries over a twin functional material has been applied previously to syngas (CO and H_2) production, which is a useful reactant for the synthesis of methanol. In methane tri-reforming, a combination (synergetic) of steam and CO_2 reforming and methane partial oxidation arises in a single reactor with supported nickel catalysts at 850 °C [178, 179]. Hydrogenation of CO_2 is another example of combined capturing and reduction reported widely, and simultaneously capturing and mineralization of flue gases (burning of coal) is another hybrid process reported at pilot scale in the USA [180, 181].

These types of integrated systems could offer a solution to current worldwide energy and environmental problems. However, modified technologies in-situ capture conversion techniques require the amendment in material science side as well as processing engineering side. Table 8.3 discusses the various challenges and opportunities of different utilization techniques. Assuming the diverse nature of catalysis and adsorption, the ultimate features of hybrid catalyst/adsorbent characteristics in conjunction with operation conditions and process considerations should be considered wisely for achieving cost-effective and highly efficient technology. Furthermore, for direct use of waste gas stream the resistivity of material to impurities impose a challenge and efforts should be take on for ensuring consistency of materials for a long-life period.

8.5 Current Challenges and Future Prospects

The effective utilization and storage of CO_2 should be evaluated from the perceptive of cost and efficiency. In the recent past, significant improvements in the design and development of several carbon storage and utilization (CSU) technologies and few implementations on an industrial level have been observed. However, the majority

Table 8.3 Challenges and opportunities of different utilization techniques

Utilization technology	Challenges	Opportunities
Chemical conversion	• Regeneration of catalyst • Higher operating condition • Lower yield rate • Catalyst stability to coke formation • High selective catalyst development	• Methane dry reforming • Reduction of catalyst to formic acid (and its derivatives) • Biological paths to synthetic fuels
Enhanced oil/comparison	• CO_2 transportation • Instability in oil prices	• Water alternating gas system • Compensated neutron log
Water treatment	• Corrosion of equipment • Expensive process • Waste of high amount of brine	• Provide portable water to municipal and residential customers • Possible to implement in various regions • Improved Solvay process
Mineralization	• High temperature and pressure operation • Expensive implementation • Relatively slow kinetics	• Carbonation (indirectly) • Organic waste utilization

of opportunities are at the laboratory scale. Hence, in both circumstances, the viable employment of advanced materials that accomplish the existing state-of-the-art materials will assuredly reduce the energy requirements of both capture and utilization. In Table 8.3, challenges and opportunities of different utilization techniques are comprehensively discussed. However, Research and Development on material concepts should be coupled with consideration of process performance for their valuation of potential under real conditions. Taking such a holistic view of both processes' materials and sharing communication between engineers and material scientists will help the scale-up of CSU technologies tremendously.

Furthermore, to lessen the uncertainties in assessing the associated costs, small-scale estimation of processes and materials must be taken into account to fulfill the implementation requirements of large-scale. Cost estimation is basically the ultimate factor which determines the practicability of implementation of emerging carbon storage and utilization (CSU) technologies. The stability of material over an extensive period in the carbon capturing and utilization (CCU) method is also a significant consideration that directly affects the system performance as well as process economics. Hybrid processes which link CO_2 capturing subsystems and offer instantaneous capturing and utilization must be focus of future study as thermodynamic investigation of such systems has highlighted their cost effectiveness and energy efficiency (by reducing operating and capital costs). The developing hybrid systems seem to have many characteristics of future generation CCU technologies. However, more research and development are required for commercial availability of these technologies in near future.

8.6 Conclusion

The CCU technology is efficiently developing and heading towards sustainable carbon capture and utilization procedures. Since, CCU technology displays excellent CO_2 conversion efficiency, overcomes the cost of CO_2 compression and transportation, and is energy efficient, it should be improved in terms of specific capturing of CO_2. More materials should be designed and fabricated that can capture CO_2 more efficiently and have enough stability to store CO_2 for longer durations. This will enhance the sustainability of industrial procedures and simultaneously remove massive amounts of CO_2 from the environment. Material-based storage can play a better role in industrial utilization of CO_2 than geological storage methods. Moreover, more utilization routes of CO_2 should be explored to increase its industrial applications and to make it a valuable industrial gas.

References

1. Bala G, Bopp L, Brovkin V, Canadell J, Chhabra A (2013) Working group I contribution to the IPCC fifth assessment report (AR5), climate change 2013: the physical basis, working group I-twelfth session. Intergovernmental Panel on Climate Change (IPCC). Geneva, Switzerland
2. IPCC (2013) Climate change 2013: the physical science basis (trans: IPCC)
3. Liu S, Ji C, Wang C, Chen J, Jin Y, Zou Z, Li S, Niu S, Zou J (2018) Climatic role of terrestrial ecosystem under elevated CO_2: a bottom-up greenhouse gases budget. Ecol Lett 21(7):1108–1118. https://doi.org/10.1111/ele.13078
4. Metz B, Davidson O, De Coninck H, Loos M, Meyer L (2005) IPCC special report on carbon dioxide capture and storage. Cambridge University Press, Cambridge
5. Najafi-Marghmaleki A, Barati-Harooni A, Mohammadi AH (2017) Impact of gas impurities on CO_2 mole fraction: application in carbon capture and storage (CCS) processes. Int J Greenhouse Gas Control 57:173–184. https://doi.org/10.1016/j.ijggc.2016.12.008
6. Ren B, Delaney JM, Lake LW, Bryant SL (2018) Interplay between permeability retardation and capillary trapping of rising carbon dioxide in storage reservoirs. SPE J 23(05):1866–1879. https://doi.org/10.2118/187356-pa
7. IPCC (2019) Global warming of 1.5 °C
8. Xie H, Li X, Fang Z, Wang Y, Li Q, Shi L, Bai B, Wei N, Hou Z (2014) Carbon geological utilization and storage in China: current status and perspectives. Acta Geotech 9(1):7–27. https://doi.org/10.1007/s11440-013-0277-9
9. Cao C, Liu H, Hou Z, Mehmood F, Liao J, Feng W (2020) A review of CO_2 storage in view of safety and cost-effectiveness. Energies 13(3):600
10. Bui M, Adjiman CS, Bardow A, Anthony EJ, Boston A, Brown S, Fennell PS, Fuss S, Galindo A, Hackett LA, Hallett JP, Herzog HJ, Jackson G, Kemper J, Krevor S, Maitland GC, Matuszewski M, Metcalfe IS, Petit C, Puxty G, Reimer J, Reiner DM, Rubin ES, Scott SA, Shah N, Smit B, Trusler JPM, Webley P, Wilcox J, Mac Dowell N (2018) Carbon capture and storage (CCS): the way forward. Energy Environ Sci 11(5):1062–1176. https://doi.org/10.1039/c7ee02342a
11. Pawar RJ, Bromhal GS, Carey JW, Foxall W, Korre A, Ringrose PS, Tucker O, Watson MN, White JA (2015) Recent advances in risk assessment and risk management of geologic CO_2 storage. Int J Greenhouse Gas Control 40:292–311. https://doi.org/10.1016/j.ijggc.2015.06.014

12. Gislason SR, Oelkers EH (2014) Carbon storage in basalt. Science 344(6182):373–374. https://doi.org/10.1126/science.1250828
13. Li J-R, Kuppler RJ, Zhou H-C (2009) Selective gas adsorption and separation in metal–organic frameworks. Chem Soc Rev 38(5):1477–1504. https://doi.org/10.1039/b802426j
14. Sevilla M, Fuertes AB (2011) Sustainable porous carbons with a superior performance for CO_2 capture. Energy Environ Sci 4(5):1765–1771. https://doi.org/10.1039/c0ee00784f
15. Wei H, Deng S, Hu B, Chen Z, Wang B, Huang J, Yu G (2012) Granular bamboo-derived activated carbon for high CO_2 adsorption: the dominant role of narrow micropores. ChemSusChem 5(12):2354–2360. https://doi.org/10.1002/cssc.201200570
16. Coromina HM, Walsh DA, Mokaya R (2016) Biomass-derived activated carbon with simultaneously enhanced CO_2 uptake for both pre and post combustion capture applications. J Mater Chem A 4(1):280–289. https://doi.org/10.1039/c5ta09202g
17. Balahmar N, Al-Jumialy AS, Mokaya R (2017) Biomass to porous carbon in one step: directly activated biomass for high performance CO_2 storage. J Mater Chem A 5(24):12330–12339. https://doi.org/10.1039/c7ta01722g
18. Serafin J, Narkiewicz U, Morawski AW, Wróbel RJ, Michalkiewicz B (2017) Highly microporous activated carbons from biomass for CO_2 capture and effective micropores at different conditions. J CO_2 Utilization 18:73–79. https://doi.org/10.1016/j.jcou.2017.01.006
19. Sevilla M, Valle-Vigón P, Fuertes AB (2011) N-doped polypyrrole-based porous carbons for CO_2 capture. Adv Func Mater 21(14):2781–2787. https://doi.org/10.1002/adfm.201100291
20. Hu M, Reboul J, Furukawa S, Torad NL, Ji Q, Srinivasu P, Ariga K, Kitagawa S, Yamauchi Y (2012) Direct carbonization of al-based porous coordination polymer for synthesis of nanoporous carbon. J Am Chem Soc 134(6):2864–2867. https://doi.org/10.1021/ja208940u
21. Introduction to Metal–Organic Frameworks (2012) Chem Rev 112(2):673–674. https://doi.org/10.1021/cr300014x
22. Farha OK, Eryazici I, Jeong NC, Hauser BG, Wilmer CE, Sarjeant AA, Snurr RQ, Nguyen ST, Yazaydın AÖ, Hupp JT (2012) Metal-organic framework materials with ultrahigh surface areas: is the sky the limit? J Am Chem Soc 134(36):15016–15021. https://doi.org/10.1021/ja3055639
23. Furukawa H, Ko N, Go YB, Aratani N, Choi SB, Choi E, Yazaydin AÖ, Snurr RQ, O'Keeffe M, Kim J, Yaghi OM (2010) Ultrahigh porosity in metal-organic frameworks. Science 329(5990):424–428. https://doi.org/10.1126/science.1192160
24. Yuan D, Zhao D, Sun D, Zhou H-C (2010) An isoreticular series of metal-organic frameworks with dendritic hexacarboxylate ligands and exceptionally high gas-uptake capacity. Angew Chem Int Ed 49(31):5357–5361. https://doi.org/10.1002/anie.201001009
25. Farha OK, Özgür Yazaydın A, Eryazici I, Malliakas CD, Hauser BG, Kanatzidis MG, Nguyen ST, Snurr RQ, Hupp JT (2010) De novo synthesis of a metal–organic framework material featuring ultrahigh surface area and gas storage capacities. Nat Chem 2(11):944–948. https://doi.org/10.1038/nchem.834
26. Alezi D, Belmabkhout Y, Suyetin M, Bhatt PM, Weseliński ŁJ, Solovyeva V, Adil K, Spanopoulos I, Trikalitis PN, Emwas A-H, Eddaoudi M (2015) MOF crystal chemistry paving the way to gas storage needs: aluminum-based soc-MOF for CH_4, O_2, and CO_2 storage. J Am Chem Soc 137(41):13308–13318. https://doi.org/10.1021/jacs.5b07053
27. Llewellyn PL, Bourrelly S, Serre C, Vimont A, Daturi M, Hamon L, De Weireld G, Chang J-S, Hong D-Y, Kyu Hwang Y, Hwa Jhung S, Férey G (2008) High uptakes of CO_2 and CH_4 in mesoporous metal—organic frameworks MIL-100 and MIL-101. Langmuir 24(14):7245–7250. https://doi.org/10.1021/la800227x
28. Dietzel PDC, Besikiotis V, Blom R (2009) Application of metal–organic frameworks with coordinatively unsaturated metal sites in storage and separation of methane and carbon dioxide. J Mater Chem 19(39):7362–7370. https://doi.org/10.1039/b911242a
29. Bourrelly S, Llewellyn PL, Serre C, Millange F, Loiseau T, Férey G (2005) Different adsorption behaviors of methane and carbon dioxide in the isotypic nanoporous metal terephthalates MIL-53 and MIL-47. J Am Chem Soc 127(39):13519–13521. https://doi.org/10.1021/ja054668v

30. Herm ZR, Swisher JA, Smit B, Krishna R, Long JR (2011) Metal−organic frameworks as adsorbents for hydrogen purification and precombustion carbon dioxide capture. J Am Chem Soc 133(15):5664–5667. https://doi.org/10.1021/ja111411q
31. Mu B, Schoenecker PM, Walton KS (2010) Gas adsorption study on mesoporous metal−organic framework UMCM-1. J Phys Chem C 114(14):6464–6471. https://doi.org/10.1021/jp906417z
32. Arthur W, Eric G (2009) Zeolite characterization and catalysis. Springer, London
33. Chue KT, Kim JN, Yoo YJ, Cho SH, Yang RT (1995) Comparison of activated carbon and zeolite 13X for CO_2 recovery from flue gas by pressure swing adsorption. Ind Eng Chem Res 34(2):591–598. https://doi.org/10.1021/ie00041a020
34. Siriwardane RV, Shen M-S, Fisher EP (2003) Adsorption of CO_2, N_2, and O_2 on natural zeolites. Energy Fuels 17(3):571–576
35. Thakkar H, Eastman S, Hajari A, Rownaghi AA, Knox JC, Rezaei F (2016) 3D-printed zeolite monoliths for CO_2 removal from enclosed environments. ACS Appl Mater Interfaces 8(41):27753–27761. https://doi.org/10.1021/acsami.6b09647
36. Cheung O, Hedin N (2014) Zeolites and related sorbents with narrow pores for CO_2 separation from flue gas. RSC Adv 4(28):14480–14494. https://doi.org/10.1039/c3ra48052f
37. Li S, Falconer JL, Noble RD (2004) SAPO-34 membranes for CO_2/CH_4 separation. J Membr Sci 241(1):121–135. https://doi.org/10.1016/j.memsci.2004.04.027
38. Bhatta LKG, Subramanyam S, Chengala MD, Olivera S, Venkatesh K (2015) Progress in hydrotalcite like compounds and metal-based oxides for CO_2 capture: a review. J Clean Prod 103:171–196. https://doi.org/10.1016/j.jclepro.2014.12.059
39. Belmabkhout Y, Serna-Guerrero R, Sayari A (2009) Adsorption of CO_2 from dry gases on MCM-41 silica at ambient temperature and high pressure. 1: pure CO_2 adsorption. Chem Eng Sci 64(17):3721–3728. https://doi.org/10.1016/j.ces.2009.03.017
40. Jahandar Lashaki M, Khiavi S, Sayari A (2019) Stability of amine-functionalized CO_2 adsorbents: a multifaceted puzzle. Chem Soc Rev 48(12):3320–3405. https://doi.org/10.1039/c8cs00877a
41. Belmabkhout Y, Sayari A (2009) Effect of pore expansion and amine functionalization of mesoporous silica on CO_2 adsorption over a wide range of conditions. Adsorption 15(3):318–328. https://doi.org/10.1007/s10450-009-9185-6
42. Suekane T, Nobuso T, Hirai S, Kiyota M (2008) Geological storage of carbon dioxide by residual gas and solubility trapping. Int J Greenhouse Gas Control 2(1):58–64. https://doi.org/10.1016/S1750-5836(07)00096-5
43. Lokhorst A, Wildenborg T (2005) Introduction on CO_2 geological storage—classification of storage options. Oil Gas Sci Technol Rev IFP 60(3):513–515
44. Gale J (2004) Geological storage of CO_2: what do we know, where are the gaps and what more needs to be done? Energy 29(9–10):1329–1338. https://doi.org/10.1016/j.energy.2004.03.068
45. Bachu S, Stewart S (2002) Geological sequestration of anthropogenic carbon dioxide in the western Canada sedimentary basin: suitability analysis. J Can Petrol Technol 41(02). https://doi.org/10.2118/02-02-01
46. Boot-Handford ME, Abanades JC, Anthony EJ, Blunt MJ, Brandani S, Mac Dowell N, Fernández JR, Ferrari M-C, Gross R, Hallett JP, Haszeldine RS, Heptonstall P, Lyngfelt A, Makuch Z, Mangano E, Porter RTJ, Pourkashanian M, Rochelle GT, Shah N, Yao JG, Fennell PS (2014) Carbon capture and storage update. Energy Environ Sci 7(1):130–189. https://doi.org/10.1039/C3EE42350F
47. Gaurina-Međimurec, Mavar N, Novak K (2019) Carbon capture and storage (CCS): geological sequestration of CO_2. In: CO_2 sequestration. IntechOpen
48. Jiang X (2011) A review of physical modelling and numerical simulation of long-term geological storage of CO_2. Appl Energy 88(11):3557–3566. https://doi.org/10.1016/j.apenergy.2011.05.004
49. Bradshaw J, Bachu S, Bonijoly D, Burruss R, Holloway S, Christensen NP, Mathiassen OM (2007) CO_2 storage capacity estimation: issues and development of standards. Int J Greenhouse Gas Control 1(1):62–68. https://doi.org/10.1016/s1750-5836(07)00027-8

50. Bachu S, Gunter WD, Perkins EH (1994) Aquifer disposal of CO_2: hydrodynamic and mineral trapping. Energy Convers Manage 35(4):269–279. https://doi.org/10.1016/0196-8904(94)90060-4
51. Zhang D, Song J (2014) Mechanisms for geological carbon sequestration. Procedia IUTAM 10:319–327. https://doi.org/10.1016/j.piutam.2014.01.027
52. Altman SJ, Aminzadeh B, Balhoff MT, Bennett PC, Bryant SL, Cardenas MB, Chaudhary K, Cygan RT, Deng W, Dewers T, DiCarlo DA, Eichhubl P, Hesse MA, Huh C, Matteo EN, Mehmani Y, Tenney CM, Yoon H (2014) Chemical and hydrodynamic mechanisms for long-term geological carbon storage. J Phys Chem C 118(28):15103–15113. https://doi.org/10.1021/jp5006764
53. Gunter WD, Bachu S, Benson S (2004) The role of hydrogeological and geochemical trapping in sedimentary basins for secure geological storage of carbon dioxide. Geol Soc Lond Spec Publ 233(1):129. https://doi.org/10.1144/GSL.SP.2004.233.01.09
54. Wilberforce T, Olabi AG, Sayed ET, Elsaid K, Abdelkareem MA (2021) Progress in carbon capture technologies. Sci Total Environ 761:143203. https://doi.org/10.1016/j.scitotenv.2020.143203
55. Pires JCM, Martins FG, Alvim-Ferraz MCM, Simões M (2011) Recent developments on carbon capture and storage: an overview. Chem Eng Res Des 89(9):1446–1460. https://doi.org/10.1016/j.cherd.2011.01.028
56. De Silva GPD, Ranjith PG, Perera MSA (2015) Geochemical aspects of CO_2 sequestration in deep saline aquifers: a review. Fuel 155:128–143. https://doi.org/10.1016/j.fuel.2015.03.045
57. Pan Y, Hui D, Luo P, Zhang Y, Sun L, Wang K (2018) Experimental investigation of the geochemical interactions between supercritical CO_2 and shale: implications for CO_2 storage in gas-bearing shale formations. Energy Fuels 32(2):1963–1978. https://doi.org/10.1021/acs.energyfuels.7b03074
58. Gaus I, Audigane P, André L, Lions J, Jacquemet N, Durst P, Czernichowski-Lauriol I, Azaroual M (2008) Geochemical and solute transport modelling for CO_2 storage, what to expect from it? Int J Greenhouse Gas Control 2(4):605–625. https://doi.org/10.1016/j.ijggc.2008.02.011
59. Ranjith PG, Perera MSA, Khan E (2013) A study of safe CO_2 storage capacity in saline aquifers: a numerical study. Int J Energy Res 37(3):189–199. https://doi.org/10.1002/er.2954
60. Michael K, Arnot M, Cook P, Ennis-King J, Funnell R, Kaldi J, Kirste D, Paterson L (2009) CO_2 storage in saline aquifers I—current state of scientific knowledge. Energy Procedia 1(1):3197–3204. https://doi.org/10.1016/j.egypro.2009.02.103
61. Bachu S, Gunter WD (2004) Acid-gas injection in the Alberta basin, Canada: a CO_2-storage experience. Geol Soc Lond Spec Publ 233(1):225. https://doi.org/10.1144/GSL.SP.2004.233.01.15
62. Arts R, Winthaegen P (2005) Chapter 19—monitoring options for CO_2 storage. In: Thomas DC (ed) Carbon dioxide capture for storage in deep geologic formations. Elsevier, Amsterdam, pp 1001–1013. https://doi.org/10.1016/B978-008044570-0/50147-1
63. Bentham M, Kirby M (2005) CO_2 Storage in saline aquifers. Oil Gas Sci Technol Rev IFP 60(3):559–567
64. Kumar A, Noh M, Pope GA, Sepehrnoori K, Bryant S, Lake LW (2004) Reservoir simulation of CO_2 storage in deep saline aquifers. In: Paper presented at the SPE/DOE symposium on improved oil recovery
65. Shi JQ, Durucan S (2005) CO_2 storage in deep unminable coal seams. Oil Gas Sci Technol Rev IFP 60(3):547–558
66. Ross HE, Hagin P, Zoback MD (2009) CO_2 storage and enhanced coalbed methane recovery: reservoir characterization and fluid flow simulations of the big george coal, powder river basin, wyoming, USA. Int J Greenhouse Gas Control 3(6):773–786. https://doi.org/10.1016/j.ijggc.2009.06.002
67. Pan Z, Ye J, Zhou F, Tan Y, Connell LD, Fan J (2018) CO_2 storage in coal to enhance coalbed methane recovery: a review of field experiments in China. Int Geol Rev 60(5–6):754–776. https://doi.org/10.1080/00206814.2017.1373607

68. Gibbins J, Chalmers H (2008) Carbon capture and storage. Energy Policy 36(12):4317–4322. https://doi.org/10.1016/j.enpol.2008.09.058
69. Mazzotti M, Pini R, Storti G (2009) Enhanced coalbed methane recovery. J Supercrit Fluids 47(3):619–627. https://doi.org/10.1016/j.supflu.2008.08.013
70. Jessen K, Kovscek AR, Orr FM (2005) Increasing CO_2 storage in oil recovery. Energy Convers Manage 46(2):293–311. https://doi.org/10.1016/j.enconman.2004.02.019
71. Obdam A, van der Meer L, May F, Kervevan C, Bech N, Wildenborg A (2003) Effective CO_2 storage capacity in aquifers, gas fields, oil fields and coal fields. In: Gale J, Kaya Y (eds) Greenhouse gas control technologies—6th international conference. Pergamon, Oxford, pp 339–344. https://doi.org/10.1016/B978-008044276-1/50055-6
72. Agartan E, Gaddipati M, Yip Y, Savage B, Ozgen C (2018) CO_2 storage in depleted oil and gas fields in the Gulf of Mexico. Int J Greenhouse Gas Control 72:38–48. https://doi.org/10.1016/j.ijggc.2018.02.022
73. Godec M, Kuuskraa V, Van Leeuwen T, Stephen Melzer L, Wildgust N (2011) CO_2 storage in depleted oil fields: the worldwide potential for carbon dioxide enhanced oil recovery. Energy Procedia 4:2162–2169. https://doi.org/10.1016/j.egypro.2011.02.102
74. Bordenave S, Chatterjee I, Voordouw G (2013) Microbial community structure and microbial activities related to CO_2 storage capacities of a salt cavern. Int Biodeterior Biodegradation 81:82–87. https://doi.org/10.1016/j.ibiod.2012.08.001
75. Grobe M (2000) Distribution and thickness of salt within the devonian Elk Point group, western Canada sedimentary basin. Earth Sci Rep 2:1–12
76. Bachu S (2000) Sequestration of CO_2 in geological media: criteria and approach for site selection in response to climate change. Energy Convers Manage 41(9):953–970. https://doi.org/10.1016/S0196-8904(99)00149-1
77. Crossley NG (1998) Conversion of LPG salt caverns to natural gas storage "a transgas experience". J Can Pet Technol 37(12). https://doi.org/10.2118/98-12-03
78. Lux KH (2009) Design of salt caverns for the storage of natural gas, crude oil and compressed air: geomechanical aspects of construction, operation and abandonment. Geol Soc Lond Spec Publ 313(1):93–128. https://doi.org/10.1144/sp313.7
79. Tayari F, Blumsack S, Dilmore R, Mohaghegh SD (2015) Techno-economic assessment of industrial CO_2 storage in depleted shale gas reservoirs. J Unconventional Oil Gas Resour 11:82–94. https://doi.org/10.1016/j.juogr.2015.05.001
80. Louk K, Ripepi N, Luxbacher K, Gilliland E, Tang X, Keles C, Schlosser C, Diminick E, Keim S, Amante J, Michael K (2017) Monitoring CO_2 storage and enhanced gas recovery in unconventional shale reservoirs: results from the morgan county, tennessee injection test. J Nat Gas Sci Eng 45:11–25. https://doi.org/10.1016/j.jngse.2017.03.025
81. Boosari SSH, Aybar U, Eshkalak MO (2015) Carbon dioxide storage and sequestration in unconventional shale reservoirs. J Geosci Environ Prot 03(01):9. https://doi.org/10.4236/gep.2015.31002
82. Mahzari P, Jones AP, Oelkers EH (2020) Impact of in-situ gas liberation for enhanced oil recovery and CO_2 storage in liquid-rich shale reservoirs. Energy Sour Part A Recovery Utilization Environ Effects: 1–21. https://doi.org/10.1080/15567036.2020.1815907
83. Mikaloff Fletcher SE, Gruber N, Jacobson AR, Doney SC, Dutkiewicz S, Gerber M, Follows M, Joos F, Lindsay K, Menemenlis D, Mouchet A, Müller SA, Sarmiento JL (2006) Inverse estimates of anthropogenic CO_2 uptake, transport, and storage by the ocean. Glob Biogeochem Cycles 20(2). https://doi.org/10.1029/2005GB002530
84. Adams EE, Caldeira K (2008) Ocean storage of CO_2. Elements 4(5):319–324. https://doi.org/10.2113/gselements.4.5.319
85. Li Q, Wu Z, Li X (2009) Prediction of CO_2 leakage during sequestration into marine sedimentary strata. Energy Convers Manage 50(3):503–509. https://doi.org/10.1016/j.enconman.2008.11.011
86. Singh U (2013) Carbon capture and storage: an effective way to mitigate global warming. Curr Sci 105(7):914–922

87. Wong CS, Matear RJ (1998) Ocean disposal of CO_2 in the North Pacific Ocean: assessment of CO_2 chemistry and circulation on storage and return to the atmosphere. Waste Manage 17(5):329–335. https://doi.org/10.1016/S0956-053X(97)10049-6
88. Freund P, Ormerod WG (1997) Progress toward storage of carbon dioxide. Energy Convers Manage 38:S199–S204. https://doi.org/10.1016/S0196-8904(96)00269-5
89. Styring P (2015) Chapter 7—Novel sorbent materials for carbon capture. In: Shi F, Morreale B (eds) Novel materials for carbon dioxide mitigation technology. Elsevier, Amsterdam, pp 207–229. https://doi.org/10.1016/B978-0-444-63259-3.00007-0
90. Huang C-H, Tan C-S (2014) A review: CO_2 utilization. Aerosol Air Qual Res 14(2):480–499. https://doi.org/10.4209/aaqr.2013.10.0326
91. Kanjilal B, Masoumi A, Nabavinia M, Krishnadoss V, Savelski M, Noshadi I (2020) Chapter 8—CO_2 adsorption by carbonaceous materials and nanomaterials. In: Rahimpour MR, Farsi M, Makarem MA (eds) Advances in carbon capture. Woodhead Publishing, pp 173–192. https://doi.org/10.1016/B978-0-12-819657-1.00008-6
92. Aresta M, Dibenedetto A (2007) Utilisation of CO_2 as a chemical feedstock: opportunities and challenges. Dalton Trans 28:2975–2992. https://doi.org/10.1039/B700658F
93. Boot-Handford ME, Abanades JC, Anthony EJ, Blunt MJ, Brandani S, Mac Dowell N, Fernández JR, Ferrari M-C, Gross R, Hallett JP (2014) Carbon capture and storage update. Energy Environ Sci 7(1):130–189
94. Obrien K, Lu Y, Patel V, Greenberg S, Locke R, Larson M, Krishnamurthy K, Byron M, Naumovitz J, Guth D, Bennett S (2017) Creating markets for captured carbon: retrofit of abbott power plant and future utilization of captured CO_2. Energy Procedia 114:1263–1272. https://doi.org/10.1016/j.egypro.2017.03.1380
95. Poliakoff M, Leitner W, Streng ES (2015) The twelve principles of CO_2 chemistry. Faraday Discuss 183:9–17. https://doi.org/10.1039/c5fd90078f
96. Rozzi E, Minuto FD, Lanzini A, Leone P (2020) Green synthetic fuels: renewable routes for the conversion of non-fossil feedstocks into gaseous fuels and their end uses. Energies 13(2):420
97. Al-mamoori AAA (2020) Development of novel materials and processes for direct conversion of CO_2 flue gas to chemicals and fuels. Ph.D., Missouri University of Science and Technology, Ann Arbor
98. MacDowell N, Florin N, Buchard A, Hallett J, Galindo A, Jackson G, Adjiman CS, Williams CK, Shah N, Fennell P (2010) An overview of CO_2 capture technologies. Energy Environ Sci 3(11):1645–1669. https://doi.org/10.1039/C004106H
99. Van Niekerk D, Markley B, Li Y, Rodriguez-Santiago V, Thompson D, Mohan AR, Elsworth D, Mathews JP, Pisupati SV, Song C (2006) Utilization of carbon dioxide from coalfired power plant for the production of value-added products
100. Aresta M, Dibenedetto A, Angelini A (2014) Catalysis for the valorization of exhaust carbon: from CO_2 to chemicals, materials, and fuels. Technological use of CO_2. Chem Rev 114(3):1709–1742. https://doi.org/10.1021/cr4002758
101. Homs N, Toyir J, Piscina P (2013) Catalytic processes for activation of CO_2. In: pp 1–26. https://doi.org/10.1016/b978-0-444-53882-6.00001-2
102. Angunn E, Nada A, Gaëlle B-C (2014) Evaluation of carbon dioxide utilisation concepts: a quick and complete methodology. Energy Procedia 63:8010–8016. https://doi.org/10.1016/j.egypro.2014.11.837
103. Meylan FD, Moreau V, Erkman S (2015) CO_2 utilization in the perspective of industrial ecology, an overview. J CO_2 Utilization 12:101–108. https://doi.org/10.1016/j.jcou.2015.05.003
104. Styring P (2019) Carbon dioxide utilization as a mitigation tool. In: pp 527–551. https://doi.org/10.1016/b978-0-12-814104-5.00018-1
105. Ampelli C, Perathoner S, Centi G (2015) CO_2 utilization: an enabling element to move to a resource-and energy-efficient chemical and fuel production. Philos Trans Roy Soc A Math Phys Eng Sci 373(2037):20140177

106. Hu B, Guild C, Suib SL (2013) Thermal, electrochemical, and photochemical conversion of CO_2 to fuels and value-added products. J CO_2 Utilization 1:18–27. https://doi.org/10.1016/j.jcou.2013.03.004
107. Song C (2006) Global challenges and strategies for control, conversion and utilization of CO_2 for sustainable development involving energy, catalysis, adsorption and chemical processing. Catal Today 115(1–4):2–32. https://doi.org/10.1016/j.cattod.2006.02.029
108. Fan MS, Abdullah AZ, Bhatia S (2009) Catalytic technology for carbon dioxide reforming of methane to synthesis gas. Chem Cat Chem 1(2):192–208
109. He C-S, Gong L, Zhang J, He P-P, Mu Y (2017) Highly selective hydrogenation of CO_2 into formic acid on a nano-Ni catalyst at ambient temperature: process, mechanisms and catalyst stability. J CO_2 Utilization 19:157–164. https://doi.org/10.1016/j.jcou.2017.03.012
110. Xiong S, Lian Y, Xie H, Liu B (2019) Hydrogenation of CO_2 to methanol over Cu/ZnCr catalyst. Fuel 256:115975. https://doi.org/10.1016/j.fuel.2019.115975
111. Li N, Zou X, Liu M, Wei L, Shen Q, Bibi R, Xu C, Ma Q, Zhou J (2017) Enhanced visible light photocatalytic hydrogenation of CO_2 into methane over a Pd/Ce-TiO_2 nanocomposition. J Phys Chem C 121(46):25795–25804. https://doi.org/10.1021/acs.jpcc.7b07298
112. Al-Mamoori A, Lawson S, Rownaghi AA, Rezaei F (2020) Oxidative dehydrogenation of ethane to ethylene in an integrated CO_2 capture-utilization process. Appl Catal B 278:119329. https://doi.org/10.1016/j.apcatb.2020.119329
113. Kuld S, Conradsen C, Moses PG, Chorkendorff I, Sehested (2014). Quantification of zinc atoms in a surface alloy on copper in an industrial-type methanol synthesis catalyst. Angew Chem Int Ed 53:5941–5945
114. Pendry LF, Salvatore J (2015) Individual and social benefits of online discussion forums. Comput Hum Behav 50:211–220. https://doi.org/10.1016/j.chb.2015.03.067
115. Dayton DC, Turk B, Gupta R (2019) Syngas cleanup, conditioning, and utilization thermochemical processing of biomass: conversion into fuels, chemicals and power
116. Li Y, Markley B, Mohan AR, Rodriguez-Santiago V, Thompson D, Niekerk DV (2006) Utilization of carbon dioxide from coal-fired power plant for the production of value-added products. Design engineering of energy and geo-environmental systems course (EGEE 580) internal document for a course college of earth and mineral science
117. Ofélia de Queiroz FA, de Medeiros JL (2014) CO_2 utilization: a process systems engineering vision. Intech, CO_2 separation and valorisation
118. Liu H, Zhang Z, Li H, Huang Q (2011) Intrinsic kinetics of oxidative dehydrogenation of propane in the presence of CO_2 over Cr/MSU-1 catalyst. J Nat Gas Chem 20(3):311–317. https://doi.org/10.1016/s1003-9953(10)60180-6
119. Estes DP, Coperet C (2015) The role of proton transfer in heterogeneous transformations of hydrocarbons. Chimia (Aarau) 69(6):321–326. https://doi.org/10.2533/chimia.2015.321
120. Cheng Y, Zhang F, Zhang Y, Miao C, Hua W, Yue Y, Gao Z (2015) Oxidative dehydrogenation of ethane with CO_2 over Cr supported on submicron ZSM-5 zeolite. Chin J Catal 36(8):1242–1248. https://doi.org/10.1016/s1872-2067(15)60893-2
121. Koirala R, Buechel R, Krumeich F, Pratsinis SE, Baiker A (2015) Oxidative dehydrogenation of ethane with CO_2 over flame-made ga-loaded TiO_2. ACS Catal 5(2):690–702. https://doi.org/10.1021/cs500685d
122. Senboku H (2014) Electrochemical fixation of carbon dioxide. In: Transformation and utilization of carbon dioxide. Green chemistry and sustainable technology, pp 245–262. https://doi.org/10.1007/978-3-642-44988-8_10
123. Aresta M, Dibenedetto A (2010) 14—Industrial utilization of carbon dioxide (CO_2). In: Maroto-Valer MM (ed) Developments and innovation in carbon dioxide (CO_2) capture and storage technology, vol 2. Woodhead Publishing, pp 377–410. https://doi.org/10.1533/9781845699581.4.377
124. Ito K, Ikeda S, Yamauchi N, Iida T, Takagi T (1985) Electrochemical reduction products of carbon dioxide at some metallic electrodes in nonaqueous electrolytes. Bull Chem Soc Jpn 58(10):3027–3028

125. Kim T, Logan BE, Gorski CA (2017) A pH-gradient flow cell for converting waste CO_2 into electricity. Environ Sci Technol Lett 4(2):49–53. https://doi.org/10.1021/acs.estlett.6b00467
126. Quadrelli EA, Armstrong K, Styring P (2015) Chapter 16—Potential CO_2 utilisation contributions to a more carbon-sober future: a 2050 vision. In: Styring P, Quadrelli EA, Armstrong K (eds) Carbon dioxide utilisation. Elsevier, Amsterdam, pp 285–302. https://doi.org/10.1016/B978-0-444-62746-9.00016-5
127. Fan WK, Tahir M (2022) Recent developments in photothermal reactors with understanding on the role of light/heat for CO_2 hydrogenation to fuels: a review. Chem Eng J 427:131617. https://doi.org/10.1016/j.cej.2021.131617
128. Olah GA, Goeppert A, Prakash GS (2009) Chemical recycling of carbon dioxide to methanol and dimethyl ether: from greenhouse gas to renewable, environmentally carbon neutral fuels and synthetic hydrocarbons. J Org Chem 74(2):487–498
129. Demirbas A (2008) Biofuels sources, biofuel policy, biofuel economy and global biofuel projections. Energy Convers Manage 49(8):2106–2116. https://doi.org/10.1016/j.enconman.2008.02.020
130. Larson ED (2006) A review of life-cycle analysis studies on liquid biofuel systems for the transport sector. Energy Sustain Dev 10(2):109–126. https://doi.org/10.1016/S0973-0826(08)60536-0
131. Centi G, Quadrelli EA, Perathoner S (2013) Catalysis for CO_2 conversion: a key technology for rapid introduction of renewable energy in the value chain of chemical industries. Energy Environ Sci 6(6):1711–1731. https://doi.org/10.1039/c3ee00056g
132. Saito M, Murata K (2004) Development of high performance Cu/ZnO-based catalysts for methanol synthesis and the water-gas shift reaction. Catal Surv Asia 8(4):285–294. https://doi.org/10.1007/s10563-004-9119-y
133. Semelsberger TA, Borup RL, Greene HL (2006) Dimethyl ether (DME) as an alternative fuel. J Power Sour 156(2):497–511. https://doi.org/10.1016/j.jpowsour.2005.05.082
134. Vafajoo L, Afshar SHA, Firouzbakht B (2009) Developing a mathematical model for the complete kinetic cycle of direct synthesis of DME from Syngas through the CFD technique. In: Jeżowski J, Thullie J (eds) Computer aided chemical engineering, vol 26. Elsevier, pp 689–694. https://doi.org/10.1016/S1570-7946(09)70115-4
135. Lim H-W, Park M-J, Kang S-H, Chae H-J, Bae JW, Jun K-W (2009) Modeling of the kinetics for methanol synthesis using $Cu/ZnO/Al_2O_3/ZrO_2$ catalyst: influence of carbon dioxide during hydrogenation. Ind Eng Chem Res 48(23):10448–10455. https://doi.org/10.1021/ie901081f
136. Corma A, Garcia H (2013) Photocatalytic reduction of CO_2 for fuel production: possibilities and challenges. J Catal 308:168–175. https://doi.org/10.1016/j.jcat.2013.06.008
137. Li H, Opgenorth PH, Wernick DG, Rogers S, Wu T-Y, Higashide W, Malati P, Huo Y-X, Cho KM, Liao JC (2012) Integrated electromicrobial conversion of CO_2 to higher alcohols. Science 335(6076):1596–1596. https://doi.org/10.1126/science.1217643
138. Omodolor IS, Otor HO, Andonegui JA, Allen BJ, Alba-Rubio AC (2020) Dual-function materials for CO_2 capture and conversion: a review. Ind Eng Chem Res 59(40):17612–17631. https://doi.org/10.1021/acs.iecr.0c02218
139. Tian S, Yan F, Zhang Z, Jiang J (2019) Calcium-looping reforming of methane realizes in situ CO_2 utilization with improved energy efficiency. Sci Adv 5(4):eaav5077. https://doi.org/10.1126/sciadv.aav5077
140. Kim SM, Abdala PM, Broda M, Hosseini D, Copéret C, Müller C (2018) Integrated CO_2 capture and conversion as an efficient process for fuels from greenhouse gases. ACS Catal 8(4):2815–2823. https://doi.org/10.1021/acscatal.7b03063
141. Al-Mamoori A, Rownaghi AA, Rezaei F (2018) Combined capture and utilization of CO_2 for syngas production over dual-function materials. ACS Sustain Chem Eng 6(10):13551–13561. https://doi.org/10.1021/acssuschemeng.8b03769
142. Kim A, Sanchez C, Haye B, Boissière C, Sassoye C, Debecker DP (2019) Mesoporous TiO_2 support materials for Ru-based CO_2 methanation catalysts. ACS Appl Nano Mater 2(5):3220–3230. https://doi.org/10.1021/acsanm.9b00518

143. Duyar MS, Treviño MAA, Farrauto RJ (2015) Dual function materials for CO_2 capture and conversion using renewable H_2. Appl Catal B 168–169:370–376. https://doi.org/10.1016/j.apcatb.2014.12.025
144. Duyar MS, Wang S, Arellano-Treviño MA, Farrauto RJ (2016) CO_2 utilization with a novel dual function material (DFM) for capture and catalytic conversion to synthetic natural gas: an update. J CO_2 Utilization 15:65–71. https://doi.org/10.1016/j.jcou.2016.05.003
145. Myint M, Yan B, Wan J, Zhao S, Chen JG (2016) Reforming and oxidative dehydrogenation of ethane with CO_2 as a soft oxidant over bimetallic catalysts. J Catal 343:168–177. https://doi.org/10.1016/j.jcat.2016.02.004
146. Hyakutake T, van Beek W, Urakawa A (2016) Unravelling the nature, evolution and spatial gradients of active species and active sites in the catalyst bed of unpromoted and K/Ba-promoted Cu/Al_2O_3 during CO_2 capture-reduction. J Mater Chem A 4(18):6878–6885. https://doi.org/10.1039/c5ta09461e
147. Yao ZT, Ji XS, Sarker PK, Tang JH, Ge LQ, Xia MS, Xi YQ (2015) A comprehensive review on the applications of coal fly ash. Earth Sci Rev 141:105–121. https://doi.org/10.1016/j.earscirev.2014.11.016
148. Ahmaruzzaman M (2010) A review on the utilization of fly ash. Prog Energy Combust Sci 36(3):327–363. https://doi.org/10.1016/j.pecs.2009.11.003
149. Dindi A, Quang DV, Vega LF, Nashef E, Abu-Zahra MRM (2019) Applications of fly ash for CO_2 capture, utilization, and storage. J CO_2 Utilization 29:82–102. https://doi.org/10.1016/j.jcou.2018.11.011
150. Shawabkeh RA, Aslam Z, Hussien IA (2015) Thermochemical treatment of fly ash for synthesis of mesoporous activated carbon. J Therm Anal Calorim 122(3):1191–1201. https://doi.org/10.1007/s10973-015-4964-7
151. Siriruang C, Toochinda P, Julnipitawong P, Tangtermsirikul S (2016) CO_2 capture using fly ash from coal fired power plant and applications of CO_2-captured fly ash as a mineral admixture for concrete. J Environ Manage 170:70–78. https://doi.org/10.1016/j.jenvman.2016.01.010
152. Wei Z, Wang B, Falzone G, La Plante EC, Okoronkwo MU, She Z, Oey T, Balonis M, Neithalath N, Pilon L, Sant G (2018) Clinkering-free cementation by fly ash carbonation. J CO_2 Utilization 23:117–127. https://doi.org/10.1016/j.jcou.2017.11.005
153. Peters M, Köhler B, Kuckshinrichs W, Leitner W, Markewitz P, Müller TE (2011) Chemical technologies for exploiting and recycling carbon dioxide into the value chain. ChemSusChem 4(9):1216–1240. https://doi.org/10.1002/cssc.201000447
154. Zhou J, Tian S, Yang K, Dong Z, Cai J (2022) Chapter Eleven—Enhanced gas recovery technologies aimed at exploiting captured carbon dioxide. In: Wood DA, Cai J (eds) Sustainable natural gas reservoir and production engineering, vol 1. Gulf Professional Publishing, pp 305–347. https://doi.org/10.1016/B978-0-12-824495-1.00010-3
155. Melzer LS (2012) Carbon dioxide enhanced oil recovery (CO_2 EOR): factors involved in adding carbon capture, utilization and storage (CCUS) to enhanced oil recovery. Center Clim Energy Solutions:1–17
156. Saleem Qadir Tunio AHT, Ghirano NA, El Adawy ZM (2011) Comparison of different enchanced oil recovery techniques for better oil productivity. Int J Appl Sci Technol 1(5)
157. Cho J, Kim TH, Lee KS (2016) Modeling of CO_2 EOR process combined with intermediate hydrocarbon solvents for higher recovery efficiency. In: SPE EOR conference at oil and gas West Asia. OnePetro
158. Perera M, Gamage R, Rathnaweera T, Ranathunga A, Koay A, Choi X (2016) A review of CO_2-enhanced oil recovery with a simulated sensitivity analysis. Energies 9(7). https://doi.org/10.3390/en9070481
159. Miller BG (2011) 10—CO_2 capture and storage. In: Miller BG (ed) Clean coal engineering technology. Butterworth-Heinemann, Boston, pp 483–511. https://doi.org/10.1016/B978-1-85617-710-8.00010-8
160. Cuéllar-Franca RM, Azapagic A (2015) Carbon capture, storage and utilisation technologies: a critical analysis and comparison of their life cycle environmental impacts. J CO_2 Utilization 9:82–102. https://doi.org/10.1016/j.jcou.2014.12.001

161. Cuéllar Franca RM, Azapagic A (2017) Life cycle environmental impacts of carbon capture, storage, and utilization. In: Abraham MA (ed) Encyclopedia of sustainable technologies. Elsevier, Oxford, pp 447–459. https://doi.org/10.1016/B978-0-12-409548-9.10123-X
162. Rackley SA (2017) 2—Overview of carbon capture and storage. In: Rackley SA (ed) Carbon capture and storage (2nd edn). Butterworth-Heinemann, Boston, pp 23–36. https://doi.org/10.1016/B978-0-12-812041-5.00002-7
163. Bijl DL, Bogaart PW, Kram T, de Vries BJM, van Vuuren DP (2016) Long-term water demand for electricity, industry and households. Environ Sci Policy 55:75–86. https://doi.org/10.1016/j.envsci.2015.09.005
164. Kheshgi H, de Coninck H, Kessels J (2012) Carbon dioxide capture and storage: seven years after the IPCC special report. Mitig Adapt Strat Glob Change 17(6):563–567. https://doi.org/10.1007/s11027-012-9391-5
165. Max MD, Tatro SR, Osegovic JP, Brazel LA, Sheps KM (2006) New method for extraction of water from natural gas. Marine Desalination Systems LLC
166. McCutcheon JL, McGinnis RL, Elimelech M (2006) The ammonia-carbon dioxide forward osmosis desalination process. Water Conditioning & Purification
167. Dawoud MA (2012) Environmental impacts of seawater desalination: Arabian Gulf case study. Int J Environ Sustain 1(3):1927–9566
168. Latteman S (2010) Development of an environmental impact assessment and decision support system for seawater desalination plants. CRC Press
169. Dindi A, Quang DV, Abu-Zahra MRM (2015) Simultaneous carbon dioxide capture and utilization using thermal desalination reject brine. Appl Energy 154:298–308. https://doi.org/10.1016/j.apenergy.2015.05.010
170. Buonomenna MG, Golemme G (2012) Advanced materials for membrane preparation. Bentham Science Publishers
171. Chen H, Zhang H, Yan Y (2013) Catalytic combustion of volatile organic compounds over a structured zeolite membrane reactor. Ind Eng Chem Res 52(36):12819–12826. https://doi.org/10.1021/ie401882w
172. Avila AM, Yu Z, Fazli S, Sawada JA, Kuznicki SM (2014) Hydrogen-selective natural mordenite in a membrane reactor for ethane dehydrogenation. Microporous Mesoporous Mater 190:301–308. https://doi.org/10.1016/j.micromeso.2014.02.024
173. Carvill BT, Hufton JR, Anand M, Sircar S (1996) Sorption-enhanced reaction process. AIChE J 42(10):2765–2772
174. van Selow ER, Cobden PD, Verbraeken PA, Hufton JR, van den Brink RW (2009) Carbon capture by sorption-enhanced water−gas shift reaction process using hydrotalcite-based material. Ind Eng Chem Res 48(9):4184–4193. https://doi.org/10.1021/ie801713a
175. Hufton JR, Mayorga S, Sircar S (1999) Sorption-enhanced reaction process for hydrogen production. AIChE J 45(2):248–256
176. Jang HM, Lee KB, Caram HS, Sircar S (2012) High-purity hydrogen production through sorption enhanced water gas shift reaction using K_2CO_3-promoted hydrotalcite. Chem Eng Sci 73:431–438. https://doi.org/10.1016/j.ces.2012.02.015
177. Al-Mamoori A, Krishnamurthy A, Rownaghi AA, Rezaei F (2017) Carbon capture and utilization update. Energ Technol 5(6):834–849. https://doi.org/10.1002/ente.201600747
178. Minutillo M, Perna A (2009) A novel approach for treatment of CO_2 from fossil fired power plants, part A: the integrated systems ITRPP. Int J Hydrogen Energy 34(9):4014–4020. https://doi.org/10.1016/j.ijhydene.2009.02.069
179. Song C, Pan W (2004) Tri-reforming of methane: a novel concept for catalytic production of industrially useful synthesis gas with desired H_2/CO ratios. Catal Today 98(4):463–484. https://doi.org/10.1016/j.cattod.2004.09.054
180. Reddy KJ, John S, Weber H, Argyle MD, Bhattacharyya P, Taylor DT, Christensen M, Foulke T, Fahlsing P (2011) Simultaneous capture and mineralization of coal combustion flue gas carbon dioxide (CO_2). Energy Procedia 4:1574–1583. https://doi.org/10.1016/j.egypro.2011.02.027

181. Lao DB, Galan BR, Linehan JC, Heldebrant DJ (2016) The steps of activating a prospective CO_2 hydrogenation catalyst with combined CO_2 capture and reduction. Green Chem 18(18):4871–4874

Chapter 9
Techno-economic Analysis of Facilitated Transport Membranes (FTMs) Based CO_2 Separation Processes

Muhammad Ahsan, Arshad Hussain, Syed Shujaat Karim, and Sarah Farrukh

Abstract Facilitated transport membranes (FTMs) for carbon dioxide (CO_2) separation have emerged as one of the most promising technologies due to their high gas separation performance, lower energy consumption, compactness, ease of operation, and eco-friendliness. However, the industrial implementation of this technology requires a technical and economical analysis to predict its performance and cost. Therefore, in this chapter, the techno-economic analysis of FTMs for the removal of CO_2 from different gas mixture streams of flue gas, natural gas, and syn-gas processing has been comprehensively discussed. For the process design system investigation, a simulation analysis based on Aspen Hysys has also been described in order to study the impact of process parameters and membrane stage configurations on the cost and energy demand of CO_2 separation from different gas mixture streams. The results of the techno-economic analysis showed that FTMs are the most feasible and comparatively low-cost technology, which implies the potential of FTMs as a viable technology for the separation of CO_2 from major industrial sectors such as flue gas, syn-gas, and natural gas processing.

Keywords Facilitated transport membranes (FTMs) · CO_2 capture · Flue gas separation · Natural gas separation · Syngas purification · Techno-economic analysis

---

M. Ahsan · S. S. Karim · S. Farrukh (✉)
School of Chemical and Materials Engineering (SCME), National University of Sciences and Technology (NUST), Sector H-12, Islamabad, Pakistan
e-mail: sarah.farrukh@scme.nust.edu.pk

M. Ahsan
e-mail: ahsan@scme.nust.edu.pk

S. S. Karim
e-mail: sshujaat.che6scme@student.nust.edu.pk

A. Hussain
Department of Chemical, Mechanical, Materials, and Mining Engineering, Pak-Austria Fachhochschule, Institute of Applied Sciences and Technology, Haripur, KPK, Pakistan
e-mail: arshad.hussain@fcm3.paf-iast.edu.pk

S. Farrukh et al. (eds.), *Facilitated Transport Membranes (FTMs) for CO_2 Capture: Overview and Future Trends*, Green Energy and Technology,
https://doi.org/10.1007/978-3-031-21444-8_9

9.1 Process Design Analysis of FTMs Based CO_2 Capture from Flue Gas and Natural Gas

This CO_2 emissions remedial approach is multidimensional and focuses on enhancing energy conservation, energy efficiency, renewable and clean energy alternatives, and developing/employing conventional and new technologies to capture the current CO_2 emissions from point sources. Membrane-based CO_2 capture processes offer great potential for CO_2 capture, despite a few limitations in their sensitivity towards high temperature/pressure and severe process conditions. However, membrane processes benefit from low capital investment, small footprints, ease of scale-up, and low energy requirements [1]. Two types of facilitated transport membranes (FTMs), developed by the research group MEMFO at NTNU, have been considered for the CO_2 capture processes [2].

The membrane used for CO_2 removal from natural gas is made of a thin PVAm and PVA composite membrane layer on a polysulfone support. While the membrane used for CO_2 capture from flue gas is made of a thin PVAm layer on polysulfone support. Both membranes exhibited high selectivity and permeability in extensive experimental work [1]. The successful experimental results showed that these fixed site carrier membranes could play a promising role in CO_2 capture from major industrial sectors like natural gas processing and flue gas from power plants.

Membrane-based process design for CO_2 capture and cost evaluation has been done with a membrane module interfaced within Aspen-Hysys to calculate the required membrane, process energy requirement, and the parameters influencing the energy and membrane area requirements for a specific application. This in-house developed program calculates the permeate and retentate for a system with any number of components, given the permeances of the components and the membrane area [3]. The membrane is divided into several compartments with the equal area and perfectly mixed stages in this model [4]. The process design analysis and cost evaluation for CO_2 capture and natural gas are discussed separately in the subsequent paragraphs [5].

9.1.1 Process Design Analysis of FTM Based CO_2 Capture from Natural Gas

PVAm has a high density of fixed amino groups in this blend membrane that act as CO_2 transport carriers. A mechanically robust polymer, PVA enhances the polymeric network with good membrane formation properties [6]. This ultrathin PVAm/PVA blend layer supported on a polysulfone membrane has been evaluated, and the experimental results revealed its suitability for CO_2/CH_4 separations. Pressure at 2 bar has been recorded, but the experiments at higher pressures are ongoing to validate its reliability for natural gas sweetening. However, this analysis has a selectivity of 40 and a permeance of 0.1 m^3(STP)/m^2.bar.h [3].

The technical viability of the membrane-based CO_2 capture process has been investigated using a two-stage membrane process to meet natural gas pipeline requirements and confine the methane loss in permeate below 2% [3]. First, though, the CO_2 content in the feed has been set to 3.8 mol%, and the feed pressure is about 15 bar [4]. The influence of CO_2 composition in feed and feed pressure on methane loss, membrane area, the energy required, and gas processing cost has been investigated [1].

The feed gas at 15 bar and 55 °C is fed to the membrane. As the FSC membrane performs well when swollen with water, water vapors are introduced as sweep gas on the permeate side. This enhances the membrane performance and reduces the partial pressure of CO_2 on the permeate side, which correspondingly increases the separation efficiency of the membrane. Heat exchangers are used to economizing energy consumption [4].

The overhead products of both flash vessels and the distillation column contain lighter hydrocarbons, predominantly methane, and are fed to the mixer. The gas stream from the mixer is fed to the membrane module to remove CO_2 from natural gas and other lighter hydrocarbons. The process conditions for the simulation analysis are the process conditions of the gas coming out of the mixer. The discrete area in the process flow diagram demarcated by broken lines shows the pre-treatment section for removing heavier hydrocarbons from natural gas before entering the membrane separation unit.

In this process design, most of the separation is accomplished in 1st stage, while the 2nd stage acts as a CO_2 enrichment stage. Therefore, water vapors as a sweep in the 2nd membrane stage are avoided to reduce energy consumption. A one-dimensional isothermal model has been implemented to evaluate the performance of a hollow-fiber membrane module composed of the described membrane. In this work, a feasibility analysis has been carried out with an in-house membrane program interfaced within the process simulation program to investigate the influence of process parameters and membrane stage configurations on the energy demand and gas processing cost [4]. This program can utilize Hysys capability to calculate and couple energy balances in the process model [3].

This simulation analysis is started with the influence of CO_2 concentration in feed gas on the required energy and membrane area. As mentioned earlier, for all results, the CO_2 concentration in retentate has been set constant at 2% and the CO_2 purity at the permeate side was maintained from 92 to 98% depending on the processing conditions. Figure 9.1 shows that both process energy and membrane area increase as the CO_2 concentration in the feed rises. More membrane area is required if a considerable amount of CO_2 must be removed from the natural gas. This also means that there will be more CO_2 permeate, and hence more energy required for the permeate to be compressed before entering the 2nd membrane stage.

Although both membrane area and required process energy increase with increasing CO_2 concentration in the feed, both curves show different rising trends [7]. The rise in energy is steeper than a rise in membrane area because the rise in CO_2 concentration dampens the increase in membrane area. Figure 9.2 shows the effect of CO_2 concentration in feed on gas processing cost and the methane lost through

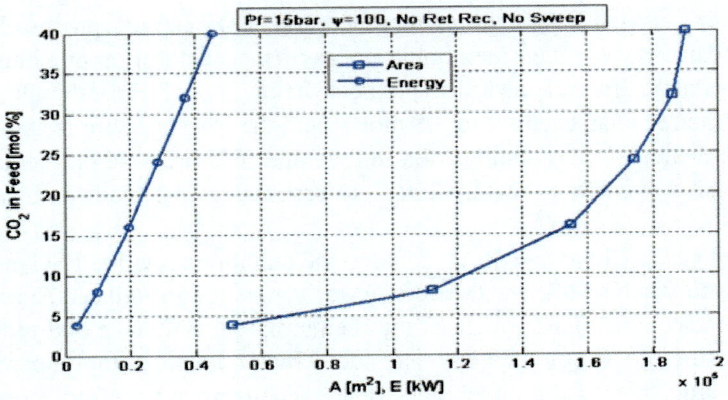

Fig. 9.1 Effect of area and energy on CO_2 in feed

the permeate [8]. The gas processing cost is strongly influenced by the required membrane area, process energy, and CH_4 loss in the permeate [3]. Therefore, a cumulative influence of membrane area, process energy, and CH_4 loss in permeate is evident in the rise in gas processing cost. It is also evident from Fig. 9.4 that the rise in gas processing cost is significantly low when the CO_2 concentration in feed reaches 30 mol.% [7]. This shift in GPC's rising trend can be seen as replicating the rising trend of membrane area with increasing concentration of CO_2 in feed [7].

Figure 9.3 shows that the required membrane area decreases considerably as the sweep flow rate increases. However, the required energy decreases at first but increases as the sweep flow rate increases [9]. Any additional sweep flow rate can result in less membrane area.

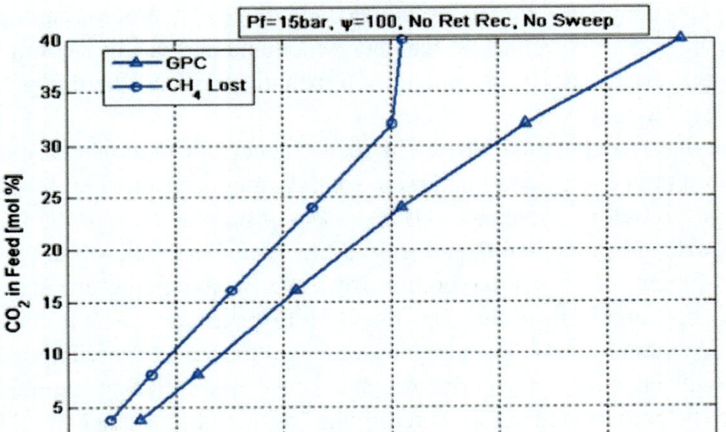

Fig. 9.2 Effect of CO_2 concentration in feed on gas processing cost and the methane lost through the permeate

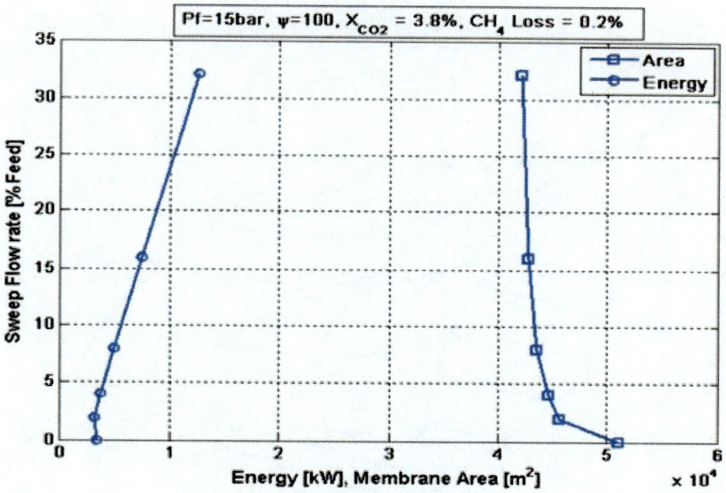

Fig. 9.3 Effect of membrane area on sweep flow rate

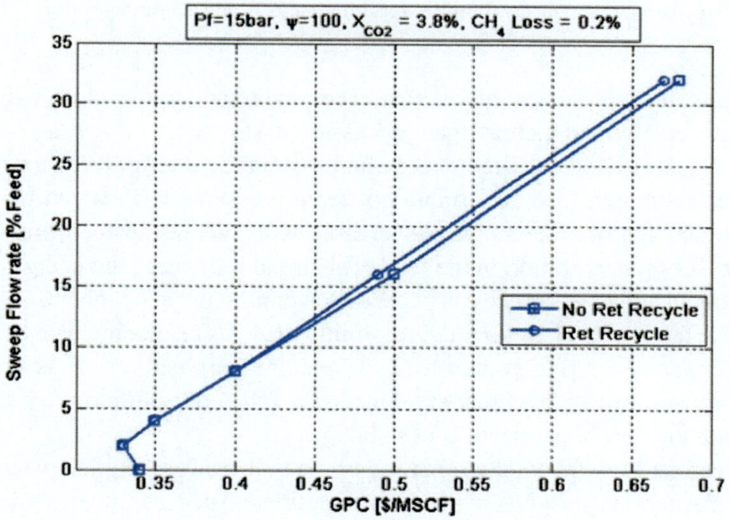

Fig. 9.4 Effect of process conditions on sweep flow rate

However, more process energy is required to separate water vapors from permeate and recycle these, ultimately raising the gas processing cost, as evident in Fig. 9.4. Depending on the feed composition and process conditions, recycling the retentate can increase the driving force across the membrane, improving the separation efficiency of a membrane system [4].

It can be deduced that recycling the retentate decreases the gas processing cost compared to the case when no retentate is being re-circulated. The optimum amount of retentate recycling might differ depending upon the process conditions and feed

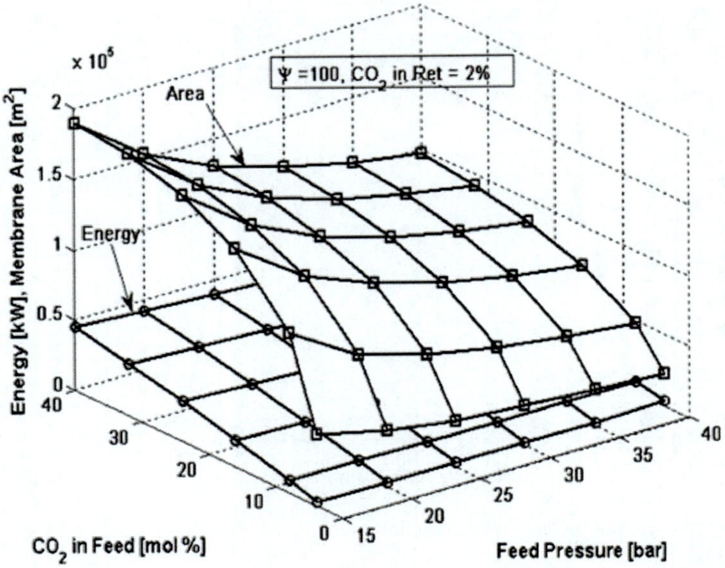

Fig. 9.5 Energy and membrane area profiles for different CO_2 concentrations

compositions. It is important to note that excessive retentate recycling will require extra process energy and increase gas processing costs.

Figure 9.5 shows the required energy and membrane area profiles for different CO_2 concentrations in feed at variable upstream pressures [10]. It can be inferred from the figure that membrane area decreases with increasing feed pressure and decreasing CO_2 concentration in the feed. This trend influences the energy profiles, thus resulting in higher gas processing costs at lower pressure and higher CO_2 concentration in the feed [11]. Therefore, the feasibility analysis results indicate that a two-stage membrane separation system for CO_2 capture from natural gas is viable [3]. Furthermore, the results show that attaining a 2% CO_2 concentration in retentate at a lower methane loss in permeate is possible.

The sweeping flow favors the separation process by increasing the driving force across the membrane. However, it needs to be optimized because an excessive sweep flow rate requires more compressional energy for permeation and thus results in a higher gas processing cost. The gas processing cost increases with decreasing feed pressure but increases with increasing CO_2 concentration in the feed [7]. Recycling retentate from the 2nd membrane stage slightly decreases the gas processing cost compared to the case without recycling. Montazer-Rahmati, Optimization of membrane-based CO_2-removal from natural gas using simple models considering the pressure and temperature effects. In this analysis, the natural gas mixture contains carbon dioxide, methane, ethane, and propane. The rest of the mixture comprises small molar fractions of butane, pentane, and hexane. The membrane process is designed to reduce the concentration of carbon dioxide in natural gas by up to 1 mol% [12]. This preliminary simulation analysis is conducted to study the optimum process configuration to identify the required membrane area by varying process parameters. In this analysis, feed temperatures, feed pressures, sweep flow, and membrane selectivity have been varied to see their influence on the required membrane area.

The simulations are based on the SRK equation of state, and the input data are feed compositions, process conditions, and membrane permeance values of the gaseous mixture. The membrane selectivity values have been deduced from the permeation experiments conducted within the membrane research group. The CO_2/CH_4 selectivity value of 40 has been measured at around 15 bar for a relatively humid facilitated membrane. For this analysis, the gas flow rate has been taken at 8200 Nm^3/h at 21 °C [4]. Heat exchangers are used to economizing energy consumption. The overhead products of flash vessels V100, V101, and the distillation column contain lighter hydrocarbons, predominantly methane, and are fed to the mixer. The gas stream from the mixer is fed to the membrane module to remove CO_2 from natural gas and other lighter hydrocarbons. The process conditions for the simulation analysis are the process conditions of the gas coming out of the mixer. As depicted in Fig. 9.6, a rise in the feed pressure leads to a higher driving force for gas separation, requiring a smaller membrane area [13].

Figure 9.7 shows that a particular sweep contributes to better membrane performance, but once it acquires its optimum value, it does not contribute further to the separation process.

The desired concentration of CO_2 in the retentate stream was varied. The analysis depicted in Fig. 9.8 shows that a smaller membrane area will be required if a higher concentration of CO_2 is maintained in the retentate stream.

It has been found that there is no considerable difference between the results from both types of flow. Feasibility analysis of CO_2 capture from natural gas by employing Amine absorption and Membrane technology. Technical analysis of CO_2

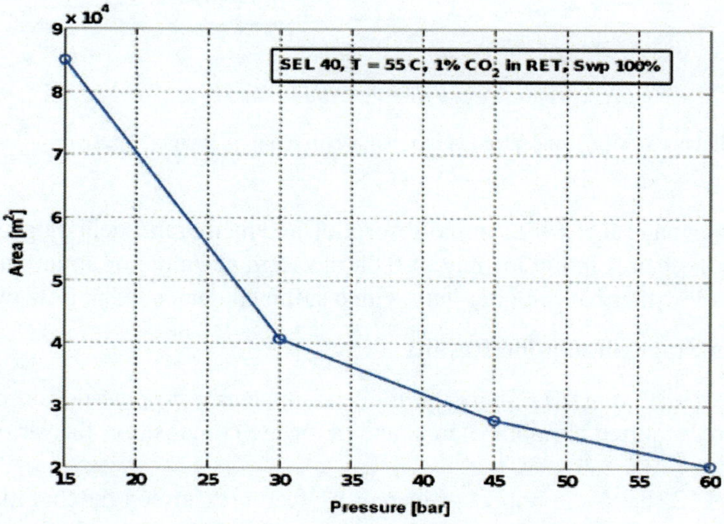

Fig. 9.6 Influence of pressure on required membrane area

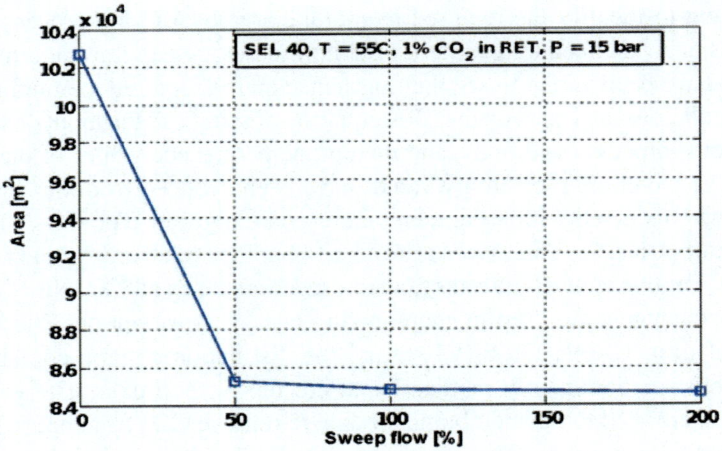

Fig. 9.7 Influence of sweep flow on required membrane area

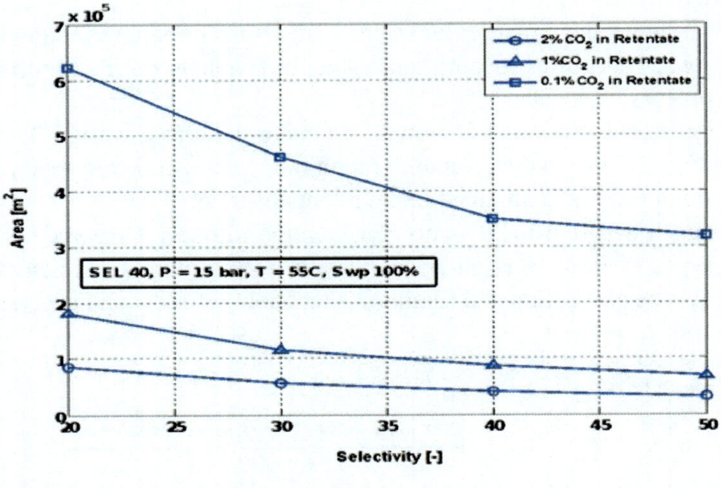

Fig. 9.8 Influence of CO_2 concentration in retentate on required membrane area

removal from natural gas with amine absorption and membrane technology has been conducted. A lean solution loading of 0.08 mol acid gas/mole of amine fulfills the desire for a low thermal re-boiler duty with a sufficient lean solvent flow rate [14].

(a) **Single-Stage Simulation Results**

Figures 9.9, 9.10, and 9.11 show the influence of permeate pressure on permeate temperature, required membrane area, and permeate composition for various cases [14].

Figure 9.12 shows the impact of increasing feed pressure on the 2nd membrane stage. These simulation results show that gas stream purity is depending on different separation techniques [14]. The higher purity of solid and vent gas in the amine

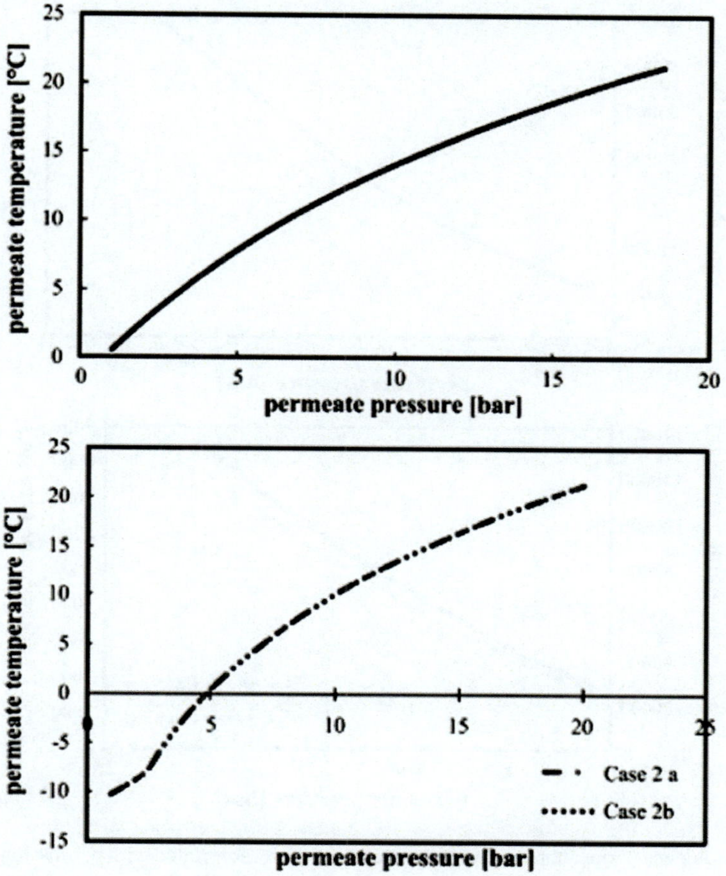

Fig. 9.9 Influence permeate pressure on the permeate temperature. Reprinted with permission from Ref. [14], copyright © 2022 Elsevier

process might justify the high total capital investment and harmful environmental penalties [14]. On the other hand, membrane processes yield lower purity but are environmentally friendly, easy to scale up, have low energy requirements, and have smaller footprints. Moreover, as reported by other authors, amine absorption yields high capital investment and is too complicated to handle for low flow rates. Therefore, membrane applications are recommended for flow rates less than 5 MMSCFD.

The feed gas flow rates are lower than existing membrane systems for natural gas sweetening. It is worth mentioning that membrane systems for natural gas sweetening are typically favored for small-sized and remote applications. This supports employing a membrane system for natural gas sweetening in this analysis. Membrane and amine absorption systems become competitive for a size of 5–50 MMSCFD [3].

Initially, the feed gas at 90 bar and 50 °C is mixed with retentate from stage 3 and then fed to the first membrane stage. The first membrane stage enables 50% trimming of CO_2, and thus the remaining CO_2-containing stream is fed to stage 2. Pressure on the permeate side varies depending on its influence on the required

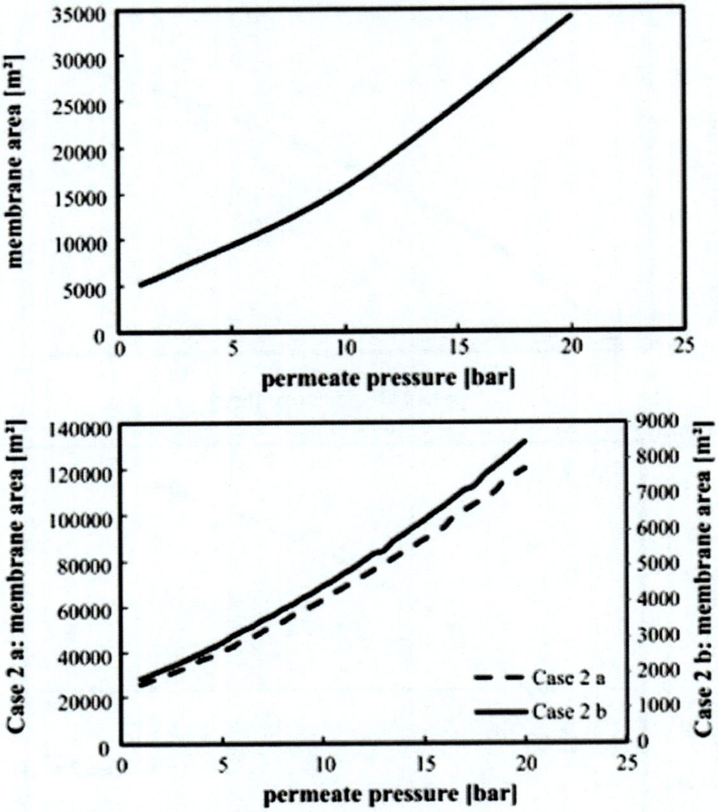

Fig. 9.10 Influence permeate pressure on the membrane area. Reprinted with permission from Ref. [14] copyright © 2022 Elsevier

membrane area and energy. The temperature of permeate from stage 1 and stage 2 decreases due to a high-pressure drop across the membrane and a fair number of higher hydrocarbons in the permeate side liquefy. Heat exchangers are used to reduce energy consumption. A one-dimensional isothermal model has been implemented to evaluate the performance of a hollow-fiber membrane module composed of the described membrane. In this work, a feasibility analysis has been carried out with an in-house membrane program interfaced within the process simulation program to investigate the influence of permeate pressure on the energy demand and gas processing cost. This program can utilize Hysys capability to calculate and couple energy balances in the process model [3].

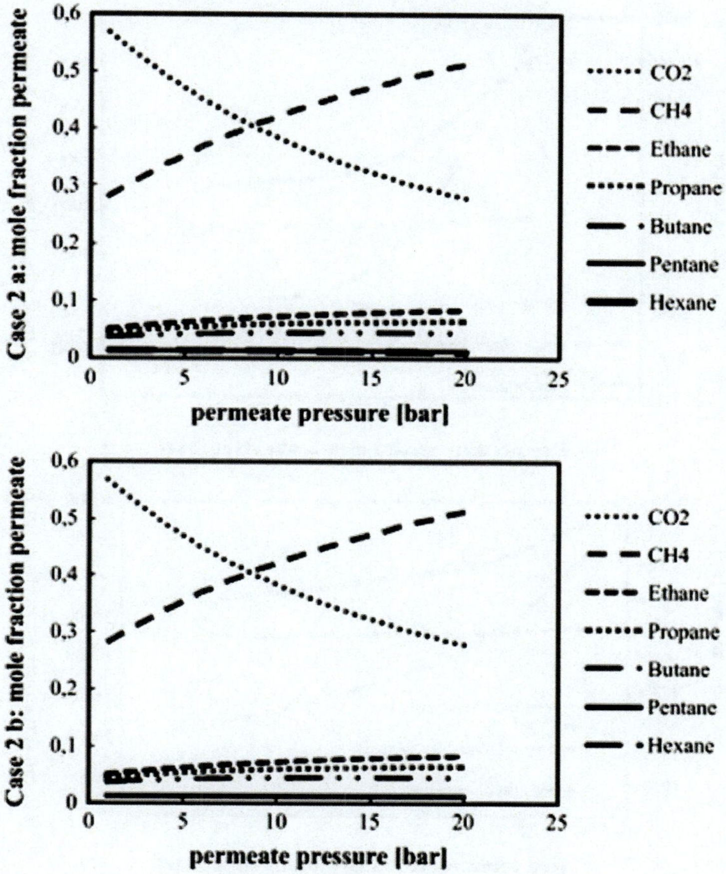

Fig. 9.11 Influence permeate pressure on the mole fraction. Reprinted with permission from Ref. [14] copyright © 2022 Elsevier

9.1.2 *Process Design Analysis of FTM Based CO_2 Capture from Flue Gas*

The ultrathin PVAm membrane supported on polysulfone is considered in this work, which exhibits a facilitated transport mechanism to capture CO_2 from the flue gas mixtures. The preliminary results of selectivity ~ 200, permeance ~ 1 m^3(STP)/bar.h.m^3 of gas separation studies show a promising application in CO_2 capture from flue gas [15]. The technical viability of a membrane-based CO_2 capture process has been investigated in this work by using a two-stage membrane process with an efficient process design utilizing water vapors as the sweep gas [4].

The separation target of this simulation analysis is to achieve 90% CO_2 recovery and 90% purity in permeate, which is considered a requisite for the applicability of the membrane-based CO_2 capture process [4]. In addition, the CO_2 content in flue gas has been set at 10% to show that the membrane process can still be an attractive

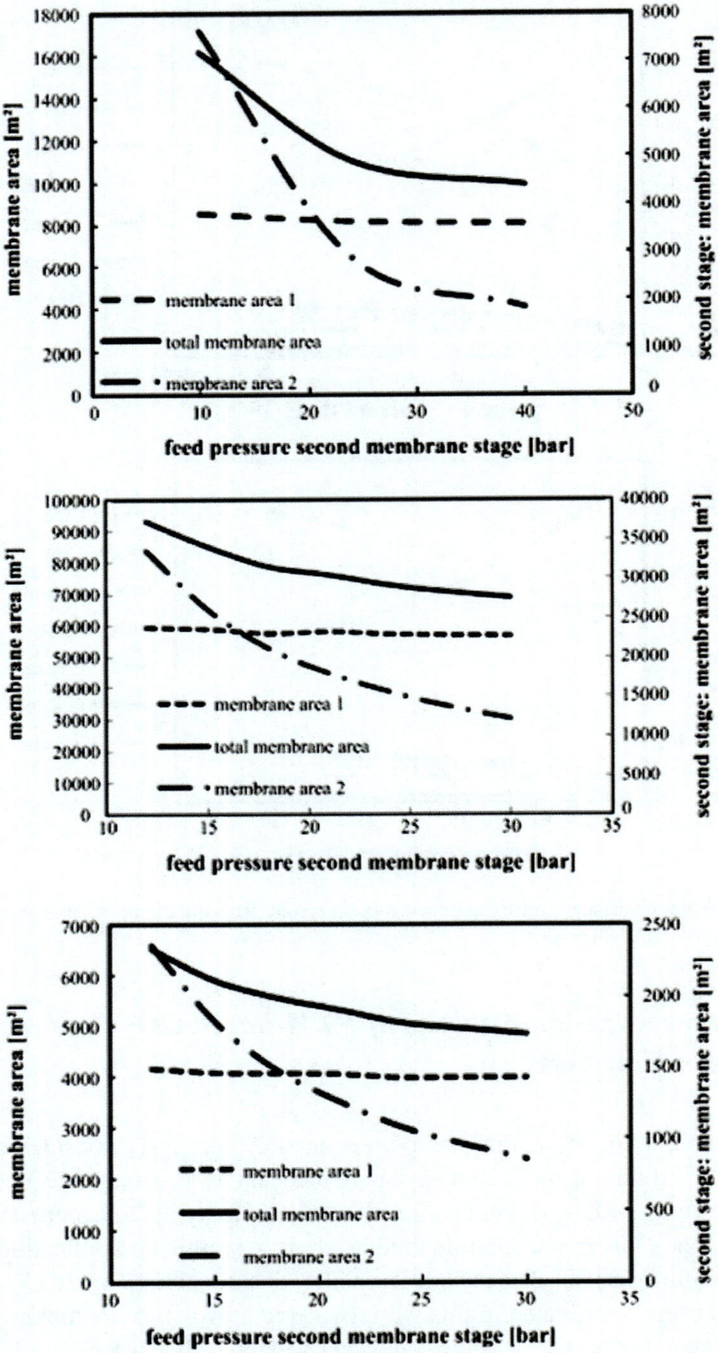

Fig. 9.12 Impact of increasing feed pressure on membrane area. Reprinted with permission from Ref. [14] copyright © 2022 Elsevier

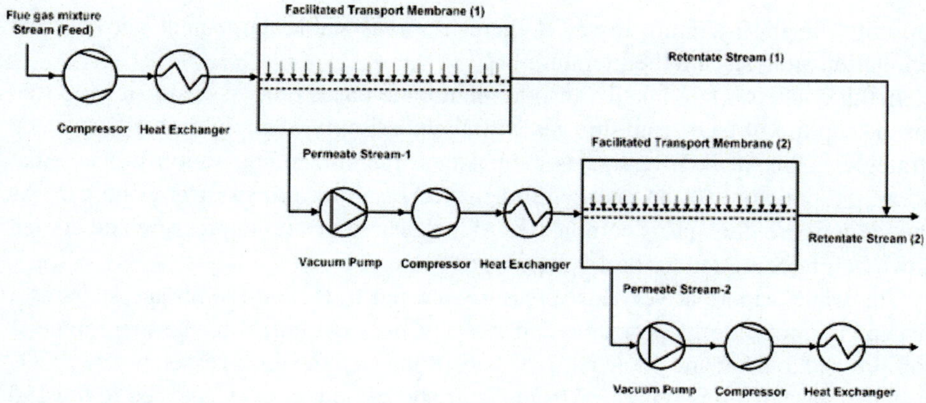

Fig. 9.13 Two-stage membrane separation process [16]

CO_2 capture technique. However, the higher content of CO_2 in flue gas favors the feasibility of the membrane separation process.

Apart from flow rates, temperature, and compositions, other process variables when using polymeric membranes are the pressure ratio, between the upstream and downstream pressures over the membrane, and stage cut, which is the ratio of the permeate flow rate to the feed flow rate [15]. A process flow diagram of a two-stage membrane separation process is depicted in Fig. 9.13. In a post-combustion scenario, flue gas flow rates can be huge depending on the emissions source. The feed gas pressure and temperature are 1 bar and 40 °C, respectively.

In the current study, the feed gas is compressed before being fed to the membrane to maximize the driving force across the membrane. Simulations are performed with varying permeate pressures. The facilitated membrane performs well when wetted with water. Therefore, water vapors are suggested as sweep gas on the permeate side. This enhances the membrane performance and reduces the partial pressure of CO_2 on the permeate side, which correspondingly increases the separation efficiency of the membrane [4].

The membrane selectivity values have been deduced from the permeation experiments conducted within the membrane research group. The CO_2/N_2 selectivity value of 200 has been measured up to 5 bars for a relatively humid, facilitated membrane. In addition, a one-dimensional isothermal model has been implemented to evaluate the performance of a hollow-fiber membrane module composed of the described membrane. In this work, a feasibility analysis has been carried out with an in-house membrane program interfaced within the process simulation program to investigate the influence of process parameters and membrane stage configurations on the energy demand and flue gas processing cost [4].

This program can utilize Hysys capability to calculate and couple energy balances in the process model. Other input data are the feed compositions, process conditions, and membrane permeance values of the flue gas mixture. For better comprehension, the discussion on this feasibility study is started by comparing the simulation results presented in the literature to see if there is a reasonable agreement between the simulation analysis and the literature [4].

In this context, a facilitated transport membrane-based process design is proposed for the capture of CO_2 from flue gas with the possibility of applying various sweep strategies. This work investigates a two-stage membrane separation process that offers the potential for an energy-efficient CO_2 capture process from flue gas. As shown in figure, flue gas is compressed to the desired upstream pressure and cooled down before being fed to the 1st membrane stage [4].

The water vapors at very low pressure, are fed to the 1st membrane stage as a sweep. The mixture of permeate and water vapors exiting from the membrane is compressed to 2 bar and cooled at 3 °C before entering the flash vessel, where a CO_2 enriched gas stream is separated from water and heated to 25 °C and fed to the 2nd membrane stage for further CO_2 separation. The water separated from permeate is heated by passing through a series of heat exchangers to acquire the vaporization temperature and then expanded to the respective downstream pressure for its re-use as sweep.

The permeate from 1st membrane stage contains 42–55% of CO_2 when using water vapors as sweep, while the CO_2 concentration in the permeate from 1st membrane stage varies from 42 to 82% when using no sweep or the permeate fraction as sweep. The 2nd membrane stage contributes toward further enrichment of the permeate stream. For a specific sweep strategy, at different pressure ratios, the membrane area of the 1st stage is varied to attain the desired recovery and purity. In contrast, the membrane area and pressure ratio are kept constant for 2nd membrane stage. 0.5 bar intervals varied the upstream pressure, while the downstream pressure was varied by 0.02 bar intervals.

Using the permeate fraction as a sweep contributes to the efficient removal of permeate from the membrane stage. However, on the contrary, it slightly reduces the CO_2 driving force across the membrane, which depends on the difference in partial pressures of CO_2 on the upstream and downstream sides. Therefore, 70% of the permeate from the 1st membrane stage is fed to the 2nd membrane stage as feed, and the remaining 30% of the permeate is re-circulated as a sweep to the first membrane stage, whereas 5% of the permeate from the 2nd membrane stage is re-circulated as a sweep to 2nd membrane stage. This distribution of sweep corresponds to roughly 5% of the feed flow rate in the respective membrane stage.

9.2 Process Design Analysis of FTM Based CO_2 Separation from Syngas

A single-stage membrane process has been designed using facilitated transport membranes (FTMs) to decarbonize the coal-derived syngas from an integrated gasification combined cycle (IGCC) power plant. The necessary process model and costing method have also been developed to assess the technical feasibility and process

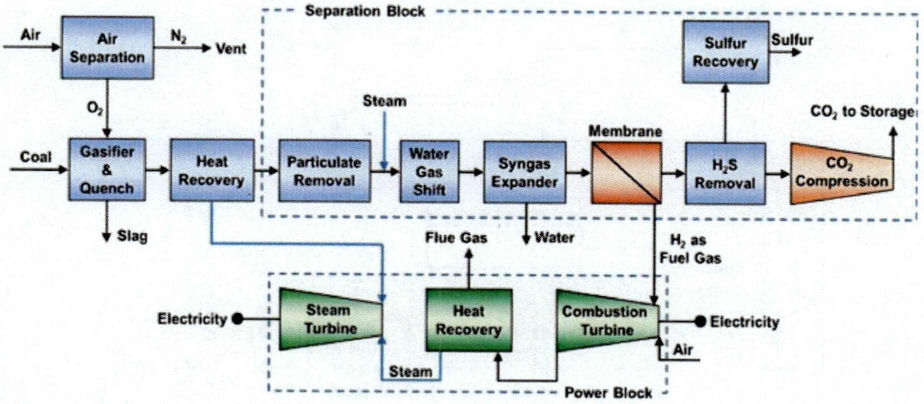

Fig. 9.14 Location of the membrane separation unit in an IGCC power plant. Reprinted with permission from Ref. [17] copyright © 2022 Elsevier

economics. The techno-economic study reveals that the mitigated carrier saturation upon bulk CO_2 removal can lead to appreciable increases in the CO_2 permeance and CO_2/H_2 selectivity, which can be utilized to achieve 95% CO_2 purity and 95% H_2 recovery with a CO_2/H_2 selectivity of 50 at complete carrier saturation. Furthermore, FTMs with different facilitated transport characteristics can also be arranged in a hybrid membrane configuration to render a H_2 recovery of 99% and a cost of electricity of $118 [17].

In this work, a single-stage membrane process has been designed to capture 90% CO_2 from an IGCC power plant while delivering a H_2 purity of > 99%. In addition, a single-stage membrane process has been designed for using FTMs to decarbonize the syngas from an IGCC power plant. The location of the membrane process in an IGCC power plant is shown in Fig. 9.14. A detailed process flow diagram of the single-stage membrane process is shown in Fig. 9.15 [17].

The process model and costing method have also been developed to assess the technical feasibility and process economics. The mitigated carrier saturation upon bulk CO_2 removal is the enabling factor for using FTM in the single-stage process. In order to achieve the separation specifications, the FTM should at least possess an onset fugacity for carrier saturation of 10 bar and a CO_2/H_2 selectivity of 50 at complete carrier saturation. FTMs with different facilitated transport characteristics can be arranged in a hybrid membrane configuration to achieve a H_2 recovery of 99% and a cost of electricity (COE) increase of 15.5% compared to the no capture scenario [17].

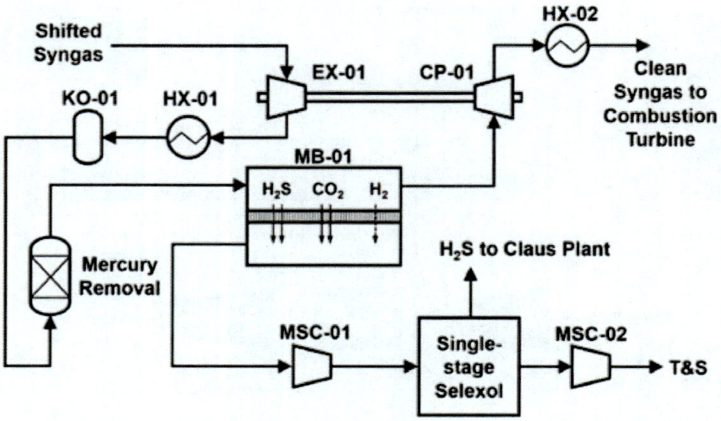

Fig. 9.15 Flowsheet of single-stage membrane process for the integration in an IGCC power plant. Abbreviations: *CP* compressor, *EX* turbo expander, *HX* heat exchanger, *MB* membrane, *MSC* multistage compressor, *T&S* transportation and sequestration. Reprinted with permission from Ref. [17] copyright © 2022 Elsevier

9.3 Conclusion

This chapter focuses on a comprehensive discussion regarding the techno-economic feasibility of facilitated transport membranes (FTMs)-based CO_2 capture processes for industrial sectors such as for natural gas sweetening process (CO_2/CH_4), flue gas separation (CO_2/N_2), and syn-gas purification process (CO_2/H_2). A simulation analysis based on Aspen Hysys has also been discussed for the process design system analysis in order to investigate the effects of process parameters and membrane stage configurations on the cost and energy requirement for the CO_2 capture processes from industrial gas mixture streams. According to the simulation results, the conventional technology, such as amine absorption, produced a higher purity of gas, however, it required large area, high energy, high capital and operating costs and also had an effect on the environment. Whereas, the FTMs processes yield a lower purity but are environmentally friendly, have a smaller footprint, are easy to operate and scale up, and require low energy for the CO_2 separation processes. The CO_2 separation systems of amine absorption and membrane are competitive for gas separation operation sizes in the range of 5 to 50 MMSCFD. Because amine absorption process requires a significant capital investment and is difficult to manage at low flow rates; therefore, membrane applications are recommended for flow rates below 5 MMSCFD. Furthermore, FTM systems also provide the flexibility to form hybrid systems with other separation systems to achieve better CO_2 separation efficiency. The overall discussion implies that FTMs have the potential to be a viable alternate technology for the separation of CO_2 from flue gas, syn-gas, and natural gas streams.

References

1. Hussain A (2017) Three stage membrane process for CO_2 capture from natural gas. AA 50:1
2. Hussain A, Farrukh S, Minhas FT (2015) Two-stage membrane system for post-combustion CO_2 capture application. Ener Fuels 29(10):6664–6669
3. Hussain A, Nasir H, Ahsan M (2014) Process design analyses of CO_2 capture from natural gas by polymer membrane. J Chem Soc Pakistan 36(3)
4. Hussain A, Hägg M-B (2010) A feasibility study of CO_2 capture from flue gas by a facilitated transport membrane. J Membr Sci 359(1–2):140–148
5. Lock SSM, Lau KK, Shariff A, Yeong YF, Bustam M, Jusoh N, Ahmad F (2018) An atomistic simulation towards elucidation of operating temperature effect in CO_2 swelling of polysulfone polymeric membranes. J Nat Gas Sci Eng 57:135–154
6. Deng L, Kim T-J, Sandru M, Hägg M-B (2009) PVA/PVAm blend FSC membrane for natural gas sweetening. In: Proceedings of the 1st annual gas processing symposium, Elsevier, pp 247–255
7. Hao J, Rice P, Stern S (2002) Upgrading low-quality natural gas with H_2S-and CO_2-selective polymer membranes: Part I. Process design and economics of membrane stages without recycle streams. J Membr Sci 209(1):177–206
8. Zhang X, Zhu A, Li X, Gong W (2004) Oxidative dehydrogenation of ethane with CO_2 over catalyst under pulse corona plasma. Catal Today 89(1–2):97–102
9. Nemitallah MA, Abdelhafez AA, Habib MA (2020) Hybrid membrane and porous-plates reactors for gas turbine applications. In: Approaches for clean combustion in gas turbines. Springer, pp 321–417
10. Ahmad F, Lau KK, Lock SSM, Rafiq S, Khan AU, Lee M (2015) Hollow fiber membrane model for gas separation: process simulation, experimental validation and module characteristics study. J Ind Eng Chem 21:1246–1257
11. Niu MW, Rangaiah G (2014) Retrofitting amine absorption process for natural gas sweetening via hybridization with membrane separation. Int J Greenhouse Gas Control 29:221–230
12. Chuah CY, Goh K, Yang Y, Gong H, Li W, Karahan HE, Guiver MD, Wang R, Bae T-H (2018) Harnessing filler materials for enhancing biogas separation membranes. Chem Rev 118(18):8655–8769
13. Micari M, Dakhchoune M, Agrawal KV (2021) Techno-economic assessment of postcombustion carbon capture using high-performance nanoporous single-layer graphene membranes. J Membr Sci 624:119103
14. Peters L, Hussain A, Follmann M, Melin T, Hägg M-B (2011) CO_2 removal from natural gas by employing amine absorption and membrane technology—a technical and economical analysis. Chem Eng J 172(2–3):952–960
15. Hussain A (2012) A single stage membrane process for CO_2 capture from flue gas by a facilitated transport membrane. Sep Sci Technol 47(13):1857–1865
16. Hussain A, Ijaz H, Peter L (2013) Techno-economic analysis of CO_2 capture from flue gas by amine absorption and membrane technology. Int J Ener Clean Environ 14(4)
17. Han Y, Ho WW (2021) Facilitated transport membranes for H_2 purification from coal-derived syngas: a techno-economic analysis. J Membr Sci 636:119549

Chapter 10
Conclusions and Future Trends

Zarrar Salahuddin, Syed Shujaat Karim, and Sarah Farrukh

Carbon capture and storage (CCS) is a key component of scientific and applied research in this age of industrial development. This subject has now developed into a complete scientific field of its own, owing to the vast amounts of carbon dioxide emissions into the atmosphere which need to be tackled to reduce the risk of climate change. The first step in CCS is the capture of carbon dioxide from various sources including stationary power plants operated on fossil fuels, or mobile applications such as automobiles. The main aim in this step is to reduce the net anthropogenic carbon dioxide emissions. The next step is to make use of the captured carbon dioxide to make useful products to improve human life and to benefit the society. For the capture stage of the process, various conventional technologies have been in use for decades, if not centuries. These processes include liquid absorption, solid adsorption and cryogenic separation. However, as is evident, these processes are both highly energy-intensive and may pose environmental hazards of their own. Therefore, there was a dire need to introduce new and improved processes with much less environmental impact and greater degree of energy efficiency. This led to the development of membrane technology which quickly gained the interest of researchers, academics and industrialists alike. Initially, membrane technology made use of pristine materials for CO_2 sequestration. However, it was quickly realized that this was not worthwhile, as every individual material type was hindered by its own inherent shortcomings. Therefore, new methods of incorporating seemingly incompatible materials to form

Z. Salahuddin · S. S. Karim · S. Farrukh (✉)
School of Chemical and Materials Engineering (SCME), National University of Sciences and Technology (NUST), Sector H-12, Islamabad, Pakistan
e-mail: sarah.farrukh@scme.nust.edu.pk

Z. Salahuddin
e-mail: zarrar.phdscme@student.nust.edu.pk

S. S. Karim
e-mail: sshujaat.che6scme@student.nust.edu.pk

S. Farrukh et al. (eds.), *Facilitated Transport Membranes (FTMs) for CO_2 Capture: Overview and Future Trends*, Green Energy and Technology,
https://doi.org/10.1007/978-3-031-21444-8_10

mixed matrix membranes were devised. Further development on both experimental and simulation aspects led to the development of an innovative type of membrane known as facilitated transport membranes. These membranes offer multiple advantages, including the combination of the superior properties of both polymers and inorganic fillers and also the incorporation of reversible chemical reaction mechanisms to improve the efficiency and selectivity of separation of CO_2 over other gases.

In this book, focus has mainly been laid on the capture of carbon dioxide from various industrial processes, with a section focusing the utilization of carbon dioxide after capture. Chapter 1 aims at introducing the reader to the field of membrane technology. However, before delving into the discussion on membrane technology, the different conventional technologies for carbon capture have been described with an emphasis on the reason for the development of an advanced process like membrane technology. The different types of membranes that have been used for CO_2 separation have been discussed, including liquid and solid membranes. Moreover, various membrane morphologies such as isotropic dense and porous membranes as well as anisotropic membranes as well as the mechanisms of gas flow through these different morphologies have been laid out, describing the models for gas permeation such as solution-diffusion, molecular sieving, Poiseuille flow and Knudsen diffusion. Furthermore, this chapter also discusses the configurations of membranes during separation processes, such as dead-end filtration and cross-flow filtration. Then, it goes on to elaborate on the different membrane-based industrial gas separation processes such as air separation, separation of hydrogen from hydrocarbons and CO_2 capture. Then, the focus is shifted completely to the different types of carbon capture processes, such as pre-combustion, post-combustion and oxy-fuel carbon capture processes. Finally, a detailed discussion of facilitated transport membranes has been given, including its types and their mechanisms of gas separation, such as fixed-site carrier and mobile carrier based facilitated transport membranes. Distinction has been made between membranes comprising amine-containing fillers and membranes without amine-based fillers. As a whole, this chapter lays the groundwork for the next chapters to build upon.

To fully understand and appreciate the performance enhancement brought about by the use of facilitated transport membranes. In Chap. 2, different reaction mechanisms of the two main types of facilitated transport membranes, namely fixed-site carrier and mobile carrier membranes have been discussed. In addition to reaction mechanisms, subsequent proposed gas transport mechanisms have also be laid out, including vehicle permeation model as well as carrier hopping mechanism. Approximate analytical solutions of the various reaction mechanisms emphasizing on reaction chemistries as well as the different gas transport models have been discussed in detail. In conclusion, in this chapter, it has been said that for improvement of the selectivity of facilitated transport membranes (FTMs), more advanced simulations need to be developed with better understanding of the chemistries involved in FTM reactions. Inter-layer diffusion in composite FTMs also needs to be studied as well as developing a better understanding of carrier stability needs to be worked on. In Chap. 3, the different methods of carrier particle as well as membrane synthesis have been discussed, followed by the conventional and innovative characterization techniques used to properly understand the structure and working principle of FTMs have also been discussed. In the end, it has been concluded that the research is focused on the right track; however there is great need to develop new methods of fabricating FTMs which can truly make membrane technology a cost-effective method to gain a competitive advantage over conventional processes. Furthermore, there is also a need to devise innovative characterization techniques to properly understand the working of FTMs so that they can be made more efficient in future. Chapters 4–7 have focused on the various industrial processes utilizing carbon capture technologies based on membranes. Nowadays, membrane technology is gradually and progressively being used in industrial gas separation such as natural gas processing, biogas upgrading and syngas purification as well as flue gas separation, due to the environmental and energy-efficiency advantages. Facilitated transport membranes, mainly supported ionic liquid membranes (SILMs), hybrid filler based membranes and thin film composite (TFC) membranes have been thoroughly discussed in Chap. 4 and are proclaimed to be among the most promising candidates for future exploration. Moreover, there has been extensive work on the performance of FTMs based on both their permeability as well as selectivity. Discussions have also been laid out on the state-of-the-art technology in natural gas processing in Chap. 4. Certain FTMs have also been developed which claim to separate both CO_2 and H_2S gases simultaneously. This is a great advantage of FTMs over conventional membrane materials. These membranes are also far more resilient than conventional membrane materials, thus making them highly suitable for use in harsh industrial conditions. However, to make FTMs a suitable material for use in industry, cost of production of the materials and carriers on large scale needs to be reduced. Moreover, there is need to understand and study the process through extensive modelling and simulation. Finally, new methods of surface modification of FTMs as well as different fillers need to be used to improve their lifetime in natural gas processing. In Chap. 5, a detailed discussion has been laid out on the use of FTMs for biogas upgrading. The performance of various FTMs has shown that they are exceptional candidates for use in biogas upgrading. The biggest advantage of FTMs is that their performance is enhanced by the presence of water vapour in biogas, which is otherwise detrimental to the performance of conventional membranes. FTMs can therefore be used for the separation of gases in biogas. However, for the proper upscaling of the technology to the industry, there is still need to understand the process and to test the reliability of FTMs in industrial applications. For this purpose, large-scale modelling and simulations need to be carried out as well as field tests to check the performance of FTMs in real time. Furthermore, in Chap. 6, the various production and purification methods for the efficient supply of syngas are discussed. Syngas is mostly produced through reforming and gasification processes coupled with water gas shift reactions. This produces an output stream comprised mainly of hydrogen and carbon dioxide. FTMs are considered the most suitable for the separation of CO_2/H_2 gas mixtures due to their high throughput combined with excellent separation performance. Also, various types of FTMs are discussed which have widely been employed to effect this gas separation. Lastly, this chapter emphasizes the need to develop further understanding of the performance of the membranes by using them in industrial operating conditions. It is also realized that hybrid filler based membranes are the best candidate among FTMs for optimized CO_2/H_2 gas separation. Also, emphasis is laid on the need to improve the adhesion between the composite layers and the functionalizing membrane materials so that the advantages of FTMs can be fully exploited for exceptional gas separation performance at high permeability. In Chap. 7, the use of FTMs in flue gas treatment is extensively reviewed. It is found that FTMs are excellent candidates for use in flue gas separation due to their high permeability for CO_2 and high CO_2/N_2 selectivity. However, these membranes are still far from industrial application, due to the lack of comprehensive field tests which need to be extensively conducted. Moreover, there is also need for better dispersion and compatibility between filler particles and polymeric matrices. This can be achieved by using functionalization techniques as well as using interfacial polymerization. Therefore, there is urgent need to carry out more extended research works using these techniques to improve compatibility of inorganic fillers and organic polymers, as well as the performance of FTMs.

In Chap. 8, the second step of the carbon capture and storage cycle is studied in detail that is the storage and utilization of CO_2. In the recent past, the exceptional capture of carbon dioxide has led to the development of technologies at both small-scale and industrial scale, for the storage and utilization of CO_2. Carbon dioxide storage using advanced materials is being developed, as advanced materials based CO_2 storage will improve the energy efficiency of the industrial process, also it will eliminate the cost of carbon dioxide transportation and storage in other formations. Therefore, there are extensive advantages for the storage of carbon dioxide in advanced materials instead of geological formations. Therefore, there is need to develop new materials to achieve this goal. However, the main blockade in the path of such implementation at the industrial scale is the cost involved in using such advanced materials and their development, synthesis and fabrication. There is great need to develop such materials in a cost-effective manner on a large scale. An innovative and emerging technology in this field is the hybrid processes which link carbon capture subsystems, thus providing instantaneous carbon capture and utilization. Such systems are the focus of future research but there is great need for more extensive research and development in this field. Lastly, in Chap. 9, a case study is presented for the use of FTMs in natural gas processing, flue gas separation, and syn-gas purification. In these three cases, different membrane specimen are used and this study has been carried out extensively in conditions very similar to those in the industry. Furthermore, cost evaluation and process design parameters are critically evaluated and analyzed as the next step in the utilization of FTMs in the industry is the proper and efficient design of modules and processes to execute the gas separation processes. Various FT membranes working on different mechanisms are studied and it is found that to achieve an excellent recovery of 99%, different configurations and mechanisms need to be utilized in a hybrid system and work needs to be done in that regard.

In conclusion of the contents of this book, it can be said that facilitated transport membrane (FTM) technology offers a high-quality and innovative solution to the problems associated with membrane technology, such as the trade-off between selectivity and permeability, as well as the lack of stability on a large scale. Thus, FTMs can play a major rule in the future of carbon capture and storage (CCS) processes. This is mainly due to the fact that many of the problems encountered by conventional membrane materials are greatly reduced by the use of FTMs due to their exceptional physicochemical properties. However, there is still a need to develop more cost-effective methods of developing such materials on a large scale. There is also a need to understand the chemistries involved in facilitated transport mechanism to be able to exploit it more effectively. This can be done by developing more advanced and specific simulation models to study the effects of various physical parameters on the performance of FTMs. Finally, there is also an urgent need to carry out comprehensive field tests to test the performance, durability and stability of FTMs under industrial simulated conditions of pressure, temperature and gas concentrations. Moreover, there is also a necessity of developing carbon dioxide storage and utilization techniques so that the process can be formed into a closed-loop system and we can begin to see the tangible benefits of improved carbon capture performance. For this, new and improved materials need to be developed at a reasonable cost to store carbon dioxide for long periods of time in a stable environment. Moreover, there is also a need to develop hybrid processes linking several carbon capture subsystems and simultaneously carry out carbon dioxide capture and utilization. Development of such a process will bear fruit for all the tireless efforts being carried out to improve carbon capture performance efficiency. Focus also needs to be laid on the process design and cost evaluation of implementing FTMs in the industry as FTMs are only advantageous if they are competitive in terms of cost efficiency. More research needs to be carried out not only to improve the performance of FTMs and to understand the phenomenon of facilitated transport, but also to cut costs and make membrane technology a feasible and reliable alternative to current energy-intensive and environmentally unfriendly carbon capture processes.